111 Calculation Exercises in the Field of Chemical Technology

Günter Jüptner

111 Calculation Exercises in the Field of Chemical Technology

Springer

Günter Jüptner
Hammah, Germany

ISBN 978-3-662-66919-8 ISBN 978-3-662-66920-4 (eBook)
https://doi.org/10.1007/978-3-662-66920-4

This Springer imprint is published by the registered company Springer-Verlag GmbH, DE, part of Springer Nature.
The registered company address is: Heidelberger Platz 3, 14197 Berlin, Germany

In grateful retrospect of about five decades of fulfilled, challenging and satisfied activity in the field of chemistry and process engineering

"Learning is like rowing against the current, whoever stops, drifts back."

Xunzi (Chinese philosopher, 300 BC)

Preface

The impetus for this book came from the author's review of the practice problems in chemical engineering that were created during more than a decade of teaching for aspiring industrial chemists. From my many years of experience, I have found that there is a need for such practical exercises during the training of industrial chemists, but also in the everyday work of a chemical production plant.

In addition, during my approximately 50-year career in the chemical industry, from an apprenticeship as a laboratory assistant in a laboratory of a chemical production plant to "Global Technology Leader" in an internationally active company, I have experienced that people directly involved in a chemical production plant are often overwhelmed by relatively simple technical calculations—especially at the interface between chemistry and production technology. This often applies to plant engineers who have partially received their university education in mechanical engineering or a similar field. Even chemists who did not come into contact with technical chemistry during their studies often find such calculations common and necessary in a chemical plant to be foreign.

The book created here is intended to remedy this. In addition to the target groups already mentioned, chemical laboratory assistants, chemical engineers, foremen and chemical operators may be named who want to develop professionally.

Only the most important topics prevailing in most chemical plants are dealt with in the resulting collection of tasks, but not specialties of chemical technology. Only a very brief mention is made of the basics of technical chemistry and process engineering, as the collection of tasks does not and cannot replace a corresponding textbook. It is designed as a practice book for the application and deepening of relatively simple calculations in a chemical plant. Therefore, only the formulas required for this purpose are given below with a very brief description. For the acquisition of the background knowledge and the derivation of the relationships mentioned, reference is made to the corresponding textbooks.

I wish the reader and practitioner much success and enjoyment in the acquisition of knowledge through the processing and solving of tasks.

July 2020 Günter Jüptner

Symbol Index, Constants & Conversion Factors

i. Symbol Index

The symbols used in this book correspond to the letters and characters mostly used in the relevant literature. Since the chemical industry is global, the naming of such data is mainly based on terms from the English language. This is taken into account here. However, this sometimes results in a double assignment of the symbols used. As a hint for the reader, it should also be mentioned that in the textbooks of chemical technology and process engineering, unfortunately, often still an inconsistent naming of the used quantities exists.

Symbol	Unit	Meaning
A	m^2	Area
c_i	mol/L	Molar concentration
C_i	kg/L	Mass concentration based on volume
C_i	wt%	Mass-related concentration
cp_i	J/(mol * °C) = J/(mol * K)	Heat capacity of substance i (molar-based)
cp_i	kJ/(kg * °C) = kJ/(kg * K)	Heat capacity of substance i (mass-based)
d_a	m	Outer diameter
d_i	m	Inner diameter
E	kJ	Energy, heat, work (Joule) Electrical energy → kWh
E_A	kJ/mol	Activation energy
F_T		Factor of reaction rates at temperature change
g	m/s^2	Gravitational acceleration, acceleration of gravity
h_{geo}	m	Geodetic height pipe system, pumping of liquids

Symbol	Unit	Meaning
h_p	m	Equivalent height pressure difference pipe system, pumping of liquids
h_f	m	Equivalent height friction losses pipe system, pumping of liquids
H	m	Total height pipe system, pumping of liquids
I	A	Electrical current (amperes)
$\Delta_f H_{i0}$	kJ/mol	Formation enthalpy substance i under standard conditions (25 ° C; 1 bar)
$\Delta_L H$	kJ/mol	Dissolution enthalpy
$\Delta_S H$	kJ/mol	Specific melting enthalpy (molar)
$\Delta_S H$	kJ/kg	Specific heat of melting (mass)
$\Delta_R H$	kJ/mol	Reaction enthalpy
$\Delta_V H$	kJ/mol	Specific enthalpy of vaporization (molar-based)
$\Delta_V H$	kJ/kg	Specific heat of vaporization (mass-based)
k_o	s^{-1} or L/(mol * s)	Maximum rate constant; action constant
k_1	s^{-1}	Rate constant of a first-order reaction
k_2	L/(mol * s)	Rate constant of a second-order reaction
k_x		Equilibrium constant
Kw	W/(m^2 * °C)	Heat transfer coefficient
L	m	Length
L_P	$(mol/L)^{\Phi}$	Solubility product $\left(\phi = \sum v_i\right)$
m_i	kg or t	Mass of substance i *(t=metric ton)*
$\dot{m}_i$	kg/s or t/h	Mass flow of substance i *(t=metric ton)*
M_i	g/mol	Molar mass of substance i
n		Number
n_i	mol	Molar amount of substance i
$\dot{n}_i$	mol/s	Molar flow of substance i
n_{Eq}	mol	Equivalent amount
p	Pa = kg/(m * s^2)	Pressure *(pressure → Pascal)*
P	W = kg * m^2/s^3	Power *(Power → J/s = Watt)*
Q	J = kg * m^2/s^2	Energy, heat, work (Joule) Electrical energy → KWh
$\dot{Q}$	W = kg * m^2/s^3	Power or heat flow *(J/s = Watt)*
r	mol/s	Reaction rate
s	m	Thickness (e.g. pipe wall)
ScF		Scale-up factor

Symbol	Unit	Meaning
$S_{P/A}$		Selectivity of product *P* with respect to reactant *A*
t	s	Time
T	°C	Temperature when calculating with temperature differences
T	K	Absolute temperature (thermodynamics & gas laws)
U	V	Voltage electrolysis cell (volt)
V	m^3 or L	Volume
$\dot{V}$	m^3/s	Volume flow
V_R	m^3 or L	Reactor volume
w	m/s	Flow rate
x		Mole fraction
X_A		Conversion of reactant *A*
$Y_{P/A}$		Yield of product *P* with respect to reactant *A*
α_a, α_1	$W/(m^2 * °C) = W/(m^2 * K)$	Outer heat transfer coefficient
α_i, α_2	$W/(m^2 * °C) = W/(m^2 * K)$	Inner heat transfer coefficient
α	1/°C	Linear expansion coefficient
γ	1/°C	Cubic expansion coefficient
ρ_i	kg/m^3	Density of substance *i*
Δ		Difference e.g. ΔT = temperature difference or Δp = pressure difference
η_E		Electric power efficiency
η_P		Pump efficiency
η		Efficiency
η	Pa * s = kg/(m*s)	Dynamic viscosity
λ	W/(m * °C) = W/(m * K)	Thermal conductivity
τ	s	Residence time
ν_i		Stoichiometric factor
ν_e		Number of electrons gained or lost in an electrolytic reaction according to the reaction equation
ν	m^2/s	Kinematic viscosity

For mass, volume, concentration, temperature, pressure, and other quantities, the index o stands for the conditions at the beginning of a reaction or the inflow to a reactor.

ii. Constants & Conversion Factors

The natural constants frequently used in chemical and engineering calculations are listed below:

Universal gas constant $R = 8.3146$ J/(mol * K)
= 8.3146 kg * m^2/(s^2 * mol * K) = 8.3146 Pa * m^3/(mol * K)
= 0.083146 bar * L/(mol * K) = 8.315 * 10^{-5} bar * m^3/(mol * K)
Gravitational acceleration, acceleration of gravity $g = 9.80665$ m/s^2 ≅ 9.81 m/s^2
Euler's number, Napier's constant $e = 2.718282 \cong 2.718$
Faraday's constant $F = 96{,}485.33$ A * s/Eq ≅ 96,485 A * s/Eq
(Eq = equivalent = quotient of molar amount and electrons carried in or out in the reaction formula)
Circular number $\pi = 3.1416 \cong 3.14$
Loschmidt's number $N = 6.023*10^{23}$ molecules/mol

In older tabular works, diagrams or reports, often outdated non-SI compliant units are used. Unfortunately, in some few non-European countries, especially in the USA, such units are still in use. A special case are pipeline diameters. Here, too, inches are still mostly used in European countries (1″ = 25.4 mm).

Only five sizes with their conversion factors are listed here, which are relevant for the areas covered in this book:

British Thermal Unit → 1 BTU = 1.055 kJ
Calorie → 1 cal = 4.19 J
Atmosphere → 1 atm = 1.013 bar = 101,325 Pa
mmHg → 1 Torr = 1.332 mbar = 1332 Pa
Pounds per Square Inch → 1 PSI ≅ 0.07 bar = 700 Pa

Contents

1 Introduction

A chemical production plant is a complex system in which chemical reactions are carried out specifically and under well-controlled conditions. For this purpose, chemical engineering is required as a combination of different and diverse natural scientific and engineering knowledge. The phenomenological understanding of the processes in a production plant is the prerequisite for its safe and optimal operation. A second, equally important condition for this is the quantitative consideration of the operating conditions and processes. This is where this book comes in. Based on formulas, examples and exercise problems, calculation methods for the most important basic operations in a chemical plant are presented. These are the ideal gas law, the law of mass action (reaction equilibria, pH value, solubility product), material balances (conversion, yield, selectivity), reaction rate, heat (reaction heat, heat capacity, heat transfer, heat balances, reactor stability), electrochemical equivalent, liquid transport and scale-up. In addition, tasks are set from the combination of these different areas. These are relatively simple problems from plant practice that do not require any knowledge of differential or integral calculus for their solution.

Special fields of process engineering, such as rectification, extraction, crystallization, centrifugation, filtration, grinding and many others, have to be dispensed with in this book because they would go beyond its scope. Reference is made to the corresponding specialized textbooks.

At the beginning of the book, general relationships are described, such as magnitudes, numerical values, units as well as some basic mathematical rules, relationships between mechanical magnitudes (force, work, power, pressure, etc.) and selected methods of averaging. This is followed by the formula collection for the different, already mentioned above subareas. In doing so, we refrain from deriving the formulas—reference is also made to the corresponding textbooks. However, in some cases, phenomenological explanations or problem examples are given for better understanding. Finally, the exercise problems follow. First, the text of the

G. Jüptner, *111 Calculation Exercises in the Field of Chemical Technology*,
https://doi.org/10.1007/978-3-662-66920-4_1

task is given, followed by a description of the solution strategy, the detailed solution and, if necessary, an explanation of the result.

1.1 Sizes, Numerical Values, Units

As a rule, a size consists of a numerical value and a unit of measurement (dimension). The indication of production plant parameters, material quantities or corresponding calculation results without the associated unit is meaningless and can lead to expensive or even dangerous misunderstandings. The consistent use of the associated units in the equations for solving a task should become "second nature" for the trainee, since a result with a false dimension is a good indicator of an error when entering the data into the corresponding equations or an error in the calculation. The data with their units can be entered directly into the corresponding formulas or a separate units consideration may be carried out. In the solution examples, the author prefers the former. In the present collection of tasks, SI ("Standard International") units are used consistently in the solution paths. These units were also listed in the previous table of the symbol index. As already mentioned, in tables of older books, data collections and internal company records, often outdated units such as calories, atmospheres, pounds-per-square-inch (PSI), gallons and the like can be found. In a few countries with a highly developed chemical industry, such as the USA, these outdated units are unfortunately still common. Also in Europe, pipe systems are often made of inch pipes and fittings. Such data should be converted consistently into SI units before the start of a calculation in order to avoid the subsequent errors from a confusion of dimensions. The corresponding conversion factors were given in section ii.

A often asked question is how exactly a calculation result must be represented. In general, it can be said that one should not give more digits of a calculated number than can be deduced from the accuracy of the measured data used, otherwise a greater accuracy of the calculated value is implied than corresponds to reality. As a rule of thumb, the representation of a calculation result in chemical engineering with three digits is usually accurate enough, since, for example, the usual deviation of a measurement of a mass flow is about $\pm 1\%$. In process engineering calculations, often approximations are used, which also lead to a slight deviation of the calculated result from reality. In complex calculations, differences occur due to the rounding of the intermediate results. Such slight deviations will also be noticed by the trainee when comparing his calculation results with the numerical values of the solutions given in the book. The author recommends the representation of a calculated numerical value with four to six digits, then possibly a reduction to the "safe" digits, accordingly, for example, a volume $V = 12{,}345\,\mathrm{m}^3 \cong 12{,}3\,\mathrm{m}^3$.

1.2 Important Relationships

In the training for a profession with a scientific focus, the most important mathematical rules, the basic knowledge of geometry and the relationships between various mechanical variables are imparted. They are listed again here as a reminder, in addition to other laws.

1.2.1 Mathematical Rules

In the context of exercise problems, the operators of summation and product formation are used. Knowledge of exponentiation is also required. In some calculations, areas and volumes are calculated. The corresponding basic formulas are summarized here:

Summation sign:

$$\sum_i A_i = \text{Sum of } A_1 + A_2 + A_3 + A_4 \ldots$$

Product sign:

$$\prod_i A_i = \text{Product of } A_1 * A_2 * A_3 * A_4 \ldots$$

Exponentiation:

$$A^a * A^b = A^{a+b}$$

Examples: $10^5 * 10^3 = 10^8 \quad 10^8 * 10^{-2} = 10^6 \quad 10^4 * 10^{-7} = 10^{-3}$

$$A^a / A^b = A^{a-b}$$

Examples: $10^5/10^3 = 10^5 * 10^{-3} = 10^2 \quad 10^{-6}/10^{-8} = 10^{-6} * 10^8 = 10^2 \quad 10^4/10^9 = 10^4 * 10^{-9} = 10^{-5}$

Circular area:	$A = d^2 * \pi/4$	e.g. pipe cross-section
Cylinder shell area:	$A = L * d * \pi$	e.g. heat exchange area of pipe
Volume of cylinder:	$V = L * d^2 * \pi/4$	
Surface of sphere:	$A = d^2 * \pi$	
Volume of sphere:	$V = d^3 * \pi/6$	

It should be noted at this point that the outer diameter of pipes or containers is usually given, while for the calculation of material flows through pipes or for the calculation of the reactor content, the inner diameter is decisive. For this purpose, the double wall thickness must be subtracted from the outer diameter.

1.2.2 Force, Pressure, Work, Heat, Energy and Power

The use of slightly more complex units such as N or J is facilitated if the relationships between the individual variables are understood. Their explanation as a "word equation" may serve as a "mnemonic". The relationships between force, pressure, work, heat, energy and power are given below with the corresponding units:

Force = mass * acceleration

$$F = \text{kg} * \frac{\text{m}}{\text{s}^2} = \text{N(Newton)}$$

Work, energy, heat = force * distance

$$E = Q = \frac{\text{kg} * \text{m}}{\text{s}^2} * \text{m} = \frac{\text{kg} * \text{m}^2}{\text{s}^2} = \text{J (Joule)}$$ electrical energy is also given as kWh

1 kWh = 3600 kJ

Performance = work or energy or heat per unit of time

$$P = \frac{\text{kg} * \text{m}^2}{\text{s}^2} * \frac{1}{\text{s}} = \frac{\text{kg} * \text{m}^2}{\text{s}^3} = \text{W(Watt)}$$

For electrical power:

$$P = \text{voltage} * \text{current} = V * A = W$$

Pressure = force / area

$$p = \frac{\text{kg} * \text{m}}{\text{s}^2} * \frac{1}{\text{m}^2} = \frac{\text{kg}}{\text{s}^2 * \text{m}} = \text{Pa(Pascal)} = 10^{-5}\ \text{bar}$$

1.2.3 More Relationships

Flow rate:

Volume flow per cross-sectional area → $w = \frac{\dot{V}}{A}$
for pipe flow → $w = \frac{4 * \dot{V}}{di^2 * \pi}$
Residence time in a reactor → $\tau = \frac{V_R}{\dot{V}}$

1.2.4 Efficiency

The efficiency is the ratio of useful energy to supplied energy or useful power to supplied power. As an example, the operation of an electric motor is mentioned.

η = useful energy/supplied energy

An electric motor will always perform less mechanical work than it consumes in electrical energy.
Another definition is the proportion of useful energy to total energy generated.

η = useful energy/generated energy

The operation of an electric generator or a steam boiler is given here as two examples: A power generator will always generate less electrical energy than the turbine driving it would expect in terms of the energy generated. The power of a pump will always be less than that of the electric motor driving it. A combustion plant will always provide less useful heat than the mass and calorific value of the fuel would theoretically expect. Many such examples could be added. The efficiency will therefore be in the range of 0 to 1. You will also find percentage figures for the efficiency, which represent the aforementioned ratio multiplied by 100%.

If several devices are connected in series, e.g. an electric motor drives a pump, the efficiencies multiply:

$$\eta_{\text{Total}} = \eta_{\text{Motor}} * \eta_{\text{Pump}}$$

1.2.5 Averaging

The arithmetic mean of a property, e.g. a material stream, whose temperature fluctuates, can be formed if only one substance is present and its mass or the mass stream is constant:

$$\bar{T} = \frac{T_1 + T_2}{2}$$

If several substances are present, they must be weighted with the proportion of each substance. As an example, the averaging of the heat capacity is mentioned:

$$\text{mass related: } \overline{cp} = \frac{\sum_i (cp_i * m_i)}{\sum_i m_i} \text{ or mole related: } \overline{cp} = \frac{\sum_i (cp_i * n_i)}{\sum_i n_i}$$

Mean Pipe Diameter
For the calculation of the heat transfer through pipe walls, the area of the pipe wall is needed. In general, the accuracy of the arithmetic mean of the pipe diameter is sufficient for this purpose:

$$\bar{d}_m = \frac{d_a + d_i}{2}$$

In the case of pipes, in which the total wall thickness (e.g. pipe wall + heat insulation layer) is of a similar size as the inner pipe diameter or more, the arithmetic mean results in too small values. This is the case, for example, with thick stone or glass wool layers for heat insulation. If the total thickness of the pipe wall is equal to the inner pipe diameter, this deviation is about 4%. In such cases, the logarithmic mean of the pipe diameter should be used:

$$\bar{d}_m = \frac{d_a - d_i}{\ln \frac{d_a}{d_i}}$$

Mean Temperature Difference

In heat exchangers, e.g. a double-pipe heat exchanger (see Fig. 1.1), the temperature of the fluid to be cooled decreases along the length of the pipe, while the temperature of the cooling medium increases at the same time. In order to calculate the heat transfer according to the later mentioned formula 30, a mean temperature difference between fluid stream and cooling medium must be determined. Since the temperature profiles of both streams are not linear along the heat exchanger, the mean logarithmic temperature difference is calculated according to the following formula:

$$\Delta\bar{T}_m = \frac{\Delta T_1 - \Delta T_2}{\ln \frac{\Delta T_1}{\Delta T_2}}$$

This is demonstrated in the following example with a double-pipe heat exchanger running in countercurrent operation:

Inner pipe:
Fluid to be cooled
Inlet = 90 °C
Outlet = 30 °C

Outer pipe:
Cooling fluid
Inlet = 10 °C
Outlet = 55 °C

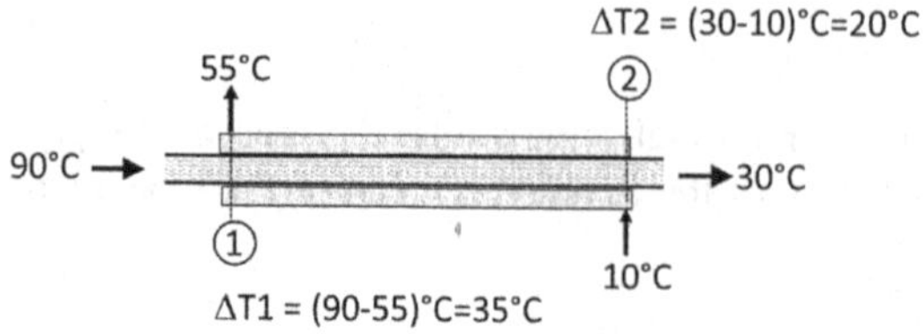

$$\Delta\bar{T}_m = \frac{35°C - 20°C}{ln\frac{35°C}{20°C}} = 26.8°C$$

Fig. 1.1 Calculation of the mean logarithmic temperature difference using the example of a double-pipe heat exchanger in countercurrent operation

Basics and Formula Collection

2

The following formula collection provides the essential relationships that are important for calculations in a chemical plant. As already mentioned, this work does not claim to be a textbook of process engineering, so the derivation of such relationships is omitted. No knowledge of differential or integral calculus is required to solve the exercises, but the mastery of exponents and logarithms is. In some cases, the scientific background is described phenomenologically in order to provide an understanding of the formulas through a certain level of intuition.

The focus of this formula collection is on the general gas law, the law of mass action, kinetics, material balances, heat balances and fluid transport. In addition, simple relationships of scale-up and electrochemistry are also treated.

2.1 Ideal Gas Law

The ideal gas law describes the relationship between volume, pressure and temperature of an ideal gas. However, in reality there are slight deviations from the ideal behavior due to intermolecular forces. However, since the accuracy of the ideal gas law generally meets the needs of calculations in a chemical production plant, it is used exclusively within the scope of this exercise book. The temperature must always be entered in the unit K (Kelvin) here. It should be noted that often the overpressure of a gas is given in operational data. However, the absolute pressure must be used in the gas law.

1 mol of an ideal gas has a volume of 22.414 liters under standard conditions (0 °C = 273.15 K and 1.013 bar). These conditions also apply to the definition of the "standard cubic meter" (Std-m^3). The relationship between volume, pressure and temperature of a gas is given by the following formulas, where instead of the volume the volume flow may be used:

G. Jüptner, *111 Calculation Exercises in the Field of Chemical Technology*, https://doi.org/10.1007/978-3-662-66920-4_2

Formula 1: $\frac{V_0 * p_0}{T_0} = \frac{V_1 * p_1}{T_1} = \frac{V_2 * p_2}{T_2} = \ldots$

or for volume flows $\frac{\dot{V}_0 * p_0}{T_0} = \frac{\dot{V}_1 * p_1}{T_1} = \frac{\dot{V}_2 * p_2}{T_2} = \ldots$

This results in the general gas law for ideal gases according to formula 2. The volume can also be replaced by the volume flow if the molar flow is used instead of the molar number:

Formula 2: $p * V = n * R * T$
or for volume flows $p * \dot{V} = \dot{n} * R * T$
with the general gas constant $R = 8.315 * 10^{-5}$ bar * m³/(mol * K)
→Further units of R are given in Sect. 1.2.

2.2 Law of Mass Action

Many reactions do not proceed to completion, but there is an equilibrium between the forward and backward reactions. Ammonia synthesis, which is an important chemical process, is given as an example here:

$$N_2 + 3H_2 \rightleftarrows 2\,NH_3$$

The position of such equilibrium reactions is described by equilibrium constants. From this, the pressures or concentrations or the molar fractions of the reactants and products in equilibrium can be calculated.

The law of mass action also finds application in a modified form in the solubility product and in the calculation of pH values.

2.2.1 Equilibrium Reactions

The equilibrium constants describe the pressures, concentrations and molar fractions of the reactants and products in equilibrium as follows. Their calculation is carried out using thermodynamic quantities by means of the van't Hoff equation, but this is not elaborated on any further in this book.

Equilibrium constants are pressure- and temperature-dependent according to LeChatelier's "principle of least constraint": In a reaction in which the molar number decreases in the reaction equation, that is, the molar number of the reactants is greater than that of the products, a high reactor pressure pushes the equilibrium in the direction of the products. In an exothermic reaction, the equilibrium is shifted in the direction of the reactants by a high reaction temperature. The definition of the pressure-related equilibrium constant k_p, the concentration-related k_c and the molar fraction-related k_x is given below in general form and using the example of the reaction $3A + B \rightleftarrows 2C + 4D$:

Formula 3a: $k_P = \prod_i p_i^{\upsilon_i}$

Example: $k_p = \frac{p_C^2 * p_D^4}{p_A^3 * p_B^1}$

If the molar number of the reactants is equal to that of the products in the reaction equation, the sum of v_i is equal to 0. It follows from this that k_p is not associated with a unit in such a case. However, if the molar number of the reactants is not equal to that of the products in the reaction equation, k_p acquires a unit of pressure. In the case of the above example, six product molecules are produced from four reactant molecules. The equilibrium constant k_p would thus have the unit bar^2 or Pa^2. Something similar applies to the concentration-related equilibrium constant k_c. For the given example, k_c would have the unit (mol/L)^2. It should be noted that the corresponding concentrations must be used as molar quantities.

Formula 3b: $k_c = \prod_i c_i^{\upsilon_i}$

$$\text{Example:}\quad k_c = \frac{c_C^2 * c_D^4}{c_A^3 * c_B^1}$$

The equilibrium constant of the molar fraction k_x is in principle not associated with a unit.

Formula 3c: $k_x = \prod_i x_i^{\upsilon_i}$

Example: $k_x = \frac{x_C^2 * x_D^4}{x_A^3 * x_B^1}$

The equilibrium constants k_p, k_c and k_x can be converted into each other using the following formula:

Formula 3d: $k_p = k_c * (R * T)^{\sum \upsilon_i} = k_x * p^{\sum \upsilon_i}$

As already mentioned, the equilibrium constant is temperature-dependent. The temperature dependence of the equilibrium constant k_p is given by the reaction enthalpy according to the van't Hoff equation.

Formula 3e: $\frac{k_{p2}}{k_{p1}} = e^{\frac{\Delta_R H}{R} * \left(\frac{1}{T_1} - \frac{1}{T_2}\right)}$

2.2.2 pH and pK$_a$ Value, Buffer Solutions

The pH value is the negative decadic logarithm of the hydrogen ion concentration (in mol/L) of a solution, while the less frequently used pOH value is the negative decadic logarithm of the hydroxyl ion concentration (mol/L):

Formula 4a: $c_{H^+} = 10^{-pH}$ $\quad c_{OH^-} = 10^{-pOH}$ with the concentrations as mol/L

Via the law of mass action, one arrives at the ion product of water:

Formula 4b: $c_{H^+} * c_{OH^-} = 10^{-14}\left(\frac{mol}{L}\right)^2$ at 25 °C

This means that the sum of pH and pOH is 14:

Formula 4c: $pH + pOH = 14$

Strictly speaking, these relationships only apply to complete dissociation of an acid or a base, which is largely true for "strong" acids and bases in diluted form. The "strength" of an acid or a base is described by the acid constant pK_a, which can be taken from the relevant tables. The lower this value is, the "stronger" an acid is, the higher it is, the "stronger" a base is.

Formula 5a: $k_a = \frac{c_{H^+} * c_{A^-}}{c_{HA}} \quad k_a = 10^{-pK_a}$

The pH value of an acid of the type $HX \rightleftarrows H^+X^-$ is calculated according to

Formula 5b: $pH = (pk_a - lg[c_A])/2$ (with the acid- or base-concentration c_A in mol/L).

The pH value of a base of the type $YOH \rightleftarrows Y^+OH^-$ is calculated according to

Formula 5c: $pH = (pk_a + lg[c_B] + 14)/2$ (with the base-concentration c_B as mole/L).

The pH value of buffer solutions of weak acids and their salts with strong bases, e.g. acetic acid and sodium acetate, is calculated according to the Henderson-Hasselbalch equation as follows:

Formula 5d: $pH = pk_a + lg\frac{c_{salt}}{c_{acid}}$

Analogously, the pH value of buffer solutions of weak bases and their salts with strong acids, e.g. ammonium hydroxide and ammonium chloride, follows:

Formula 5e: $pH = pk_a + lg\frac{c_{base}}{c_{salt}}$

2.2.3 Solubility Product

The solubility product is based on the law of mass action and describes concentrations of anions and cations in the saturated solution of a sparingly soluble substance:

Formula 6: $L_P = \prod_i c_i^{\upsilon_i}$

This is described by the following examples:

Example 1: $A^+ + B^- \rightarrow AB \downarrow$

$$L_{P-AB} = c_{A^+}^1 * c_{B^-}^1 = c_{A^+} * c_{B^-}$$

Example 2: $A^+ + 2B^- \rightarrow AB_2 \downarrow$

$$L_{P-AB} = c_{A^+}^1 * c_{B^-}^2 = c_{A^+} * c_{B^-}^2$$

Data for the solubility products of various substances can be found in relevant tabular works.

2.3 Material Balances

In chemical production operations, calculations are usually made in terms of mass, mass flow or weight percent (wt%). This is partly due to the fact that raw materials are usually purchased on a mass basis, that is, in kilograms or metric tons. The same applies to the sale of products. The production capacity of a plant is often given in "annual metric tons" = t / year. However, many chemical-technical calculations have to be carried out on a molar basis. With material balances, the principle applies that the sum of the mass of material introduced into a system (e.g. a reactor) is always equal to the total mass that the system leaves again. This is different with molar balances, as the total molar number can change by chemical reactions. The relationships between mass- and molar-related quantities are presented in this section.

It should also be pointed out that the concentrations in ppm or ppb can be either mass- or volume-related. ppm stands for "parts per million", that is, one part per million parts (10^6). 1 ppm of substance A means that 1 kg of a substance contains 1 mg of substance A or 1 m^3 of a gas mixture contains 1 mL of gas A. ppb stands for "parts per billion", that is, one part per billion parts (10^9). Even smaller concentrations are also given in "parts per trillion" (10^{12}), that is, ppt.

The progress or status of a reaction is quantified by the conversion, the effectiveness by the yield and the selectivity. These three process variables are generally calculated using molar parameters (molar number, molar flow, molar concentration).

2.3.1 Mass Balances and Stoichiometric Balances

The following describes the calculation of molar and mass-based concentrations as well as molar fractions.

Molar concentration = molar amount-*A*/volume →mol/L

or molar flow-*A*/volume flow

Formula 7a: $c_A = \frac{n_A}{V} = \frac{\dot{n}_A}{\dot{V}}$

Molar fraction = molar amount substance-*A*/total molar amount *i*

Formula 7b: $x = \frac{n_A}{\sum_i n_i}$

Mass concentration = mass-*A*/volume →kg/m^3

Formula 7c: $C_A = \frac{m_A}{V}$

Mass percentage = 100 * mass-*A*/total mass →wt%

Formula 7d: $C'_A = 100 * \frac{m_A}{\sum_i m_i}$

Mass & molar amount:

Formula 7e: mass = number of moles * molar mass → $m_i = n_i * M_i$ for massflow → $\dot{m}_i = \dot{n}_i * M_i$

Mass & volume:

Formula 8a: mass = volume * density → $m = V * \rho$

Formula 8b: for massflow → $\dot{m} = \dot{V} * \rho$

2.3.2 Yield, Selectivity and Conversion

In chemical reactions, usually two or more feedstocks (→reactants, educts) are involved, and often several products are obtained, e.g. in the reaction $A+2B \rightarrow C+3D$

Thus, both for the reactant *A* and for the reactant *B* a conversion can be formulated. Similar applies to the yields and the selectivities of the substances *C* respectively *D*. Here, in addition, a reference to the feedstock is necessary. In principle, these three variables are calculated from the molar amounts or the molar flow, but never from the mass of the feedstock and the products. In the case of a constant volume reaction, the molar concentrations can be used instead of the molar amount, but this special case is not considered here. To calculate the conversion, the yield and the selectivity, the educt is usually chosen as the deficit component. In the case of a strong price difference of the reactants, the educt of the significantly higher price is chosen in rare cases, even if it is not the deficit component. When calculating the yields and selectivities, the product, which is an intermediate product and is further processed, or the most valuable product is usually considered.

It is common to represent the conversion, the yield and the selectivity as a relative share of 1 or even in percent. For example, a conversion of $X_A = 0.747$ can also be expressed as $X_A = 74.7\%$.

The *conversion* X_E indicates the molar proportion of a starting material (reactant, feedstock) that has reacted. The following applies to the conversion of a feedstock:

Calculation of the conversion for the batch operation →

Formula 9a: $X_E = \frac{n_{Eo} - n_E}{n_{Eo}}$

e.g. for the feedstock A of the reaction equation given as an example above $X_A = \frac{n_{Ao} - n_A}{n_{Ao}}$

Calculation of the conversion for the continuous operation →

Formula 9b: $X_E = \frac{\dot{n}_{Eo} - \dot{n}_E}{\dot{n}_{Eo}}$

e.g. for the feedstock A of the reaction equation given above $X_A = \frac{\dot{n}_{Ao} - \dot{n}_A}{\dot{n}_{Ao}}$

Here n_{Ao} stands for the molar amount of A at the beginning of the reaction ($t = 0$) and n_A for the molar amount of A at the end of the reaction time (t).

If the volume does not change during the reaction, the conversion can be calculated analogously by means of the concentrations →

Formula 9c: $X_E = \frac{c_{Eo} - c_E}{c_{Eo}}$

The **yield** $Y_{P/E}$ indicates how many moles of product (n_P) have been formed relative to the moles of feedstock (n_{Eo}) used. The stoichiometric numbers ν_i from the chemical equation must also be taken into account. The feedstocks are negative by definition, the products positive. As an example, for the reaction equation given above

ν_i: for reactant $A \rightarrow \nu_A = -1$
for reactant $B \rightarrow \nu_B = -2$
for product $C \rightarrow \nu_C = +1$
for product $D \rightarrow \nu_D = +3$

Often, in incomplete reactions, unreacted reactants are removed from the product stream and added to the feed of fresh reactants. Thus, such a feed mixture can also contain a certain amount of product. This is taken into account by the term n_{Po}. In the case of the use of pure reactants, $n_{Po} = 0$.

Yield calculation for the batch operation →

Formula 10a: $Y_{P/E} = \frac{\nu_E * (n_{Po} - n_P)}{\nu_p * n_{Eo}}$

For the reaction given above, in principle four yields can be defined:

For product C based on reactant $A \to Y_{C/A}$ and based on reactant $B \to Y_{C/B}$.

The same applies to product D with respect to reactant $A \to Y_{D/A}$ and with respect to reactant $B \to Y_{D/B}$.

However, as already mentioned, the most meaningful definition of the yield is chosen.

As examples, the yield of product D with respect to reactant A and C with respect to reactant B are given:

$$Y_{D/A} = \frac{\upsilon_A * (n_{Do} - n_D)}{\upsilon_D * n_{Ao}} \rightarrow Y_{D/A} = \frac{-1 * (n_{Do} - n_D)}{3 * n_{Ao}} \text{ und}$$

$$Y_{C/B} = \frac{\upsilon_B * (n_{Co} - n_C)}{\upsilon_C * n_{Bo}} \rightarrow Y_{C/B} = \frac{-2 * (n_{Co} - n_C)}{1 * n_{Bo}}$$

In a similar way, the yields $Y_{C/A}$ or $Y_{D/B}$ can be calculated.

In the case of continuous operation, the molar flows $\dot{n}_i$ are to be used instead of the molar numbers n_i analogous to the formulas of conversion:

Yield calculation for continuous operation →

Formula 10b: $Y_{P/E} = \frac{\upsilon_E * (\dot{n}_{Po} - \dot{n}_P)}{\upsilon_p * \dot{n}_{Eo}}$

If the volume does not change during the reaction, the yield can be calculated analogously by means of the concentrations →

Formula 10c: $Y_{P/E} = \frac{\upsilon_E * (c_{Po} - c_P)}{\upsilon_p * c_{Eo}}$

The **selectivity** describes the yield of a product based on the molar amount of reactant, which has reacted. While the yield does not take into account how much reactant is lost in side reactions, but only the ratio of product formed to reactant used, the selectivity describes how much reacted educt leads to the desired product. This quantifies the influence of side reactions on the yield.

Selectivity calculation for batch operation →

Formula 11a: $S_{P/E} = \frac{\upsilon_E * (n_{Po} - n_P)}{\upsilon_p * (n_{Eo} - n_E)}$

As with the yield, four selectivities can in principle be defined for the reaction given as an example above:

For product C based on reactant $A \to S_{C/A}$ and based on reactant $B \to S_{C/B}$.

The same applies to product D with respect to reactant $A \to S_{D/A}$ and with respect to reactant $B \to S_{D/B}$.

However, as mentioned above, the most meaningful definition of selectivity is chosen.

Corresponding examples of the calculation of the selectivities for the chemical equation given as an example above are as follows:

$$S_{D/A} = \frac{\upsilon_A * (n_{Do} - n_D)}{\upsilon_D * (n_{Ao} - n_A)} \rightarrow S_{D/A} = \frac{-1 * (n_{Do} - n_D)}{3 * (n_{Ao} - n_A)} \text{ und}$$

$$S_{C/B} = \frac{\upsilon_B * (n_{Co} - n_C)}{\upsilon_C * (n_{Bo} - n_B)} \rightarrow S_{C/B} = \frac{-2 * (n_{Co} - n_C)}{1 * (n_{Bo} - n_B)}$$

Selectivity calculation for continuous operation →

Formula 11b: $S_{P/E} = \frac{\upsilon_E * (\dot{n}_{Po} - \dot{n}_P)}{\upsilon_p * (\dot{n}_{Eo} - \dot{n}_E)}$

If the volume does not change during the reaction, the selectivity can be calculated using the concentrations in a similar manner →

Formula 11c: $S_{P/E} = \frac{\upsilon_E * (c_{Po} - c_P)}{\upsilon_p * (c_{Eo} - c_E)}$

2.3.3 Reaction Rate

The reaction rate is the decrease in the concentration of a reactant A or the increase in the concentration of a product P per unit of time for reactions that are volume-constant, as shown in Formula 12 as a simple differential equation.

Formula 12: $r = \frac{dc_A}{dt}$

For the reaction rate r the time law of the corresponding reaction is now inserted and the differential equation is solved. This can lead to very complicated equations, in particular for complex reactions, such as equilibrium reactions ($A + B \rightleftarrows C + D$), consecutive reactions ($A + B \rightarrow C \quad C + A \rightarrow D$) and other complex reaction processes. These and the corresponding derivatives may be taken from the corresponding textbooks. Therefore, only solutions for four simple reaction types are given here. They only apply to homogeneous systems, i.e. only in the case of one reaction phase, but not for multi-phase reactions. In this case, batch operation and continuous reactors, such as the continuous stirred tank reactor (CSTR) and the plug flow reactor (PFR) were considered. The solutions only apply, if no other reactions take place in addition to the reactions specified. If other reactions take place, such as side reactions, equilibrium reactions, consecutive reactions and others, the solutions of the differential equations are very complicated functions, some of which can no longer be solved explicitly. These cases exceed the scope of this book and are therefore not treated here.

The corresponding differential equations of the time laws were solved in the limits $t=0$ with $c_A = c_{Ao}$ and $t=t$ with $c_A = c_A$. Starting from the initial conditions $t=0$ and $c_A = c_{Ao}$, the concentration of A is calculated as a function of the reaction time t for batch operation or the residence time τ for continuous reactors. The unit of the rate constant k for reactions of 1st order is s^{-1} and for reactions of

2nd order L/(mol * s). The rate constant is temperature-dependent, as explained in more detail later in this chapter.

Reaction 1st Order: A → C

Time law →
Formula 13a: $r = k * c_A = \frac{dc_A}{dt}$
Solution:
Batch operation: **Formula 13b**: $\ln \frac{c_{Ao}}{c_A} = k * t$

Formula 13c: $c_A = c_{Ao} * e^{-k*t}$

PFR: **Formula 13d**: $c_A = c_{Ao} * e^{-k*\tau}$

CSTR: **Formula 13e**: $c_A = \frac{c_{Ao}}{1+k*\tau}$

Reaction 2nd Order: 2A → C

Time law →
Formula 14a: $r = k * c_A^2 = \frac{dc_A}{dt}$
Solution:
Batch operation: **Formula 14b**: $\frac{1}{c_A} - \frac{1}{c_{Ao}} = k * t$

Formula 14c: $c_A = \frac{c_{Ao}}{(1+c_{Ao}*k*t)}$

PFR: **Formula 14d:** $c_A = \frac{c_{Ao}}{(1+c_{Ao}*k*\tau)}$

CSTR: **Formula 14e:** $c_A = \sqrt{\frac{c_{Ao}}{k*\tau} + \frac{1}{4*k^2*\tau^2}} - \frac{1}{2*k*\tau}$

Reaction 2nd order:A + B → C (Special case of the stoichiometric ratio of $c_{Ao}=c_{Bo}$)

In the event that the initial concentrations of the reactants *A* and *B* are used in the stoichiometric ratio of the reaction equation, the equation is simplified to Formula 15a. The time-dependent concentration profiles of the reactants *A* and *B* are identical.

Time law →
Formula 15a: $r = k * c_A * c_B = k * c_A^2 = k * c_B^2$
Solution:
Batch operation: **Formula 15b:** $c_A = c_B = \frac{c_{Ao}}{(1+c_{Ao}*k*t)} = \frac{c_{Bo}}{(1+c_{Bo}*k*t)}$

PFR: **Formula 15c:** $c_A = c_B = \frac{c_{Ao}}{(1+c_{Ao}*k*\tau)} = \frac{c_{Bo}}{(1+c_{Bo}*k*\tau)}$

CSTR: **Formula 15d:** $c_A = c_B = \sqrt{\frac{c_{Ao}}{k*\tau} + \frac{1}{4*k^2*\tau^2}} - \frac{1}{2*k*\tau} = \sqrt{\frac{c_{Bo}}{k*\tau} + \frac{1}{4*k^2*\tau^2}} - \frac{1}{2*k*\tau}$

Reaction 2. Order:A + B → C (c_{Ao} and c_{Bo} are not in stoichiometric ratio)

Time law →
Formula 16a: $r = k * c_A * c_B$
Solution:
Batch: **Formula 16b:** $\frac{1}{c_{Ao}-c_{Bo}} * \ln\left\{\frac{c_{Bo} * c_A}{c_{Ao} * (c_{Bo}-c_{A0}+c_A)}\right\} = k * t$

Formula 16c: $c_A = \frac{\theta * c_{Ao} * (c_{Bo}-c_{Ao})}{c_{Bo}-\theta * c_{Ao}}$ with $\theta = e^{(c_{Ao}-c_{Bo}) * k * t}$

PFR: **Formula 16d:** $c_A = \frac{\theta * c_{Ao} * (c_{Bo}-c_{Ao})}{c_{Bo}-\theta * c_{Ao}}$ with $\theta = e^{(c_{Ao}-c_{Bo}) * k * \tau}$

A solution for the CSTR is not given at this point due to its complexity.

The reaction rate depends on the temperature. With increasing temperature, the rate constant increases exponentially. The often-cited very rough rule of thumb that the reaction rate doubles at a temperature increase of 5 °C must be avoided due to inaccuracy and uncertainty of the statement. For homogeneous reactions, that is, for non-catalyzed reactions in one phase, the law of Arrhenius applies to an increase in the rate constant with increasing temperature:

Formula 17a: $k = k_0 * e^{-E_A/R * T}$

With k_0 as the theoretically maximum rate constant, E_A as the activation energy, R as the universal gas constant and T as the absolute temperature. The higher the activation energy E_A, the more temperature-sensitive the reaction is. The following formula can be used to calculate the factor of reaction rates (F_T) if the temperature is changed from T_1 to T_2.

Formula 17b: $F_T = \frac{k_2}{k_1} = e^{\frac{E_A}{R} * \left(\frac{1}{T_1} - \frac{1}{T_2}\right)}$

2.4 Heat

To heat a substance, energy must be supplied, while heat is removed to cool it. Every material has a specific heat capacity, with which the amount of heat required for this can be calculated depending on the temperature increase or decrease.

When a material is melted, the supply of heat is required for this. If a liquid freezes, heat is released. Every meltable substance has a specific heat of melting, which is equal to the heat of solidification. Something similar happens when a liquid evaporates and condenses: The transition from liquid to gaseous takes place with the consumption of heat, while condensation releases heat. Again, this amount of heat is specific to the substance.

When a substance is dissolved, heat can be released (e.g. potassium hydroxide in water) or consumed (e.g. potassium iodide in water).

Every chemical reaction proceeds more or less with the development of heat (exothermic reaction) or the consumption of heat (endothermic reaction). Thus, a quantification of such amounts of heat is necessary in order to be able to carry out reactions in a safe, optimum and energy-saving way. The speed of heat release increases with an increasing reaction rate. If the system is not sufficiently cooled, its temperature rises. In turn, the reaction rate increases exponentially with increasing temperature, thus further increasing the temperature. This is the scenario for a runaway reaction. In a stable system, it must be ensured that the released heat is discharged by cooling. Important for this are the knowledge of the laws of heat conduction and transmission as well as the formulation of heat balances. As shown below, stability criteria such as the adiabatic temperature increase or the calculation of stable operating points can be derived from this.

In contrast to gases, solids and liquids are usually materials with an extremely low compressibility. Thus, the expansion of such substances with increasing temperature can lead to damage, if they are not given enough space to expand thermally. The same applies to falling temperatures.

The topics mentioned will be dealt with quantitatively in the following course of this chapter.

2.4.1 Heating and Cooling

The amount of heat to be supplied or the amount of heat to be dissipated for heating or cooling a substance can be calculated from the amount of the substance, the specific heat capacity (*cp*) or the molar heat capacity(*cp*) and the temperature difference before (T_0) to after (T_1). cp describes the amount of heat (kJ) required to heat or cool 1 kg or 1 mol of a substance by 1 °C (i.e. 1 K → Kelvin). The specific heat or molar heat is to be taken from the corresponding tables. Caution is advised with older tables that are partly still based on calories. But not only for this reason is it necessary to pay attention to the unit of cp: The specific heat capacity is mass-related and therefore has the unit kJ/(kg * °C), while the molar heat capacity is mole-related and therefore this unit is kJ/(mol * °C). It should be noted that cp itself is to some extent temperature-dependent, which is taken into account in some tabular works. However, this practical book does not go into this and calculates with constant values of the heat capacity in good approximation. With regard to the unit of cp, the reference to °C is sometimes replaced by a reference to K (Kelvin). Since the calculations are based on temperature differences and the expansion of the °C scale is the same as that of the K scale, both values of the heat capacity are identical.

The amounts of heat for a heating or cooling process are calculated as follows.

Mass-related (unit $cp \rightarrow \frac{\mathrm{kJ}}{\mathrm{kg} * {}^\circ\mathrm{C}} = \frac{\mathrm{kJ}}{\mathrm{kg} * \mathrm{K}}$)

For batch operation:
If only one substance is present →**Formula 18a:** $Q = m * cp * (T_1 - T_0)$
For a mixture of substances →**Formula 18b:** $Q = \sum_i [m_i * cp_i] * (T_1 - T_0)$

For continuous operation:
If only one substance is present →**Formula 19a:** $\dot{Q} = \dot{m} * cp * (T_1 - T_0)$
For a mixture of substances: →**Formula 19b:** $\dot{Q} = \sum_i [\dot{m}_i * cp_i] * (T_1 - T_0)$

Mole-related (unit $cp \rightarrow \frac{\text{kJ}}{\text{mol} * {}^\circ\text{C}} = \frac{\text{kJ}}{\text{mol} * \text{K}}$)

For batch operation:
If only one substance is present →**Formula 20a:** $Q = n * cp * (T_1 - T_0)$
For a mixture of substances →**Formula 20b:** $Q = \sum_i [n_i * cp_i] * (T_1 - T_0)$

For continuous operation:
If only one substance is present →**Formula 21a:** $\dot{Q} = \dot{n} * cp * (T_1 - T_0)$
If a mixture of several substances is present →**Formula 21b:**
$\dot{Q} = \sum_i [\dot{n}_i * cp_i] * (T_1 - T_0)$
If $T_1 > T_0$, it is a heating process.
If $T_1 < T_0$, it is a cooling process.

2.4.2 Melting and Vaporization Heat

A melting and a vaporization process have one thing in common: When the substance is heated, the temperature rises to the melting point or boiling point and remains there, even though more heat is being absorbed. This situation remains until the substance is completely melted or vaporized. Similarly, but in the reverse order, this is the case for the freezing or the condensation process. The specific heat of melting or melting enthalpy $\Delta_S H$ indicates how much heat must be supplied to melt one kg or one mole of a substance, or is released during freezing. The same applies to the specific heat of vaporization or vaporization enthalpy for the vaporization or condensation process. Therefore, attention must be paid to the unit kJ/kg or kJ/mol given in the corresponding tabular data, in order to insert the mass or the number of moles in the following equations. Due to the mass or mole reference of $\Delta_S H$ and $\Delta_V H$, the relationships for melting or freezing processes and vaporization or condensation processes result.

The **amounts of heat for a melting or freezing process** are calculated as follows, whereby for the melting process a negative value of heat Q or $\dot{Q}$ and for the freezing process a positive value results:

Based on Mass
For batch operation →**Formula 22a:** $Q = m * \Delta_S H$
For continuous operation →**Formula 22b:** $\dot{Q} = \dot{m} * \Delta_S H$

Mol Related

For the batch operation →**Formula 23a:** $Q = n * \mathbf{\Delta}_S H$

For the continuous operation →**Formula 23b:** $\dot{Q} = \dot{n} * \mathbf{\Delta}_S H$

Similarly, the heat quantities are calculated for an **evaporation** or a **condensation** process as follows, with a negative value of heat Q or $\dot{Q}$ resulting for the evaporation process and a positive value for the condensation process:

Mass Related

For the batch operation →**Formula 24a:** $Q = m * \mathbf{\Delta}_V H$

For the continuous operation →**Formula 24b:** $\dot{Q} = \dot{m} * \mathbf{\Delta}_V H$

Mol Related

For the batch operation →**Formula 25a:** $Q = n * \mathbf{\Delta}_V H$

For the continuous operation →**Formula 25b:** $\dot{Q} = \dot{n} * \mathbf{\Delta}_V H$

2.4.3 Solution Heat

The heat that is consumed or released during the dissolution of a substance is calculated from the amount of substance dissolved, multiplied by the specific heat of solution according to

Formula 26a: $Q = -n * \mathbf{\Delta}_L H$

Formula 26b: $\dot{Q} = -\dot{n} * \mathbf{\Delta}_L H$

If the enthalpy of solution is positive, the liquid cools during the dissolution process, if it is negative, it heats up. The heat of solution has the opposite mathematical sign of the enthalpy of solution. Thus, a negative specific heat of solution means cooling of the liquid during the solid's dissolution and a positive heat of solution means heating. In the literature, the enthalpy of solution is given as mol-related, while both mass-related and mole-related values can be found for the specific heat of solution.

Since the heat of solution depends, among other things, on the crystal structure of the solid, it must be ensured when working with corresponding tabular values that the corresponding state form of the solid is taken into account.

2.4.4 Heat of Reaction and Reaction Enthalpy

The reaction heat Q is calculated from the amount of moles Δn reacted and the corresponding reaction enthalpy $\Delta_R H$. Similarly, the heat output $\dot{Q}$ is calculated for continuous processes from the difference of the molar flow reactor-in to reactor-ex related to the residence time τ (which corresponds to the reaction rate) and the reaction enthalpy:

Formula 27a: $Q = -\Delta n * \Delta_R H$

Formula 27b: $\dot{Q} = -\frac{\Delta n}{\Delta \tau} * \Delta_R H$

Formula 27c: $\dot{Q} = -\Delta \dot{n} * \Delta_R H$

Formula 27d: $\dot{Q} = -r * \Delta_R H$

It is important to pay attention to which reactant the value of the reaction enthalpy refers to. A reference of the reaction enthalpy to a reaction equation is also possible. The term kJ/FC = kJ/formula conversion is often used here, which represents the unit kJ/mol. An exothermic reaction has a negative enthalpy and releases reaction heat, while an endothermic reaction is associated with a positive enthalpy and requires the supply of reaction heat.

The **reaction enthalpy** $\Delta_R H_0$ at standard conditions is calculated according to the first and second law of thermodynamics from the formation enthalpies $\Delta_f H_{i0}$ of the reactants and products under the respective standard conditions (0 °C, 1 bar) according to eq. 28 with ν_i as the corresponding stoichiometric sign in the reaction equation. The signs ν_i of the reactants are negative, those of the products positive, as shown in the following example:

$$3A + 2B = C + 4D \rightarrow \nu_A = -3$$

$$\nu_B = -2$$

$$\nu_C = +1$$

$$\nu_D = +4$$

Formula 28: $\Delta_R H_0 = \sum_i \left(\upsilon_i * \Delta_f H_{i0}\right)$

The standard formation enthalpies can be taken from richly available extensive tabular works. However, as shown in the exercise problems, it is possible to determine the formation enthalpy of a substance by the relatively simple determination of its combustion heat. The standard formation enthalpies of the elements have been set to zero by definition. If an element is present in different states (e.g. carbon as an amorphous substance, as graphite or as diamond), there are deviations from this rule.

If a reaction does not take place under standard conditions, the exact reaction enthalpy can be adjusted according to eq. 29 by means of the molar heat capacities cp_i of reactants and products:

Formula 29: $\Delta_R H_T = \sum_i [\upsilon_i * \Delta_f H_{i0} + \upsilon_i * cp_i * (T - 298,15\,\text{K})]$

A special quantitative description of the reaction heat of a combustion process is given by the **high and low heating value**. For solid and liquid fuels, the number of moles corresponding to 1 kg of fuel is used in Formula 27a: $n = \frac{\text{m}}{M_{\text{fuel}}} = \frac{1\,\text{kg}}{M_{\text{fuel}}}$

The heating value per standard cubic meter (standard-m^3) of gaseous fuels applies to ideal gases: $n = \frac{p*V}{R*T}$

Thus for 1 std-m^3 (1.013 bar; 273 °C): $n = \frac{1.013\,\text{bar} * 1\,\text{m}^3 * \text{mol} * \text{K}}{8.315 * 10^{-5}\,\text{bar} * \text{m}^3 * 273\,\text{K}} = 44.63\frac{\text{mol}}{\text{m}^3}$

The calorific value, formerly also referred to as the high heating value *HHV* or high combustion value HCV, indicates the reaction heat for 1 kg of solid fuel or 1 std-m^3 (standard conditions: 0 °C, 1.013 bar) of gaseous fuel in accordance with Formula 27a, in which all water present in the fuel and formed during combustion is condensed. To calculate it from the formation enthalpies according to Formula 28, the formation enthalpy of gaseous water must be used. In technology the low heating value *LHV* or low combustion value LCV, is used. With this definition of the heating value, it is assumed that all water leaves the combustion process in gaseous form. Thus, the formation enthalpy of water vapor must be used to calculate it according to Formula 28.

The conversion from *HHV* to *LHV* can be done by subtracting the evaporation heat of the water formed (2450 kJ/kg = 44.1 kJ/mol) based on 1 kg or 1 std-m^3 of fuel, calculated according to Formula 24a.

2.4.5 Heat Conduction

Heat flow by thermal conductivity is caused by different temperatures on both sides of a wall, that is, due to a temperature difference ΔT. Such a heat flow $\dot{Q}$ through a wall is directly proportional to this temperature difference as well as the heat transmission area and can be calculated using Formula 30. The heat transfer coefficient (heat transfer number) K_W is included in the formula as an additional factor.

Formula 30: $\dot{Q} = K_W * A * \Delta T$

In the case of pipelines, heat exchangers, reactors and others, the heat transfer coefficient is usually significantly smaller than the thermal conductivity of the wall material itself would suggest on the first view. The reason for this is the formation of thin laminar sub-layers of the gas or liquid phases inside and outside the wall (see Fig. 2.1). While there are no temperature differences in the core of the fluid phases, the laminar boundary layers show a temperature profile each and, in addition to the wall material, represent an additional resistance to heat flow, which is taken into account by the corresponding calculation of the heat transfer coefficient K_W according to formula 31a, b.

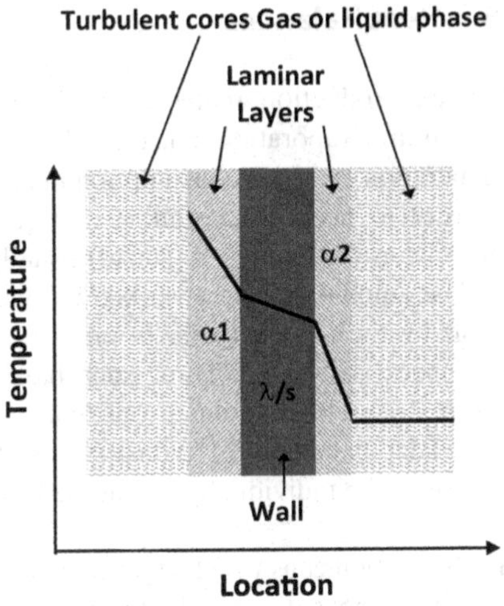

Fig. 2.1 Situation of the heat transport on a wall, which is surrounded by fluid on both sides. Formation of laminar sublayers and the corresponding temperature profile

The heat transfer coefficient K_W results from the addition of the reciprocal heat resistances, as shown in formula 31a for a wall made of a single material, or formula 31b for a multilayer wall, e.g. an insulated metal pipe. Here, s_i are the wall thicknesses, λ_i are the respective thermal conductivities of the wall materials, and α_1 and α_2 are the heat transfer coefficients of the inner and outer laminar fluid films on the walls.

Formula 31a: $\frac{1}{K_W} = \frac{1}{\alpha_1} + \frac{s}{\lambda} + \frac{1}{\alpha_2}$

Formula 31b: $\frac{1}{K_W} = \frac{1}{\alpha_1} + \sum_i \frac{s_i}{\lambda_i} + \frac{1}{\alpha_2}$

While the thermal conductivity of the wall material can be taken from various tables, the heat transfer coefficients α depend on the flow conditions of the fluids (the more turbulent the flow, the greater α), their viscosity, their thermal conductivity and their temperature. The calculation of the heat transfer coefficients in this way using dimensionless numbers such as the Nusselt, Reynolds and Prandtl numbers is complicated. Therefore, this will not be further discussed here and reference will be made to corresponding textbooks on process engineering instead.

With regard to the unit of K_W and the thermal conductivity, reference is sometimes made to °C instead of Kelvin. Since the calculations are based on temperature differences, and the expansion of the °C scale is the same as the Kelvin scale, both are identical in the case of the heat transfer coefficient and the thermal conductivity.

$$K_W \rightarrow \frac{\mathrm{W}}{\mathrm{m}^2 * {}^\circ\mathrm{C}} = \frac{\mathrm{W}}{\mathrm{m}^2 * \mathrm{K}}$$

$$\lambda \rightarrow \frac{\mathrm{W}}{\mathrm{m} * {}^\circ\mathrm{C}} = \frac{\mathrm{W}}{\mathrm{m} * \mathrm{K}}$$

2.4.6 Heat Balances

In chemical production, material streams are heated or cooled. Heat is released by condensation. Evaporation consumes heat. And there are many more examples of heat generation and heat consumption in chemical production.

In order to heat a material stream (e.g. the feed to a distillation column), another material stream (e.g. steam) usually transfers heat to it and by this cools down. The amount of heat absorbed by one stream is equal to the amount of heat given off by the other stream. In order to describe such processes quantitatively, heat balances are set up. Three such balances are developed exemplarily below. However, it should be pointed out that this only represents a small selection of possible heat balances. In industrial practice, there are manifold task settings which have to be individually considered and solved.

Example 1: By means of a heat exchanger, the liquid stream $\dot{m}_A$ of the initial temperature T_{A0} is heated to the final temperature T_A. This is done by the liquid stream $\dot{m}_B$ of an initial temperature T_{B0}, which is cooled to the final temperature T_B. The heat capacities cp_A and cp_B are known.

Heat flow which stream A absorbs: $\dot{Q}_A = \dot{m}_A * cp_A * (T_A - T_{A0})$ →see formula 19a

Heat flow which stream B gives off: $\dot{Q}_B = \dot{m}_B * cp_B * (T_{B0} - T_B)$

With $\dot{Q}_A = \dot{Q}_B$,$\dot{Q} = \dot{m}_A * cp_A * (T_A - T_{A0}) = \dot{m}_B * cp_B * (T_{B0} - T_B)$ results

This equation may be rearranged according to the value to be calculated (one of the temperatures or the mass flow of A or B).

Example 2: The mass m_A, with a heat capacity of cp_A, is heated from T_{A0} to the temperature T_A by means of saturated steam (S) with an evaporation heat (= condensation heat) of $\Delta_V H$. The steam condensate leaves the system at the temperature of the saturated steam.

Heat released during the condensation of the steam: $Q_S = m_S * \Delta_V H$ →see formula 24a

Heat absorbed by substance A: $Q_A = m_A * cp_A * (T_A - T_{A0})$ →see formula 18a

With $Q_S = Q_A$ we get $Q = m_S * \Delta_V H = m_A * cp_A * (T_A - T_{A0})$

This equation can be rearranged according to the desired value (the temperatures or the mass of the steam or the mass of substance A).

Example 3: A solid substance B at a temperature T_{B0} below its melting point of T_{BS} is to be brought to the temperature T_B above its melting point. The specific heats of the solid cp_{Bsolid} and the melt cp_{Bliq} are given, as is the heat of melting of B ($\Delta_S H_B$). The amount of heat required for this purpose is to be supplied by a certain amount of hot oil (m_H) at the initial temperature T_{H0} and the final temperature T_H . The specific heat of the hot oil cp_H is known.

Heating substance B from T_{B0} to T_B consists of three steps:

Step 1 →The solid B is heated from its initial temperature T_{B0} to the melting temperature.
$Q_{B1} = m_B * cp_{\mathrm{Bsolid}} * (T_{BS} - T_{B0})$ →see formula 18a

Step 2 →The solid B melts to
$Q_{B2} = m_B * \Delta_S H_B$ →see formula 22a

Step 3 →The melt B is heated from the melting point to the final temperature:

$$Q_{B3} = m_B * cp_{\mathrm{Bliq}} * (T_B - T_{BS})$$

The total heat required to heat substance B from T_{B0} to T_B is thus
$Q_B = Q_{B1} + Q_{B2} + Q_{B3}$.

The heat required for this, which the hot oil gives off: $Q_H = m_H * cp_H * (T_{H0} - T_H)$ see formula 18a.

With $Q_H = Q_B$ this gives:

$$Q = m_H * cp_H * (T_{H0} - T_H) = m_B * \left[cp_{\mathrm{Bsolid}} * (T_{BS} - T_{B0}) + \Delta_S H_B + cp_{\mathrm{Bliq}} * (T_B - T_{BS})\right]$$

This equation can be rearranged according to the desired variable (one temperature or the mass of the hot oil or the mass of substance B).

2.4.7 Thermal Reactor Stability Criteria

The safe operation of an exothermic reaction requires a well-planned removal of the resulting heat of reaction. Otherwise, the disaster of runaway reactor, that is, a reactor in which more heat is generated than can be removed, threatens. Such an out-of-control reaction can have serious consequences such as explosions and the like.

The method of calculation of the adiabatic temperature increase considers the case in which an exothermic reaction takes place but no cooling of the reactor takes place.

In contrast, the method of determining stable and unstable operating points is based on the same amount of heat generated by the reaction and heat removed by cooling.

2.4.7.1 Adiabatic Temperature Rise

As already mentioned, the case of the reactor content not being cooled is considered when calculating the adiabatic temperature rise. This is comparable in an intuitive way to the case of the reaction being carried out in a thermos flask and the end temperature being measured. The heat generated by the reaction (see Formula 27a–c) is used to heat the reactor content in accordance with Formulas 18a,b, 19a,b, 20a,b or 21a,b. If the resulting reaction heat is set equal to the heat absorbed by the reactor content and the temperature difference is calculated, the adiabatic temperature rise is obtained. In these formulas, the number of moles of the substance being reacted is designated as n, while the term under the fraction represents the mean heat absorption capacity of the mixture in the reactor or of the flow of the mixture supplied to the reactor.

For batch operation: → **Formula 32a:** $\Delta T_{ad} = \frac{-n * \Delta_R H}{\sum (n_i * cp_i)}$

or **Formula 32b:** $\Delta T_{ad} = \frac{-n * \Delta_R H}{\sum (m_i * cp_i)}$

For continuous operation: → **Formula 33a:** $\Delta T_{ad} = \frac{-\dot{n} * \Delta_R H}{\sum (\dot{n}_i * cp_i)}$

or **Formula 33b:** $\Delta T_{ad} = \frac{-\dot{n} * \Delta_R H}{\sum (\dot{m}_i * cp_i)}$.

If the adiabatic temperature rise remains below the maximum allowed temperature of the reaction system, the reactor can be classified as stable, since even in the event of cooling failure, no critical state will occur. If, on the other hand, the adiabatic temperature rise is above the temperature specified for the reactor system, measures must be taken which, in the event of a failure of the existing cooling, can bring the reactor into a safe state. These are usually a redundant implementation of the cooling system, additional cooling options or measures to stop the reaction.

2.4.7.2 Stable and Unstable Operating Points

The amount of heat generated by a chemical reaction per unit of time has already been described by Formula 27a–d:

$$\dot{Q} = -r * \Delta_R H = -\frac{n_{i0} - n_i}{\tau} * \Delta_R H = -\Delta \dot{n}_i * \Delta_R H$$

Where the reaction rate is exponentially dependent on the absolute temperature via the rate constant according to Formula 17a (see Section 2.3.3):

$$k = k_0 * e^{-E_a/R * T}$$

The removal of this heat output is calculated using Formula 30 according to:

$$\dot{Q} = Kw * A * \Delta T$$

With ΔT as the temperature difference between the reactor content and the coolant.

Both relationships are shown in a diagram of heat output in relation to reactor temperature in Fig. 2.2. As mentioned before, heat generation by the chemical reaction increases exponentially with temperature, while heat removal increases linearly. The reactor strives for a state in which as much heat is removed as is generated. These operating points are represented by the intersection points of the heat generation curve with the heat removal line. Point 1 is a stable operating point: If the reactor temperature decreases for any reason, e.g. fluctuations in the system, heat removal via cooling also decreases and the temperature of the operating point 1 is reached again. In the reverse case of an unwanted, random temperature increase of the reactor contents, heat transport by cooling becomes greater as the heat formation by the reaction and the reactor temperature falls back to that of the stable operating point 1. This is not the case in the unstable operating points 2

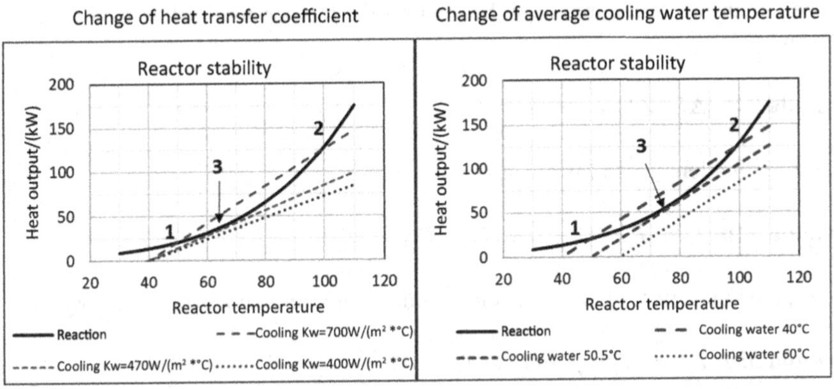

1 lower stable operating point # 2 upper unstable operating point # 3 unstable operating point

Fig. 2.2 Thermal power of a chemical reaction compared to cooling power → Stable and unstable operating points

and 3. If the temperature of the operating point is exceeded by system fluctuations here, the heat removal line always shows a lower value than the exponential curve of heat generation. Thus, the danger of a runaway reaction is given.

The influence of the heat transfer coefficient of the cooling K_W is shown in Fig. 2.2a. The slope of the cooling line decreases with a decreasing heat transfer coefficient. With a sufficiently large heat transfer coefficient (e.g. $K_W = 700$ W/m^2 * °C) there is the stable operating point 1 and the unstable upper operating point 2. The reactor is thermally controlled because it is operated at operating point 1. If the heat transfer coefficient decreases, for example due to contamination of the water-side cooling wall, to a value of $K_W = 470$ W/(m^2 * °C), only the unstable operating point 3 exists. Thus, the reactor is no longer thermally controllable. If the heat transfer coefficient decreases again to, for example, 400 W/(m^2 * °C), the heat removal line always lies below the heat generation curve - the reactor will go out of control by a runaway reaction.

Figure 2.2b shows the effect of an increasing cooling water temperature. Here, a parallel shift of the heat removal curve takes place, which has similar effects on reactor stability as described in the example of a decreasing heat transfer coefficient.

2.4.8 Thermal Expansion

In general, a body or a liquid expands in length and volume when heated. There are only a few exceptions to this general rule. The best known is water, which has its highest density at about 4 °C and thus shrinks at 0 °C to 4 °C when heated, instead of expanding. The following relationships for expansion by temperature increase are only used for solids and liquids. For gases, the gas laws are used (see Section 2.1).

The expansion of the length by heating by the temperature difference ($T_1 - T_o$) is given by:

Formula 34a: $\Delta L = L_o * \alpha * (T_1 - T_o)$

Thus, the total length at T_1 is given as

Formula 34b: $L_1 = L_o * (1 + \alpha * [T_1 - T_o])$

The expansion of the volume by a temperature increase of ΔT is described by:

Formula 35a: $\Delta V = V_o * \gamma * (T_1 - T_o)$

The volume is at T_1 given as

Formula 35b: $V_1 = V_o * (1 + \gamma * [T_1 - T_o])$

Here, ΔL and ΔV stand for the expansion or contraction of the length or volume of a body or a liquid by increasing or decreasing the temperature. L_o is the original length and V_o the original volume at the temperature T_o in °C. T_1 stands for the temperature after the body or liquid has been heated or cooled. The linear expansion coefficient α and the cubical expansion coefficient γ are associated with the unit 1/°C.

In good approximation, the cubical expansion coefficient is:

Formula 35c: $\gamma = 3\alpha$

2.5 Electrochemistry

The number of moles of a substance reduced at the cathode or oxidized at the anode per unit time during an electrolysis is described by Formula 36a, where ν_e in the reaction formula stands for the given number of electrons to be added or removed per atom or per molecule. If this molar flow is multiplied by the total duration of the electrolysis, the number of moles of the reduced or oxidized element or substance results in accordance with Formula 36b. The mass of the reduced or oxidized substance is obtained by multiplying by the molar mass or the atomic mass.

Formula 36a: $\dot{n} = \frac{I}{\nu_e * F}$

Formula 36b: $n = \frac{I * t}{\nu_e * F}$

Formula 36c: $m_i = \frac{I * M_i * t}{F * \nu_e}$

With the Faraday constant $F = 96{,}485\ \mathrm{A} * \mathrm{s}$ $\quad \mathrm{A} * \mathrm{s} = \mathrm{Coulomb}$

The electrical power required for electrolysis is calculated according to the following formula:

Formula 37a: $P = U * I$

U represents the voltage applied to the electrolysis cell, which must be higher than the decomposition voltage, i.e. the electromotive force (EMF) of the electrolysis cell. The decomposition voltage or EMF results from the voltage series of the elements and redox systems, the concentrations of the solutions in the electrolysis cell and their temperature. Such a calculation of the EMF is carried out by means of the Nernst equation, which is not dealt with in this book.

The energy required for the electrolysis process at 100% power utilization is calculated according to Formula 37b by multiplying the electrical power input by the total time of the electrolysis process, i.e. the product of the applied voltage, the Faraday constant, the number of moles of oxidized or reduced species and the electrons exchanged according to the stoichiometric formula:

Formula 37b: $E = U * I * t = U * F * n * \nu_e$

2.6 Liquid Conveying

Transporting liquids through pipelines using pumps is one of the most important basic operations in a chemical production plant. Rotary pumps, i.e. centrifugal pumps, are mostly used for this purpose. The laws for conveying liquids through a pipeline system cannot be applied directly to gases, since liquids are nearly incompressible, unlike gases.

Energy is required to transport a liquid in a pipeline system. This energy is supplied to the pipeline system by one or more pumps.

For historical reasons, these energies are still defined in terms of heights. On the one hand, this is the transportation to a higher point (geodetic height h_{geo}), for example from a ground-level storage tank to a higher located tank. On the other hand, the formation of a pressure by the pumping process is described as a pressure height (h_p) and finally the overcoming of flow resistance as a friction or loss height (h_f). According to Formula 38a, the total height H is made up of the geodetic and the pressure and friction heights.

Formula 38a: $H = h_{geo} + h_p + h_f$

- The geodetic height h_{geo} represents the height difference of the liquid transport.
- The pressure height is calculated according to Formula 38b from the difference Δp of the pressure in the tank, reactor or apparatus into which the liquid is transported, minus the pressure above the liquid at the suction point, as well as the density of the liquid ρ and the constant of gravity g.

Formula 38b: $h_p = \frac{\Delta p}{\rho * g}$

$$1\ \text{bar} = 10^5 \frac{\text{kg}}{\text{m} * \text{s}^2}$$

- The friction height is calculated analogously to Formula 38b from the pressure loss caused by the friction of the flow. The friction losses in pipeline systems depend on the flow characteristics, i.e. the Reynolds number. The calculation of friction losses in a pipeline system is quite complex and is referred to under the term "fluid dynamics". This would go beyond the scope of this book. Please refer to the appropriate textbooks.

The plot of the total height of the liquid transport against the corresponding volume flow is referred to as the plant characteristic curve (2). If the friction height increases, for example by installing a shutter or a reduced valve position, the total height increases, resulting in a higher system characteristic curve (1). The point of intersection with the characteristic curve of a rotary pump (centrifugal pump), which is dealt with in more detail below, represents the operating point of the system (see Fig. 2.3).

Characteristic curves of centrifugal pumps are provided by the manufacturer. An exemplary example is shown in Fig. 2.4: The plot of the respecting total height of the pumping volume flow is plotted for different wheel diameters at just one rotary speed. The corresponding efficiency of the pump is given by the parabolic lines.

It is also common to have a similar type of representation for just one wheel diameter, but with different rotary speeds.

The usual procedure for designing or selecting a pump is the calculation of the total height in the operating system according to formula 38a. The total height and the flow rate of this operating point are entered into the pump diagram. The wheel diameter or the rotary speed of the pump characteristic curve just above this point corresponds to the requirements of the liquid transportation task. The diagram in Fig. 2.4 also shows the corresponding efficiency of the pump for the pump task.

With the formulas 39a and 39b, the theoretically necessary power or the energy to be supplied for the pump task can be calculated:

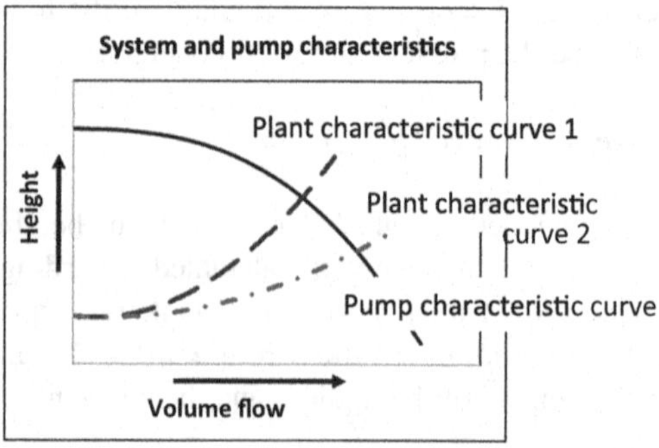

Fig. 2.3 Operating point of a liquid conveying from system and pump characteristic curve

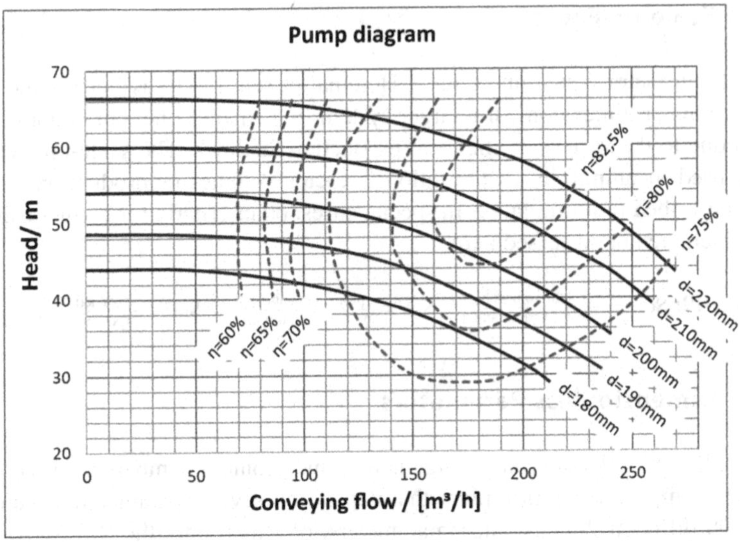

Fig. 2.4 Characteristic curves of a centrifugal pump for one speed, different impeller diameters and associated pump efficiency

Formula 39a: $P = \dot{m} * g * H$

Formula 39b: $E = \dot{m} * g * H * t = m * g * H$

In reality, one has to take into account losses in the pump itself and in the electric motor. This is done according to the formulas 39a, b by inserting the pump efficiency η_P and the efficiency of the electric motor η_E. The efficiencies are not used as a percentage, but as a decimal fraction.

Formula 39c: $P = \frac{\dot{m}*g*H}{\eta_P*\eta_E}$

Formula 39d: $E = \frac{\dot{m}*g*H*t}{\eta_P*\eta_E} = \frac{m*g*H}{\eta_P*\eta_E}$

2.7 Scale-up

The transfer of laboratory results to a large-scale chemical production facility always presents a challenging task due to the complexity of a chemical process. Understanding the entire system is an essential prerequisite for this. Therefore, the guide to the procedure of scale-up is limited to the very simple method using the scale-up factor and briefly just mentions the methodology of dimensionless numbers.

2.7.1 Scale-up Factor

The scale-up factor (scale-up factor →Hereinafter referred to as ScF) is the ratio of conditions or dimensions of a larger system to a smaller one, but is not clearly defined due to the variety of chemical production facilities. The scale-up factor is usually used meaningfully as the ratio of reactor volumes or production capacities, as described in Formula 40. In special cases, other criteria for comparison are used, which are not discussed here.

Formula 40: $ScF = \frac{\text{reactor volume big}}{\text{reactor volume small}}$ $ScF = \frac{\text{product quantity big}}{\text{product quantity small}}$ $ScF = \frac{\text{productionrate big}}{\text{produktionrate small}}$

2.7.2 Dimensionless Parameters

For complicated tasks of scale enlargement, dimensionless numbers are used. This describes complex states in a plant that depend on several parameters. There are about 100 different dimensionless parameters, of which only the Reynolds number is discussed in this chapter due to its importance, which is used to characterize flow conditions. The same value of such a dimensionless parameter in general means the same conditions. So there are very similar flow conditions with the same Reynolds number. Another example is the Nusselt number. Here, very similar heat transfer conditions from a fluid to a pipe wall are present at the same values. Another example is the Sherwood number. Here, very similar diffusion takes place between two immiscible liquids at the same size of this number, e.g. during extraction. A deeper treatment would exceed the scope of this book. As already mentioned, only the very important description of a flow state by means of the Reynold number is pointed out here, since it influences, among other things, the heat transfer from a fluid to a wall as well as the mixing and sedimentation of solids from fluids, the mass transfer from one phase to another, the dissolution of solids or the extraction and other operations of process engineering. The Reynold number is a function of the flow velocity w, a characteristic length d, the density and the viscosity of the fluid.

Formula 41: Reynolds-Number: $Re = \frac{\text{SpeedForces}}{\text{ViscousForces}} = \frac{w*d*\rho}{\eta} = \frac{w*d}{\upsilon}.$

In the frequent case of flow through pipes, the inner pipe diameter d_i is used for d. With a Reynold number of less than about 1000, there are pure laminar flow conditions. There are no eddies under these conditions, so there is only a small amount of mixing. An example is the stirring of a dispersion paint with a tinting paint for painting wallpaper. Poor mixing is observed, demonstrated by the parallel flow lines of the tinting paint. From a Reynolds number of about 2300, the flow is completely turbulent. Good mixing is only possible with such turbulent flow conditions. High turbulence, i.e. a high Reynolds number, also promotes the heat transfer from a liquid to the wall of a pipe or reactor by breaking down the laminar sublayer (see Section 2.4.5). Similar considerations apply to the faster dissolution of a solid by stirring.

3 Exercise Problems

The exercise problems are based on situations that occur in a chemical production plant. The level of difficulty and complexity within the chapters of the individual topics grows from the first to the last task. A chemical production plant is a complex system, therefore the knowledge of several areas and their combination are required in the exercises for "Combined tasks". However, this also applies to a limited extent to some tasks of the individual subject areas.

Often, a task statement from a chemical production plant looks somewhat complicated and unstructured. Instead of starting with calculations immediately, it is recommended to extract and note the relevant data, further, to understand the target variables, to develop a solution strategy and, if necessary, to record it in the form of a sequence of steps in the form of keywords. Short comments within the calculation process may also be useful. Sketches can also be helpful for understanding. Such documentation is very useful in practical operation, as the solution of a task must be traceable for colleagues and superiors. The results must be clearly highlighted, ideally in the form of a concise sentence or a short text.

The solutions of the exercise problems are presented in this form:

⊗ Solution:
→ *Strategy*
→ *Calculation*
→ *Result*

Most of the time, the solution process refers to the number of the formula to be used. Exceptions to this are formulas that describe fundamental relationships, as listed in Sect. 1.2 and the basic relationships for the conversion of concentrations, masses and moles as well as mass and volume or mass flows and volume flows (see Sect. 2.3.1 Formulas 7a–e and 8a, b).

G. Jüptner, *111 Calculation Exercises in the Field of Chemical Technology*, https://doi.org/10.1007/978-3-662-66920-4_3

Due to the complex relationships in a chemical production plant, different approaches and strategies can be used, which are all correct and lead to the goal under the motto "Many roads lead to Rome". That is exactly why the above-described procedure is very recommendable.

As already mentioned, the end and intermediate results of operating data usually consist of the numerical value and the unit. In this case, a calculation result as a pure numerical value is meaningless and useless. If the result shows the expected unit, at least a correct calculation process and the correct use of the data may be assumed.

3.1 Ideal Gas Law

Task 1

An air stream of 2 m³/s, of a temperature of 25 °C and a pressure of 1.0 bar is brought to a temperature of 120 °C and an overpressure of 1.85 bar before being introduced into a dryer. How large is the air stream under these conditions?

⊗ **Solution**

→ ***Strategy***

Formula 1a is rearranged to the volume flow $\dot{V}_2$, the corresponding data are inserted and the result is calculated.

→ ***Calculation***

$$\frac{\dot{V}_1 * p_1}{T_1} = \frac{\dot{V}_2 * p_2}{T_2} \rightarrow \dot{V}_2 = \frac{\dot{V}_1 * p_1 * T_1}{T_2 * p_2}$$

$$\text{mit } \dot{V}_1 = \frac{2\ \text{m}^3}{\text{s}} \quad T_1 = (273.15 + 25.0)\ \text{K} = 298\ \text{K} \quad T_2 = (273.15 + 120.0)\ \text{K} = 493.15\ \text{K}$$

$$p_1 = 1.0\ \text{bar} \quad p_2 = (1.0 + 1.85)\ \text{bar} = 2.85\ \text{bar}$$

$$\dot{V}_2 = \frac{\dot{V}_1 * p_1 * T_1}{T_2 * p_2} = \frac{2\ \text{m}^3 * 1.0\ \text{bar} * 393.15\ \text{K}}{\text{s} * 298.15\ \text{K} * 2.85\ \text{bar}} = \frac{0.925\ \text{m}^3}{\text{s}} = \frac{0.925\ \text{m}^3 * 3600\ \text{s}}{\text{s} * \text{h}} = \mathbf{3330\ m^3/h}$$

→ ***Result***

The air stream to the dryer is 0.925 m³/s, that is 3330 m³/h.

Task 2

The 5.5 m³ atmospheric gas space of a reactor shall be inerted with nitrogen. For this purpose, it is to be flushed completely with nitrogen four times in succession. For this purpose, 'packages' each with 6 pressure gas cylinders of a single volume of 20 L are available under a pressure of 200 bar. The temperature of the reactor with 20 °C is equal to that of the pressure gas cylinders. How many of such packages are required?

⊗ **Solution**

→ ***Strategy***

The required nitrogen volume is 4 * 5.5 m^3 at 1 bar and 20 °C, that is 22 m^3. With formula 1a this nitrogen volume is converted to a pressure of 200 bar and the number of 20 L gas cylinders is calculated. For this purpose, the relationship is rearranged to the volume at 1 bar. Since the temperature of the gas cylinders and the reactor is identical, it cancels out of formula 1a.

→ ***Calculation***

$$\frac{\dot{V}_1 * p_1}{T_1} = \frac{\dot{V}_2 * p_2}{T_2} \rightarrow \dot{V}_2 = \frac{\dot{V}_1 * p_1 * T_1}{T_2 * p_2}$$

$$\text{as } T_1 = T_2 \rightarrow \dot{V}_2 = \frac{\dot{V}_1 * p_1}{p_2} = \frac{22\,\text{m}^3 * 1\,\text{bar}}{200\,\text{bar}} = 0.11\,\text{m}^3 = 110\,\text{L}$$

1 pressure gas cylinder = 20 L → ***5 ½ gas cylinders***

→ ***Result***

One six-pack of pressure gas cylinders is sufficient to inertize the reactor.

Task 3

80 m^3 of carbon monoxide at a temperature of 150 °C and a pressure of 10 bar are cooled down to 25 °C and transported to a gasometer (cyclindrical variable gas container) with a diameter of 10 m for intermediate storage. The gasometer is at an absolute pressure of 2.0 bar at 25 °C. By how many meters is the gasometer cap lifted?

⊗ **Solution**

→ ***Strategy***

The gas volume that was supplied to the gasometer under the conditions prevailing there is calculated using the volume-resolved formula 1a. In the gasometer, this gas quantity represents a cylinder whose height is calculated from the gas volume and the gasometer cap diameter using the formula for cylinder length.

→ ***Calculation***

$$\frac{V_1 * p_1}{T_1} = \frac{V_2 * p_2}{T_2} \rightarrow V_2 = \frac{V_1 * p_1 * T_2}{T_1 * p_2}$$

$$\text{with } V_1 = 80\,\text{m}^3 \quad p_1 = 10\,\text{bar} \quad T_1 = (273.15 + 150.0)\,\text{K} = 423.15\,\text{K}$$

$$p_2 = 2\,\text{bar} \quad T_2 = (273.15 + 25.0)\,\text{K} = 278.15\,\text{K}$$

$$V_2 = \frac{80\,\text{m}^3 \, * \, 10\,\text{bar} \, * \, 298.15\,\text{K}}{423.15\,\text{K} \, * \, 2\,\text{bar}} = 281.8\,\text{m}^3$$

$$V_{Cylinder} = \frac{d^2 * \pi * L}{4} \rightarrow L = \frac{4 * V_{\text{Cyl}}}{d^2 * \pi} = \frac{4 * 281.8\,\text{m}^3}{10^2\text{m}^2 * \pi} = \mathbf{3.59\,m}$$

→ ***Result***

The supply of carbon monoxide lifts the gasometer cap by 3.59 m.

Task 4

In a chlor-alkali electrolysis plant, 16.4 kg of hydrogen are produced per hour. The hydrogen gas is compressed to 200 bar and filled into 50 L gas cylinders at 20 °C. How many gas bottles are required per hour?

⊗ **Solution**

→ ***Strategy***

Formula 2 is rearranged to volume. To solve it, the molar amount of hydrogen produced in one hour has to be calculated from the mass -to- molar mass ratio. With the respective volume of each gas cylinder of 50 L, their required number can be calculated.

→ ***Calculation***

$$p * V = n * R * T \rightarrow V = \frac{n * R * T}{p}$$

$$\text{with } n = \frac{m}{M} = \frac{16.4\,\text{kg} * \text{mol}}{0.002\,\text{kg}} = 8200\,\text{mol}$$

$$M_{\text{H}_2} = 2\frac{\text{g}}{\text{mol}} = 0.002\frac{\text{kg}}{\text{mol}}$$

$$T = (273.15 + 20.0)\,\text{K} = 293.15\,\text{K} \quad p = 200\,\text{bar}$$

$$V = \frac{8200\,\text{mol} * 8.315 * 10^{-5}\,\text{bar} * \text{m}^3 * 293.15\,\text{K}}{\text{mol} * \text{K} * 200\,\text{bar}} = 0.999\,\text{m}^3 \cong 1.00\,\text{m}^3$$

1 gas cylinder = *0.05 m³*→ **20 gas cylinders = 1.00 m³**

→ ***Result***

20 gas cylinders must be provided per hour.

Exercise 5

550 kg of propene (M = 42.1 g/mol) per hour are taken out of a liquid-gas tank, evaporated, heated up to 50 °C and led through a 4" pipe with a wall thickness of 3 mm to a compressor at a pressure of 3.5 bar. What is the flow velocity in the pipe?

⊗ Solution

→ *Strategy*

First, the volumetric flow of propene is calculated according to the rearranged formula 2. For this purpose, the mass flow of propene is converted into the molar flow. The flow velocity in the pipe results from the volumetric flow divided by the inside cross-section of the pipe.

→ *Calculation*

$$p * \dot{V} = \dot{n} * R * T \rightarrow \dot{V} = \frac{\dot{n} * R * T}{p} \text{ with } \dot{n} = \frac{\dot{m}}{M} = \frac{550\,\text{kg} * \text{mol}}{h * 0.0421\,\text{kg}} = \frac{13{,}064\,\text{mol}}{\text{h}}$$

$$= \frac{13{,}064\,\text{mol} * \text{h}}{\text{h} * 3600\,\text{s}} = 3.629\,\text{mol/s}$$

$$p = 3.5\,\text{bar},\ T = (273.15 + 50)\,\text{K} = 323.15\,\text{K}$$

$$\dot{V} = \frac{3.629\,\text{mol} * 8.315 * 10^{-5}\,\text{bar} * 323.15\,\text{K}}{\text{s} * \text{mol} * \text{K} * 3.5\,\text{bar}} = 0.02786\,\text{m}^3/\text{s}$$

Flow velocity: $w = \frac{\dot{V}}{A_{\text{Pipe}}}$

$$A_{\text{Pipe}} = \frac{d_i^2 * \pi}{4}$$

$$d_i = 4 * 25.4\,\text{mm} - 2 * 3\,\text{mm} = 95.6\,\text{mm} = 0.0956\,\text{m}$$

$$A_{\text{Pipe}} = 0.007174\,\text{m}^2$$

$$w = \frac{0.02768\,\text{m}^3}{\text{s} * 0.007174\,\text{m}^2} = \mathbf{3.86\,m/s}$$

→ *Result*

The flow velocity of the propene gas in the pipe is 3.86 m/s.

Exercise 6

A stream of methane (M = 16.0 g/mol) of a temperature of about 180–200 °C as a by-product of a process, is stored in a spherical tank with an inner diameter of 9.5 m in order to be supplied as a raw material to another production plant. The diameter of the tank can be considered constant within the scope of its operating conditions.

a) After filling, the spherical tank has a temperature of 165 °C and is under a pressure of 6.5 bar. How much methane (mass and amount of moles) does the tank contain?
b) After two days, the temperature has fallen to 90 °C. What is the pressure in the spherical tank now?

c) Now 750 standard-m^3 (0 °C, 1.013 bar) of methane are taken out of the tank. How many kg or moles of methane were taken out?
d) How much methane (mass and amount of moles) is still in the tank afterwards, and what is the pressure, if the temperature has decreased by another 25 °C?

⊗ **Solution**

→ ***Strategy***

a) The volume of the tank is calculated, using the spherical volume formula with its diameter. The number of moles of methane is calculated using the correspondingly rearranged formula 2. The mass of the stored gas results from the number of moles, multiplied by the molar mass of methane.
b) Formula 1 is rearranged to p_2=6.5bar. The tank volume is cancels off the formula because it is constant. Alternatively, the rearranged formula 2 may also be used.
c) Formula 2 is rearranged to the number of moles, and the standard pressure and standard temperature are used for the calculation. The mass of the taken methane is calculated from the thus calculated number of moles, multiplied with the molar mass of methane.
d) The number of moles of methane, taken from the tank and calculated in the previous task part, is subtracted from the initial methane number of moles in the tank. The pressure in the tank is calculated from this with the correspondingly rearranged formula 2, taking into account the temperature.

→ ***Calculation***

a) $V = \frac{d^3 * \pi}{6} = \frac{9.5^3 * \text{m}^3 * \pi}{6} = 448.7\,\text{m}^3$

$$p * V = n * R * T \rightarrow \boldsymbol{n} = \frac{p * V}{R * T} = \frac{6.5\,\text{bar} * 448.7\,\text{m}^3 * \text{mol} * \text{K}}{8.315 * 10^{-5}\,\text{bar} * \text{m}^3 * (273.15 + 165)\text{K}}$$

$$= \textbf{80,054.6 mol Methane} \cong \textbf{80,055 mol}$$

$$m = n * M = 80{,}054.6\,\text{mol} * \frac{0.016\,\text{kg}}{\text{mol}} = \textbf{1281 kg Methane} = \textbf{1.28 t Methane}$$

b) $\frac{V_1 * p_1}{T_1} = \frac{V_2 * p_2}{T_2}$

$$\text{V}_1 = \text{V}_2 \rightarrow \boldsymbol{p_2} = \frac{p_1 * T_2}{T_1} = \frac{6.5\,\text{bar} * (273.15 + 90)\text{K}}{(273.15 + 165)\text{K}} = 5.39\,\text{bar} \cong \textbf{5.4 bar}$$

Alternative:

$$\boldsymbol{p} = \frac{n * R * T}{V} = \frac{80{,}054.6\,\text{mol} * 8.315 * 10^{-5}\,\text{bar} * \text{m}^3 * (273.15 + 90)\text{K}}{\text{mol} * \text{K} * 448.7\,\text{m}^3} = 5.39\,\text{bar} \cong \textbf{5.4 bar}$$

c) *Methane removal Δn, Δm*

$$\Delta \boldsymbol{n} = \frac{p * \Delta V}{R * T} = \frac{1.013\,\text{bar} * 750\,\text{m}^3 * \text{mol} * \text{K}}{8.315 * 10^{-5}\text{bar} * \text{m}^3 * 273.15\,\text{K}} = \mathbf{33{,}451\,mol\ Methane}$$

$$\Delta \boldsymbol{m} = \Delta n * M = 33{,}451\,\text{mol} * \frac{0.016\,\text{kg}}{\text{mol}} = \mathbf{535.2\,kg\ Methane} \cong \mathbf{535\,kg\ Methane}$$

d) *Remaining methane amount*

$$\boldsymbol{n} = (80{,}054.6 - 33{,}450.9)\text{mol} = \mathbf{46{,}603.7\ molMethane}$$

$$\boldsymbol{m} = (1281 - 535)\,\text{kg} = \mathbf{746\,kg\ Methane}$$

$$\text{T} = (273.15 + 65)\text{K} = 338.15\text{K}$$

$$\boldsymbol{p} = \frac{n * R * T}{V} = \frac{46{,}603.7\,\text{mol} * 8.315 * 10^{-5}\,\text{bar} * \text{m}^3 * 338.15\text{K}}{\text{mol} * \text{K} * 448.7\,\text{m}^3} = \mathbf{2.93\,bar}$$

→ ***Result***

a) **The amount of methane in the spherical tank is 80,055 mol methane, which corresponds to a mass of 1.28 t.**
b) **The tank is under a pressure of 5.39 bar ≅ 5.4 bar.**
c) **The amount of methane withdrawn from the tank corresponds to 33,451 mol or 535 kg.**
d) **After the withdrawal, there are still 46,604 mol = 746 kg methane under a pressure of 2.93 bar in the tank.**

Task 7
In a continuous reactor, 3.6 metric t of alkaline wastewater with a sodium hydroxide content (M = 40.0 g/mol) of 0.85wt% is to be neutralized per hour. The plant design provides for the use of gaseous hydrogen chloride from another plant, which contains a certain proportion of dust. The hydrogen chloride is under a pressure of 2.5 bar and has a temperature of 15 °C. In order to avoid the deposition of dust, the flow velocity of the gas must not be less than 1.0 m/s. What is the maximum inner diameter of the supply line?

⊗ **Solution**
→ ***Strategy***
First, the mass flow of sodium hydroxide is calculated from the total flow and the percentage content. From the mass flow of sodium hydroxide its molar flow is calculated. The molar flow of sodium hydroxide corresponds to the molar flow of hydrogen chloride required for neutralization. Hence the volumetric flow of hydrogen chloride is calculated from the molar flow of sodium hydroxide using the appropriate formula 2. The flow velocity of the hydrogen chloride results from the volume flow of this gas divided by the cross-section area of the pipe.

→ ***Calculation***

$$\dot{m}_{NaOH} = \frac{3.6\,t * 0.85\%}{h * 100\%} = 0.0306\frac{t}{h} = \frac{0.0306\,t * h}{h * 3600\,s} = \frac{8.50 * 10^{-6}t}{s} = \frac{0.0085\,kg}{s}$$

$$\dot{n}_{NaOH} = \frac{\dot{m}_{NaOH}}{M_{NaOH}} = \frac{0.0085\,kg * mol}{s * 0.040\,kg} = 0.2125\frac{mol}{s} = \dot{n}_{HCl}$$

$$p * \dot{V} = \dot{n} * R * T \rightarrow \dot{V} = \frac{\dot{n} * R * T}{p} = \frac{0.2125\,mol * 8.315 * 10^{-5}\,bar * m^3 * (273.15 + 15)K}{s * 2.5\,bar * mol * K}$$

$$= 0.00204\frac{m^3}{s}$$

$$w = \frac{\dot{V}}{A}\; A = \frac{d^2 * \pi}{4} \rightarrow w = \frac{4 * \dot{V}}{d^2 * \pi}$$

$$d = \frac{\sqrt{4 * \dot{V}}}{w * \pi} = \sqrt{\frac{4 * 0.00204\,m^3 * s^3}{s * 1.0\,m * \pi}} = \mathbf{0.051\,m}$$

→ ***Result***
The inner diameter of the hydrogen chloride supply line must not be greater than 0.051 m.

3.2 Law of Mass Action

3.2.1 Equilibrium reactions

Exercise 8
A test reactor for investigations in catalysts for the synthesis of sulfur trioxide from air and sulfur dioxide is operated at 400 °C. In the temperature range of 400°C - 600°C, the enthalpy of reaction for the formula conversion $2SO_2 + O_2 \leftrightarrows 2SO_3$ is $\Delta_R H$ = -200 kJ/mol. The partial pressures of the gases leaving the reactor were measured as follows:

$$p_{N_2} = 1.70\,bar;\; p_{O_2} = 0.025\,bar;\; p_{SO_2} = 0.0035\,bar;\; p_{SO_3} = 0.110\,bar$$

It is assumed that thermodynamic equilibrium has been reached.

a) Calculate the partial pressure-related equilibrium constant k_p and the molar fraction-related equilibrium constant k_X.
b) What is the equilibrium constant k_p at 500 °C?

⊗ **Solution**
→ ***Strategy***

a) First, the stoichiometric factors ν_i are taken from the reaction equation. With formula 3a, the equilibrium constant k_p is calculated from the partial pressures of the mixture leaving the reactor with the stoichiometric factors as the corresponding exponents. The equilibrium constant of the molar fractions k_x is calculated from formula 3d, where the total pressure p is the sum of all partial pressures, so that the partial pressure of the inert nitrogen must be included here.
b) By formula 3e, the equilibrium constant determined for 400 °C is converted to 500 °C.

→ ***Calculation***

a) $2\ SO_2 + O_2 \rightleftarrows 2\ SO_3$

$$\nu_{SO_2} = -2$$

$$\nu_{O_2} = -1$$

$$\nu_{SO_3} = +2$$

$$k_p = \prod_i p_i^{\nu_i} = p_{SO_2}^{-2} * p_{O_2}^{-1} * p_{SO_3}^{+2} = \frac{p_{SO_3}^2}{p_{SO_2}^2 * p_{O_2}} = \frac{(0.110\,\text{bar})^2}{(0.0035\,\text{bar})^2 * 0.025\,\text{bar}} = \mathbf{3.95 * 10^4\,bar^{-1}}$$

$$k_p = k_x * p^{\sum \nu_i}$$

$$k_x = k_p * p^{\sum -\nu_i}$$

$$p = p_{N_2} + p_{O_2} + p_{SO_2} + p_{SO_3} = (1.70 + 0.025 + 0.0035 + 0.110)\,\text{bar} = 1.84\,\text{bar}$$

$$\sum \nu_i = -2 - 1 + 2 = -1$$

$$\sum -\nu_i = +1$$

$$k_x = 3.95 * 10^4\,\text{bar}^{-1} * 1.84\,\text{bar} = \mathbf{7.27 * 10^4}$$

b) $k_{p2} = k_{p1} * e^{\frac{\Delta_R H}{R} * \left(\frac{1}{T_1} - \frac{1}{T_2}\right)}$

$$k_{p2} = k_{p500\,°C} \quad k_{p1} = k_{p400\,°C} = 3.95 * 10^4\,\text{bar}^{-1}$$

$$T_2 = (500 + 273)\,\text{K} = 773\,\text{K} \quad T_1 = (400 + 273)\,\text{K} = 673\,\text{K}$$

$$k_{p2} = 3.95 * 10^4\,\text{bar}^{-1} * e^{\frac{-200\,\text{kJ} * \text{mol} * \text{K}}{\text{mol} * 0.008315\,\text{kJ}} * \left(\frac{1}{673\,\text{K}} - \frac{1}{773\,\text{K}}\right)} = 3.95 * 10^4\text{bar}^{-1} * e^{-4.624} = \mathbf{388\,bar^{-1}}$$

→ ***Result***

a) **The equilibrium constants at 400 °C are k_p= 3.95 * 10^4bar^{-1} and k_x= 7.27 * 10^4.**
b) **The equilibrium constant k_p is at 500 °C 3.88 * 10^2bar^{-1} and is thus about a factor of 100 higher than that at 400 °C.**

Task 9

A reactor for the production of ammonia is operated at 450 °C and 220 bar. The equilibrium conversion is approximately complete. The gas leaving the reactor in thermodynamic equilibrium has a composition of 12.3 vol% ammonia, 65.8 vol% nitrogen and 21.9 vol% hydrogen. What are the values of the equilibrium constants k_p and k_x?

⊗ **Solution**

→ ***Strategy***

First, the reaction equation is set up and the stoichiometric factors are determined from it. The equation of the equilibrium constant k_p is set up according to formula 3a. Formula 3d shows the relationship between k_p and k_x. The partial pressures of the individual components result from their volume fraction in the gas mixture and the total pressure. As an alternative, k_x can also be calculated from the relative volume fractions, which according to the ideal gas law represent the molar fraction, using formula 3c.

→ ***Calculation***

$N_2 + 3H_2 \leftrightarrows 2NH_3$

$$\nu_{N_2} = -1 \quad \nu_{H_2} = -3 \quad \nu_{NH_3} = +2 \quad \sum_i \nu_i = -1 - 3 + 2 = -2$$

$$k_p = \frac{p_{NH_3}^2}{p_{N_2} * p_{H_2}^3}$$

$$k_x = k_p * p^{-\sum \nu_i} = kp * p^{-(-2)} = k_p * p^2$$

$$p_{NH_3} = \frac{\text{vol\% NH}_3}{100\,\text{vol\%}} * 220\,\text{bar} = 0.123 * 220\,\text{bar} = 27.06\,\text{bar}$$

$$p_{N_2} = \frac{\text{vol\% N}_2}{100\,\text{vol\%}} * 220\,\text{bar} = 0.658 * 220\,\text{bar} = 144.76\,\text{bar}$$

$$p_{H_2} = \frac{\text{vol\% H}_2}{100\,\text{vol\%}} * 220\,\text{bar} = 0.219 * 220\,\text{bar} = 48.18\,\text{bar}$$

$$\boldsymbol{k_p} = \frac{27.06^2\,\text{bar}^2}{144.76\,\text{bar} * 48.18^3\,\text{bar}^3} = \mathbf{4.523 * 10^{-5}\,bar^{-2}}$$

$$k_x = 4.523 * 10^{-5}\,\text{bar}^{-2} * (220\,\text{bar})^2 = \mathbf{2.19}$$

Or alternatively:

$$k_x = \frac{0.123^2}{0.658 * 0.219^3} = \mathbf{2.19}$$

→ ***Result***
The equilibrium constants are $k_p = 4.52 * 10^{-5}/\text{bar}^2$ and $k_x = 2.19$.

Exercise 10
Ethyl acetate is produced in a batch reactor. For this purpose, 250 kg of acetic acid (M = 60.1 g/mol) and 820 kg of ethanol (M = 46.1 g/mol) are brought to reaction. There is no change in volume. The equilibrium constant is $k_c = 3.4$.

What is the composition of the mixture in mol and wt% after the reaction in equilibrium?

⊗ **Solution**
→ ***Strategy***
First, the reaction equation is set up and the stoichiometric factors are determined. From this, the formula of the equilibrium constant k_c is set up according to Formula 3b. In the reaction, there is no change in the total molarity and the volume. The concentration is the quotient of molarity and volume. Thus, the volume in the k_c-formula is eliminated. The molarity of the ethanol used and the acetic acid is calculated from the quotient of their mass and molar mass. The molarity of the resulting ester (→ Δn) is equal to the molarity of the water formed. For each mole of ester formed, one mole of ethanol and one mole of acetic acid are consumed. This balance is set in the k_c-formula and solved for the molarity Δn of the ester formed.

→ ***Calculation***

$$CH_3COOH + C_2H_5OH \rightleftarrows CH_3COO{-}C_2H_5 + H_2O$$

Stoichiometric factor ν_i	−1	−1	+1	+1
Subscripts	Ac	Et	E	W

$$k_c = \prod_i c_i^{\upsilon_i} = c_{\text{Ac}}^{-1} * c_{\text{Et}}^{-1} * c_E^{+1} * c_W^{+1}$$

$$\text{with } c = \frac{n}{V} \text{ and } V = V_{\text{Ac}} + V_{\text{Et}} + V_E + V_W$$

$$k_c = \frac{\frac{n_E}{V} * \frac{n_W}{V}}{\frac{n_{\text{Ac}}}{V} * \frac{n_{\text{Et}}}{V}}$$

Since there is no volume change, it results in $k_c = \frac{n_E * n_W}{n_{Ac} * n_{Et}}$

with $n_E = n_W = \Delta n$ und $n_{Ac} = n_{Ac_o} - \Delta n \quad n_{Et} = n_{Et_o} - \Delta n.$

The molar amount of acetic acid and ethanol are thus calculated from the respective mass fed to the reactor and the molar mass:

$$n_{Ac_o} = \frac{250\,\text{kg} * \text{mol}}{0.0601\,\text{kg}} = 4159.7\,\text{mol} \quad n_{Et_o} = \frac{820\,\text{kg} * \text{mol}}{0.0461\,\text{kg}} = 17{,}787.4\,\text{mol}$$

$$k_c = \frac{(\Delta \text{n})^2}{\left(n_{Ac_o} - \Delta \text{n}\right) * \left(n_{Et_o} - \Delta \text{n}\right)} \text{ with the current data :}$$

$$3.4 = \frac{(\Delta n)^2}{(4159.7\,\text{mol} - \Delta n) * (17{,}787.4\,\text{mol} - \Delta n)} = \frac{(\Delta n)^2}{73{,}990.248\,\text{mol}^2 - 21{,}947\,\text{mol} * \Delta n + (\Delta n)^2}$$

This is a quadratic equation of Δn*, which is solved for* Δn *by the method of quadratic completion by means of Vieta's theorem (see textbooks of algebra). For this purpose, the equation is brought into the form* $0 = x^2 + p * x + q$ *and the two solutions*

$x_{1.2} = -\frac{p}{2} \pm \sqrt{-q + \frac{p^2}{4}}$ *are calculated, of which one solution is unrealistic.*

$$251{,}566.43\,\text{mol}^2 - 74{,}620\,\text{mol} * \Delta n + 2.4 * (\Delta n)^2 = 0$$

$$(\Delta n)^2 - 31{,}092\,\text{mol} * \Delta n + 104{,}819.518\,\text{mol}^2 = 0$$

with $p = -31{,}092\,\text{mol} \quad q = 104{,}819.518\,\text{mol}^2$ the result is

$$\Delta n_{1.2} = \frac{31{,}092\,\text{mol}}{2} \pm \sqrt{\left(-104{,}819.518 + \frac{31{,}092^2}{4}\right)\text{mol}^2}$$

$$\Delta n_{1.2} = 15{,}546\,\text{mol} \pm 11{,}699\,\text{mol}$$

$$\Delta n_1 = 3847\,\text{mol}$$

$\Delta n_2 = 27{,}245\,\text{mol} \rightarrow$ *This result is irrelevant because it indicates a greater decrease in acetic acid and ethanol than was fed to the reactor.*
The mixture in equilibrium thus consists of

$$\boldsymbol{n_E = 3847\,\text{mol} \quad n_W = 3847\,\text{mol}}$$

$$n_{Ac} = (4159.7 - 3847)\,\text{mol} = \mathbf{313\,mol} \quad n_{Et} = (17{,}787.4 - 3847)\,\text{mol} = \mathbf{13{,}940\,mol}$$

With m = n * M, the mass fractions of the individual components in the final mixture result:

$$m_E = 3847\,\text{mol} * \frac{0.0881\,\text{kg}}{\text{mol}} = 339\,\text{kg} \quad m_W = 3847\,\text{mol} * \frac{0.018\,\text{kg}}{\text{mole}} = 69\,\text{kg}$$

$$m_{\text{Ac}} = 313\,\text{mol} * \frac{0.0601\,\text{kg}}{\text{mol}} = 18.8\,\text{kg} \quad m_{\text{Et}} = 13{,}940\,\text{mol} * \frac{0.0461\,\text{kg}}{\text{mol}} = 643\,\text{kg}$$

The total mass in the reactor is 1070 kg. This results in the following relative mass fractions in the final mixture for the different components:

$$\textbf{Ester} \rightarrow \frac{339\,\text{kg} * 100\,\text{wt\%}}{1070\,\text{kg}} = \mathbf{31.7\,wt\%}$$

$$\textbf{Water} \rightarrow \frac{69\,\text{kg} * 100\,\text{wt\%}}{1070\,\text{kg}} = \mathbf{6.45\,wt\%}$$

$$\textbf{AceticAcid} \rightarrow \frac{18.8\,\text{kg} * 100\,\text{wt\%}}{1070\,\text{kg}} = \mathbf{1.75\,wt\%}$$

$$\textbf{Ethanol} \rightarrow \frac{643\,\text{kg} * 100\,\text{wt\%}}{1070\,\text{kg}} = \mathbf{60.1\,wt\%}$$

→ ***Result***
The composition of the mixture after the reaction is in equilibrium:

3847 mol ethyl acetate → 31.7 wt%
3847 mol water → 6.45 wt%
313 mol acetic acid → 1.75 wt%
13,940 mol ethanol → 60.1 wt%

Task 11
In the shift reaction for the production of hydrogen, carbon monoxide is reacted with water vapor in an equilibrium reaction to carbon dioxide and hydrogen. In a tubular reactor, a gas stream of the composition 36 vol% carbon monoxide, 3 vol% carbon dioxide, 28 vol% hydrogen and 33 vol% water vapor is passed through a catalyst bed. The composition of the exiting gas stream at 250 °C corresponds to the thermodynamic equilibrium with an equilibrium constant of $k_p = 93.1$. The calculation should be based on ideal gases.

a) What is the equilibrium constant k_X?
b) What is the composition of the gas mixture leaving the reactor?

⊗ **Solution**
→ ***Strategy***

a) First, the reaction equation is set up, the stoichiometric factors are determined and the equilibrium constant k_x is formulated according to Formula 3c. Since the number of moles does not change during the reaction, according to Equation 3d $k_x = k_p$.
b) For ideal gases, the molar fractions correspond to one hundredth of the corresponding volume percentages. The composition of the inlet gas stream is given. However, the equilibrium constant describes the composition of the gas mixture leaving the reactor. For this purpose, a material balance is set up. The volume percentages of carbon monoxide and water decrease by the same amount 'Delta' (Δ), while the volume percentages of carbon dioxide and hydrogen increase by the same amount Δ. These endconcentrations, expressed as molar fractions, are inserted into the equation of the equilibrium constant (Formula 3c), the equation is solved for Δ and the numerical value of Δ is calculated. From the initial values of the composition of the gas mixture and the difference Δ, the composition of the gas mixture leaving the reactor is calculated.

→ ***Calculation***

$$\mathrm{CO} + \mathrm{H_2O} \rightleftarrows \mathrm{CO_2} + \mathrm{H_2}$$

a) *Formula 3d:* $k_p = k_x * \prod_i p^{\sum \upsilon_i}$

$$\nu_{\mathrm{CO}} = -1 \quad \nu_{\mathrm{H_2}} = -1 \quad \nu_{\mathrm{CO_2}} = +1 \quad \nu_{\mathrm{H_2}} = +1$$

$$\rightarrow \sum_i \nu_i = 0$$

$$k_p = k_x * p^0 = k_x * 1 = k_x \quad k_x = k_p = \mathbf{93.1}$$

b) Formula 3c: $k_x = \prod_i x_i^{\nu_i} = x_{\mathrm{CO}}^{-1} * x_{\mathrm{H_2O}}^{-1} * x_{\mathrm{CO_2}}^{+1} * x_{\mathrm{H_2}}^{+1} = \frac{x_{\mathrm{CO_2}} * x_{\mathrm{H_2}}}{x_{\mathrm{CO}} * x_{\mathrm{H_2O}}}$

The composition of the gas mixture leaving the reactor, expressed as molar fractions, is described by:

$$x_{\mathrm{CO}} = x_{\mathrm{CO_o}} - \Delta \quad x_{\mathrm{H_2O}} = x_{\mathrm{H_2O_o}} - \Delta$$

$$x_{\mathrm{CO_2}} = x_{\mathrm{CO_{2o}}} + \Delta \quad x_{\mathrm{H_2}} = x_{\mathrm{H_{2o}}} - \Delta$$

Subscript o →Molar fraction in the gas mixture supplied to the reactor:

$$x_{\mathrm{CO_o}} = \frac{36\ \mathrm{Vol\%}}{100\ \mathrm{Vol\%}} = 0.36 \quad x_{\mathrm{H_2O_o}} = \frac{33\ \mathrm{Vol\%}}{100\ \mathrm{Vol\%}} = 0.33$$

$$x_{\mathrm{CO_{2o}}} = \frac{3\ \mathrm{Vol\%}}{100\ \mathrm{Vol\%}} = 0.03 \quad x_{\mathrm{H_{2o}}} = \frac{28\ \mathrm{Vol\%}}{100\ \mathrm{Vol\%}} = 0.28$$

Inserted into the above Formula 3c:

$$k_x = \frac{\left(x_{CO_{2_0}}\Delta\right) * \left(x_{H_{2_0}} + \Delta\right)}{\left(x_{CO_0} - \Delta\right) * \left(x_{H_2O_0} - \Delta\right)} = \frac{(0.03 + \Delta) * (0.28 + \Delta)}{(0.36 - \Delta) * (0.33 - \Delta)}$$
$$= \frac{0.0084 + 0.03 * \Delta + 0.28 * \Delta + \Delta^2}{0.1188 - 0.36 * \Delta - 0.33 * \Delta + \Delta^2} = \frac{0.0084 + 0.31 * \Delta + \Delta^2}{0.1188 - 0.69 * \Delta + \Delta^2}$$

This is a quadratic equation of Δ, which is solved for Δ by the method of quadratic completion using Vieta's theorem (see textbooks of algebra). For this purpose, the equation is brought into the form $0 = x^2 + p * x + q$ and the two solutions

$x_{1.2} = -\frac{p}{2} \pm \sqrt{-q + \frac{p^2}{4}}$ *are calculated, of which one solution is unrealistic.*

With $k_x = 93.1$, this results in

$11.06 - 64.24 * \Delta + 93.1 * \Delta^2 = 0.0084 + 0.31 * \Delta + \Delta^2$

And from that $0 = \Delta^2 - 0.701 * \Delta + 0.120$, thus $p = -0.701$ und $q = 0.120$

With Vieta's theorem follows $\Delta_{1.2} = 0.3505 \pm \sqrt{-0.120 + 0.12281} = 0.3505 \pm 0.053$

$$\Delta_1 = 0.2975 \quad \Delta_2 = 0.4035$$

Δ_2 is unrealistic, because that would result in unrealistic negative molar fractions for CO and H_2O

($x_{CO} = -0.0435$ $x_{H_2O} = -0.0735$) for the resulting gas mixture.)

The correct value is $\Delta_1 = 0.2975$.

Thus, the composition of the gas mixture leaving the reactor is

$$x_{CO} = 0.36 - 0.2975 = \mathbf{0.0625 \rightarrow 6.25\ Vol\%}$$
$$x_{H_2O} = 0.33 - 0.2975 = \mathbf{0.0325 \rightarrow 3.25\ Vol\%}$$

$$x_{CO_2} = 0.03 + 0.2975 = \mathbf{0.3275 \rightarrow 32.75\ Vol\%}$$
$$x_{H_2} = 0.28 + 0.2975 = \mathbf{0.5775 \rightarrow 57.75\ Vol\%}$$

→ *Result*

The equilibrium constant related to the molar fractions is equal to the one related to the partial pressure and is $k_x = 93.1$.

The gas mixture, leaving the reactor consists of 6.25 Vol% carbon monoxide, 3.25 Vol% water vapor, 32.75 Vol% carbon dioxide and 57.75 Vol% hydrogen.

Exercise 12

A gas mixture of carbon dioxide (M = 44 g/mol) and tetrachlorocarbon (M = 154 g/mol) is passed through a 200 °C hot pipe. The partial pressures are $p_{CO_2} = 1.6$ bar and $p_{CCl_4} = 0.4$ bar. The total pressure is 2.0 bar. According to the following reaction equation, highly toxic phosgene (M = 99 g/mol, workplace exposure limit = 0.1 mL/m^3 = 0.1 vol-ppm) can be formed:

$$CO_2 + CCl_4 \rightleftarrows 2COCl_2$$

$$\Delta_R H = 83.7\ kJ/mol$$

a) What is the maximum expected phosgene concentration (volume-ppm and mass-ppm) in the gas mixture leaving the pipe, if the equilibrium constant at 200 °C is
$k_X = 8.0 * 10^{-10}$?
b) What would be the maximum phosgene concentration at a pipe temperature of 400 °C?

⊗ **Solution**
→ ***Strategy***

a) The reaction equation is set up and the stoichiometric factors are taken from it. There is no change in volume caused by the reaction. The sum of the stoichiometric factors is zero. This makes the exponent of the pressure in Formula 3d zero and therefore $k_p = k_x$. The equilibrium constant k_p is expressed according to Formula 3a.
Strictly speaking, the partial pressure of the formed phosgene (Δ) and the associated decrease of the carbon dioxide partial pressure ($p_{CO_2} = p_{CO_{2_0}} - \Delta/2$) and the tetrachloromethane partial pressure ($p_{CCl_4} = p_{CCl_{4_0}} - \Delta/2$) should be inserted into Formula 3a. However, at a very low equilibrium constant (<10^{-3}), the amount of reacted reactant is so small that $p_{CO_2} \cong p_{CO_{2_0}}$ and $p_{CCl_4} \cong p_{CCl_{4_0}}$ can be approximately inserted into Formula 3a. It is solved for Δ. According to the ideal gas law, the molar fraction of phosgene corresponds to the ratio of the volume of phosgene in the gas mixture to the total volume and to the ratio of the partial pressure of phosgene to the total pressure. This results in the volume-ppm. The mass-ppm results from the ratio of the mass of phosgene to the total mass. These are calculated using Formula 2 of the general gas law.
b) The equilibrium constant at 400 °C is determined by means of Formula 3e by inserting the reaction enthalpy, and further processed as already done in part a.

→ ***Calculation***

a) $CCl_4 + CO_2 \rightleftarrows 2\ COCl_2$

$$\nu_{CCl_4} = -1 \quad \nu_{CO_2} = -1 \quad \nu_{CCl_4} = +2 \quad \sum \nu_i = -1 - 1 + 2 = 0$$

$$k_x = k_p * p^{\sum -(\nu_i)} = k_p * p^0 = k_p = \frac{p^2_{COCl_2}}{p_{CCl_4} * p_{CO_2}} \cong \frac{p^2_{COCl_2}}{p_{CCl_{4_0}} * p_{CO_{2_0}}}$$

$$p_{COCl_2} = \sqrt{k_p * p_{CCl_4} * p_{CO_2}} = \sqrt{8.0 * 10^{-10} * 0.4\ bar * 1.6\ bar} = 2.26 * 10^{-5}\ bar$$

$$x_{CCl_4} = \frac{V_{CCl_4}}{V_{total}} = \frac{p_{CCl_4}}{p_{total}} = \frac{2.26 * 10^{-5}\ bar}{2\ bar} = 1.13 * 10^{-5}$$

In a million units of the total mixture
*the **volumetric phosgene content** is* $1.13 * 10^{-5} * 10^{6} = \mathbf{11.3\ vol-ppm}$.
Total mass: $m_{total} = m_{CCl_{4_0}} + m_{CO_{2_0}}$ with $n = \frac{p*V}{R*T}$ and $m = n * M$

$$m_{total} = (p_{CCl_{4_0}} * M_{CCl_4} + p_{CO_{2_0}} * M_{CO_2}) * \frac{V}{R * T}$$

Mass of phosgene: $m_{COCl_2} = p_{COCl_2} * M_{COCl_2} * \frac{V}{R*T}$
and thus

$$\frac{m_{COCl_2}}{m_{total}} = \frac{p_{COCl_2} * M_{COCl_2}}{p_{CCl_{4_0}} * M_{CCl_4} + p_{CO_{2_0}} * M_{CO_2}} = \frac{2.26 * 10^{-5}\text{bar} * 0.099\,^{g}/_{mol}}{0.4\text{bar} * 0.154\,^{g}/_{mol} + 1.6\text{bar} * 0.044\,^{g}/_{mol}}$$
$$= 1.695 * 10^{-5} \simeq 1.7 * 10^{-5}\frac{g}{g}$$

In a million units of the total mixture
*the **mass-based phosgene content*** is $1.7 * 10^{-5} * 10^{6}\frac{mg}{kg} = \mathbf{17.0\ wt-ppm}$.

b) $k_{p2} = k_{p1} * e^{\frac{\Delta_R H}{R} * \left(\frac{1}{T_1} - \frac{1}{T_2}\right)}$

$$k_{p2} = k_{p400\,°C} \quad k_{p1} = k_{p200\,°C} = 8.0 * 10^{-10}$$

$$T_2 = (400 + 273)\text{K} = 673\,\text{K} \quad T_1 = (200 + 273)\,\text{K} = 473\,\text{K}$$

$$k_{p2} = 8.0 * 10^{-10} * e^{\frac{83.7\,kJ * mol * K}{mol * 0.008315\,kJ} * \left(\frac{1}{473\,K} - \frac{1}{673\,k}\right)} = 8.0 * 10^{-10} * e^{6.3244} = 4.46 * 10^{-7}$$

$$p_{COCl_2} = \sqrt{k_p * p_{CCl_4} * p_{CO_2}} = \sqrt{4.46 * 10^{-7} * 0.4\,\text{bar} * 1.6\,\text{bar}} = 5.35 * 10^{-4}\,\text{bar}$$

$$x_{COCl_2} = \frac{V_{COCl_2}}{V_{total}} = \frac{p_{COCl_2}}{p_{total}} = \frac{5.35 * 10^{-4}\,\text{bar}}{2\,\text{bar}} = 2.675 * 10^{-4}$$

In a million units of the total mixture
*the **volumetric phosgene content*** is $2.675 * 10^{-4} * 10^{6} = \mathbf{268\ vol-ppm}$.

$$\frac{m_{COCl_2}}{m_{total}} = \frac{p_{COCl_2} * M_{COCl_2}}{p_{CCl_{4_0}} * M_{CCl_4} + p_{CO_{2_0}} * M_{CO_2}} = \frac{5.35 * 10^{-4}\,\text{bar} * 0.099\,^{g}/_{mol}}{0.4\,\text{bar} * 0.154\,^{g}/_{mol} + 1.6\,\text{bar} * 0.044\,^{g}/_{mol}}$$
$$= 4.01 * 10^{-4}\frac{g}{g}$$

In a million units of the total mixture
*the **mass-based phosgene content*** is $4.01 * 10^{-4} * 10^{6} \cong \mathbf{400\ wt-ppm}$.

→ ***Result***

a) **At a temperature of 200 °C, the gas mixture contains 11.3 vol-ppm phosgene, which corresponds to 17 wt-ppm.**

b) **At a temperature of 400 °C, the gas mixture contains 268 vol-ppm phosgene, which corresponds to 400 wt-ppm.**
According to the industial hygiene guidelines, this is a dangerous concentration of phosgene.

3.2.2 pH Value

Exercise 13
What are the pH values of the following aqueous solutions?

a) 0.01 molar hydrochloric acid solution at complete dissociation?
b) 0.01 molar sulfuric acid solution at complete dissociation?
c) 0.5 normal nitric acid solution at complete dissociation?
d) 0.015 normal formic acid, if it is 30% dissociated?
e) 0.1 molar sodium hydroxide solution at complete dissociation?
f) 0.25 molar calcium hydroxide solution, if it is 15% dissociated?

⊗ **Solution**
→ ***Strategy***
To calculate determine the pH value of the acids, the concentration of H^+ ions is calculated first. According to formula 4a, the negative decimal logarithm of the H^+ ion concentration is determined, which represents the pH value of the solution (enter the H^+ concentration into the calculator and press the "lg" or "log" button and multiply the result by −1.).

To calculate the pH value of the bases, the concentration of OH^- ions is calculated first. According to formula 4a, the negative decimal logarithm of the OH^- ion concentration is determined, which represents the pOH value of the solution (see calculator instructions above). Using the water ion product of 14.0, the pH value is calculated according to formula 4c.

→ ***Calculation***

a) $HCl \leftrightarrows H^+ + Cl^-$
In the case of complete dissociation, 0.01 molar HCl has a H^+concentration of 0.01 mol/L.
$c_{H+} = 0.01\ mole/L = 1.0 * 10^{-2} mol/L$
$lg 10^{-2} = -2.0 \rightarrow \mathbf{pH = 2.0}$
b) $H_2SO_4 \leftrightarrows 2H^+ + SO_4^{2-}$
In the case of complete dissociation, 0.01 molar H_2SO_4 has a H^+concentration of 0.02 mol/L.
$c_{H+} = 0.02\ mol/L = 2.0 * 10^{-2} mol/L$
$lg(2 * 10^{-2}) = lg(10^{-1.70}) = -1.70 \rightarrow \mathbf{pH = 1.7}$

c) $HNO_3 \leftrightarrows H^+ + NO_3^-$
In the case of complete dissociation, 0.5 molar HNO_3 has a H^+concentration of 0.5 mol/L.
$c_{H+} = 0.5\ mol/L = 5.0 * 10^{-1} mol/L$
$lg(5.0 * 10^{-1}) = lg(10^{-0.30}) = -0.30 \rightarrow \mathbf{pH = 0.30}$

d) $HCOOH \leftrightarrows HCOO^- + H^+$
*In the case of complete dissociation, 0.015 molar HCOOH has a H^+concentration of 0.015 mol/L, in the case of 30% dissociation it is 0.015 mol/L * 0.30 = 0.0045 mol/L.*
$c_{H+} = 0.0045\ mol/L = 4.5 * 10^{-3} mol/L$
$lg(4.5 * 10^{-3}) = lg(10^{-2.35}) = -2.35 \rightarrow \mathbf{pH = 2.35}$

e) $NaOH \leftrightarrows Na^+ + OH$
At complete dissociation, 0.1 molar NaOH has a OH^-concentration of 0.1 mol/L.
$c_{OH-} = 0.1\ mol/L = 1.0 * 10^{-1} mol/L$
$lg 10^{-1} = -1 \rightarrow pOH = 1.0$
$pH + pOH = 14.0$
$pH = 14.0 - pOH$
$pH = 14.0 - 1.0 = \mathbf{13.0}$

f) $Ca(OH)_2 \leftrightarrows Ca^{2+} + 2OH^-$
*At complete dissociation, 0.25 molar $Ca(OH)_2$ has a OH^-concentration of 0.5 mol/L, at 15% dissociation it is 0.5 mol/L * 0.15 = 0.075 mol/L.*
$c_{OH-} = 0.075\ mol/L = 7.5 * 10^{-2} mol/L$
$lg 10^{-1.12} = -1.12 \rightarrow pOH = 1.12$
$pH + pOH = 14.0$
$pH = 14.0 - pOH$
$pH = 14.0 - 1.12 = \mathbf{12.9}$

→ ***Result***

		%Dissociated	pH
a)	0.01 M HCl	100	2.0
b)	0.01 M H_2SO_4	100	1.7
c)	0.5 N HNO_3	100	0.30
d)	0.015 N HCOOH	30	2.35
e)	0.1 M NaOH	100	13.0
f)	0.25 M $Ca(OH)_2$	15	12.9

Task 14
What are the molarities of the following aqueous solutions?

a) A sulfuric acid solution with a pH of 1.4 (complete dissociation assumed)?
b) A potassium hydroxide solution with a pH of 11.3 (complete dissociation assumed)?

⊗ **Solution**

→ ***Strategy***

The H^+ or OH^- concentration can be calculated using Formula 4a. For part b., the pH value must first be converted to pOH using Formula 4c.

→ ***Calculation***

a) $H_2SO_4 \leftrightarrows 2H^+ + SO_4^{2-}$ $c_{H_2SO_4} = c_{SO_4-} = \frac{1}{2}c_{H+}$

$$c_{H^+} = \frac{10^{-pH}\text{mol}}{\text{L}} = \frac{10^{-1.4}\,\text{mol}}{\text{L}} = \mathbf{0.04\,mol/L} \quad c_{\mathbf{H_2SO_4}} = \mathbf{0.02\frac{mol}{L}}$$

b) $KOH \leftrightarrows K^+ + OH$ $c_{KOH} = c_{OH-} = 10^{-pOH}$

$$pOH = 14.0 - pH = 14.0 - 11.3 = 2.7$$

$$c_{\mathbf{KOH}} = c_{OH-} = \frac{10^{-2.7}\,\text{mol}}{\text{L}} = \mathbf{0.002\,mole/L}$$

→ ***Result***

a) **The sulfuric acid concentration is 0.02 mol/L.**
b) **The potassium hydroxide concentration is 0.002 mol/L.**

Exercise 15

What pH values do 0.25 molar aqueous solutions of acetic acid ($pk_S = 4.76$), chloroacetic acid ($pk_S = 2.86$), dichloroacetic acid ($pk_S = 1.48$), and trichloroacetic acid ($pk_S = 0.70$) have?

⊗ **Solution**

→ ***Strategy***

In Formula 5b, the corresponding pk_S values and the acid concentration of 0.25 mol/L are used and the pH value is calculated with this.

→ ***Calculation***

$$pH = \frac{pk_s - \lg(c_s)}{2}$$

$$\text{with } c_S = 0.25\,\text{mol/L} \rightarrow \lg(c_S) = -0.60$$

$$\textbf{Acetic Acid } pH = \frac{4.76 - (-0.60)}{2} = 2.68 \cong 2.7$$

$$\textbf{Chloroacetic Acid } pH = \frac{2.86 - (-0.60)}{2} = 1.73 \cong 1.75$$

$$\textbf{DichloroaceticAcid}\ pH = \frac{1.48 - (-0.60)}{2} = 1.04 \cong 1.05$$

$$\textbf{TrichloroaceticAcid}\ pH = \frac{0.70 - (-0.60)}{2} = 0.65$$

→ ***Result***

0.25 molar aqueous solutions of the following acids have pH values of:
Acetic acid → 2.7; Chloroacetic acid → 1.75;
Dichloroacetic acid → 1.05; Trichloroacetic acid → 0.65.

Exercise 16

a) What is the pH value of an aqueous solution of 5 g diethylamine per liter (pk_S = 9.5; M = 73.15 g/mol)?
b) What ammonia concentration (pk_S = 9.25) would have to be chosen to achieve the same pH value in an aqueous solution?

⊗ **Solution**

→ ***Strategy***

a) Amines are usually basic, as is the case here with the relatively high pk_S value. Formula 5c is used to calculate the pH value. First, the molarity of the solution is calculated and inserted into the aforementioned formula together with the pk_S value.
b) Formula 5c is rearranged to the base concentration. It is determined with the pk_S value of the ammonia and the previously calculated pH value of the diethylamine solution.

→ ***Calculation***

a) $n = \frac{m}{M} = \frac{5\,\mathrm{g} * \mathrm{mol}}{\mathrm{L} * 73.15\,\mathrm{g}} = 0.0684\,\frac{\mathrm{mol}}{\mathrm{L}}$

$$pH = \frac{(pk_s + \lg[c_B] + 14.0)}{2} = \frac{9.5 + \lg(0.0684) + 14}{2} = \frac{9.5 - 1.17 + 14.0}{2} = \mathbf{11.16}$$

b) $\lg[c_{NH_3}] = 2 * pH - 14.0 - pk_S = 22.32 - 14.0 - 9.25 = -0.93$

$$c_{NH_3} = 10^{\lg[c_{NH_3}]} = 10^{-0.93} = \mathbf{0.117\frac{mol}{L}}$$

→ ***Result***

a) **The pH value of the diethylamine solution is 11.16.**
b) **The same pH value would be achieved with a 0.117 molar ammonia solution.**

Exercise 17
Commonly used buffer systems are ammonium/ammonium chloride and acetic acid/sodium acetate.

a) What pH value does an aqueous solution have that contains 1 mol of ammonia ($pk_S = 9.25$) and 3 moles of ammonium chloride per liter?
b) What pH value does an aqueous solution have that contains 0.5 mol of acetic acid ($pk_S = 4.76$) and 1 mol of sodium acetate per liter?

⊗ **Solution**
→ ***Strategy***
The corresponding are inserted into formulas 5d and 5e and the result is calculated.

→ ***Calculation***

a) $\boldsymbol{pH} = pk_s + \lg\frac{c_{\text{base}}}{c_{\text{salt}}} = 9.25 + \lg\frac{1\,\text{mol/L}}{3\,\text{mol/L}} = 9.25 - 0.48 = \mathbf{8.77}$
b) $\boldsymbol{pH} = pk_s + \lg\frac{c_{\text{salt}}}{\text{acid}} = 4.76 + \lg\frac{1\,\text{mol/L}}{0.5\,\text{mol/L}} = 4.76 + 0.30 = \mathbf{5.06}$

→ ***Result***
The pH of the ammonium buffer is 8.77 and that of the acetate buffer is 5.06.

3.2.3 Solubility Product

Exercise 18
What is the pH of a slurry of calcium hydroxide in water at 25 °C? The corresponding solubility product is $L_P = 6.0 * 10^{-6}\ \text{mol}^3/\text{L}^3$.

⊗ **Solution**
→ ***Strategy***
The solubility product is formed from the reaction equation according to formula 6. The concentration of calcium ions is half as large as that of hydroxide ions. Thus, the value of the calcium ion concentration can be substituted by that of the hydroxide ion concentration. The equation is solved for the hydroxide ion concentration and its value is calculated. According to formulas 4a and 4c, the pH is then determined.

→ ***Calculation***
$Ca(OH)_2 \leftrightarrows Ca^{2+} + 2OH^-$
$\nu_{\text{Ca}} = 1 \quad \nu_{\text{OH}} = 2$ *(The charge of the ions was omitted in the equations.)*

$$L_P = c_{\text{Ca}} * c_{\text{OH}}^2 \text{ with } c_{\text{Ca}} = \frac{1}{2}c_{\text{OH}} \quad \rightarrow L_P = \frac{1}{2} * c_{\text{OH}}^3$$

$$c_{OH} = \sqrt[3]{2 * L_P} = \sqrt[3]{2 * 6.0 * 10^{-6}\,\text{mol}^3/\text{L}^3} = 2.29 * 10^{-2}\,\text{mol/L}$$

$$pOH = -\lg(c_{OH}) = -\lg\left(2.29 * 10^{-2}\right) = 1.64$$

$$pH + pOH = 14.0 \quad pH = 14.0 - pOH \quad \boldsymbol{pH} = 14.0 - 1.64 = \mathbf{12.36 \cong 12.4}$$

→ ***Result***

The pH of a calcium hydroxide slurry at 25 °C is 12.36 ≈ 12.4.

Task 19

To recover silver (M = 107.9 g/mol) from a wastewater stream, the concentration of silver ions should be reduced to a maximum of 0.01 ppm = 10 ppb by adding sodium chloride. How high must the minimum chloride ion concentration be for this purpose? The solubility product of silver chloride at the wastewater temperature of 20 ° C is $L_p = 1.62 * 10^{-10}$ mol ^{2}/L^2. The density of the wastewater is ρ = 1.02 kg/L.

⊗ **Solution**

→ ***Strategy***

First, the equation for the solubility product is set up according to formula 6 and transformed into the chloride concentration. The silver ion concentration, given as ppm, can be converted to the unit mol/L required for the calculation using the density of the wastewater and the atomic weight of silver.

→ ***Calculation***

$AgCl \leftrightarrows Ag^+ + Cl^-$

$L_P = \prod_i c_i^{\nu_i} = c_{Ag}^1 * c_{Cl}^1$ *(The charge of the ions was omitted in the equations.)*

$$c_{Cl} = \frac{L_P}{c_{Ag}}$$

$$1\,\text{ppm} = \frac{1\,\text{mg}}{\text{kg}}$$

$$0.01\,\text{ppm Ag+} = \frac{0.01\,\text{mg Ag+}}{\text{kg}_{\text{waste water}}} \quad \text{targeted unit:} \ \frac{\text{mol}}{\text{L}}$$

$$V = \frac{m}{\rho} \quad n = \frac{m}{M}$$

$$0.01\,\text{ppm Ag+} = \frac{0.01\,\text{mg Ag+}}{\text{kg}_{\text{waste water}}} * \frac{1.02\,\text{kg}_{\text{waste water}}}{\text{L}} * \frac{\text{mol Ag+}}{107{,}900\,\text{mg Ag+}} = 9.453 * 10^{-8}\,\frac{\text{mol Ag+}}{\text{L}}$$

$$c_{Cl^-} = \frac{1.62 * 10^{-10}\text{mol}^2 * \text{L}}{9.453 * 10^{-8}\text{mol} * \text{L}^2} = 1.72 * 10^{-3}\frac{\text{mol}}{\text{L}} \cong \mathbf{0.002}\frac{\textbf{mol}}{\textbf{L}}$$

→ ***Result***

The minimum concentration of chloride ions to achieve a silver concentration of 0.01 ppm in wastewater is 0.00172 mol/L ≅ 0.002 mol/L.

Task 20

In the past, large amounts of toxic lead fluoride have been disposed off on a landfill. What are the maximum concentrations (mol/L; ppm) of lead and fluoride in the leachates of the landfill with 20 ° C, if no other lead and fluoride salts are present? The solubility product of PbF_2 at 20 ° C is 1.74 * 10^{-8} mol^3/L^3. The density of the leachates is approximately 1.035 kg/L. (M_{Pb} = 207.2 g/mol; M_F = 19.0 g/mol).

⊗ Solution

→ ***Strategy***

First, the solubility product of PbF_2 is set up using Formula 6. The molar concentration of fluoride is twice as high as that of lead, so the lead concentration in the formula can be substituted by that of fluoride (substituting the fluoride concentration by that of lead would be a viable alternative solution path). The formula is solved for the fluoride concentration. The fluoride concentration, and hence the lead concentration, in the leachates is calculated in mol/L and converted to ppm by using the density of water.

→ ***Calculation***

$$PbF_2 \rightleftarrows Pb^{2+} + 2F^- \rightarrow \nu_{\text{pb}} = 1 \quad \nu_F = 2$$

$$L_P = \prod_i c_i^{\nu_i} = c_{\text{Pb}}^1 * c_F^2 \textit{ (The charge of the ions was left out.)}$$

$$c_F = 2 * c_{\text{Pb}} \quad \rightarrow \quad L_p = \frac{1}{2} * c_F^3 \quad \rightarrow \quad c_F = \sqrt[3]{2 * L_p}$$

$$c_F = \sqrt[3]{\frac{2 * 1.74 * 10^{-8}\text{mol}^3}{\text{L}^3}} = \mathbf{3.265 * 10^{-3}\ mol/L}$$

$$m_F = n_F * M_F \rightarrow c_F = 3.265 * 10^{-3}\frac{\text{mol}}{\text{L}} * 19.0\frac{\text{g}}{\text{mol}} = 0.0620\frac{\text{g}}{\text{L}} = 62.0\frac{\text{mg}}{\text{L}}$$

$$1\,\text{L} = 1.035\,\text{kg} \quad \rightarrow \quad c_F = \frac{62\,\text{mg}}{1.035\,\text{kg}} = 59.9\frac{\text{mg}}{\text{kg}} \quad 1\frac{\text{mg}}{\text{kg}} = 1\,\text{ppm} \quad \rightarrow \quad \boldsymbol{c_F} \quad \mathbf{60\,ppm}$$

$$c_{\text{Pb}} = \frac{3.265 * 10^{-3}\text{mol}}{2\,\text{L}} = \mathbf{1.63 * 10^{-3}mol/L}$$

$$m_{Pb} = n_{Pb} * M_{Pb} \quad \rightarrow \quad c_{Pb} = 1.63 * 10^{-3}\frac{mol}{L} * 207.2\frac{g}{mol} = 0.338\frac{g}{L} = 338\frac{mg}{L}$$

$$1\,L = 1.035\,kg \rightarrow c_{Pb} = \frac{338\,mg}{1.035\,kg} = 326.6\frac{mg}{kg} \quad 1\frac{mg}{kg} = 1\,ppm \quad \rightarrow \quad c_{Pb} \cong \mathbf{327\,ppm}$$

→ ***Result***

The leachates of the dump have a maximum concentration of fluoride ions of 3.27 * 10^{-3}mol/L or 60 ppm and a maximum concentration of lead ions of 1.63 * 10^{-3}mol/L or 327 ppm. This corresponds to 1.63 * 10^{-3}mol/L or 387 ppm of lead fluoride.

Task 21

The water of a groundwater well has an iron (II) ion content of 100 ppm. It is to be used as process water. For this purpose, the iron content must be reduced to a maximum of 0.5 ppm. This is done by oxidizing to iron (III) ions by blowing air into the water, followed by precipitation as Fe $(OH)_3$ with a flocculating agent and subsequent filtration.

a) What is the minimum pH value required for a precipitation that meets these conditions?
b) How much $Fe(OH)_3$ precipitates from a water stream of 10 m^3/h within a week?

Solubility product $Fe(OH)_3$: $L_P = 3.8 * 10^{-38}\ mol^4/L^4$
$M_{Fe} = 55.8$ g/mol
$M_{Fe(OH)_3} = 106.8$ g/mol
$\rho_{Water} = 1.00$ kg/L

⊗ **Solution**

→ ***Strategy***

a) The reaction equation is set up and from this the solubility product is formulated according to formula 6 and converted to the hydroxide ion concentration. Then the maximum allowed iron concentration of 0.5 ppm is converted to mol/L. The necessary hydroxide concentration results from using this value in the inverted relation of the solubility product. With pOH and the ion product of water (see formula 4a and 4c) the minimum pH value necessary for the precipitation of iron(III) hydroxide results.
b) The difference in the iron content of the groundwater and the purified water, multiplied by the water flow rate and the number of hours in a week, results in the amount of precipitated iron. This is converted to iron(III) hydroxide by the corresponding ratio of molar masses.

→ ***Calculation***

a) $Fe(OH)_3 \leftrightarrows Fe^{3+} + 3OH^- \rightarrow \nu_{Fe} = 1 \quad \nu_{OH} = 3$ *(The charge of the ions was left out in the equations.)*

$$L_P = c_{Fe} * c_{OH}^3 \quad c_{OH} = \sqrt[3]{\frac{L_P}{c_{Fe}}}$$

$C_{Fe} = 0.5\,\text{ppm} = \frac{0.5\,\text{mg Fe}}{1\,\text{kg Water}}$ *with V = m/ρ and n = m/M*

$$c_{Fe} = \frac{0.5\,\text{mg Fe}}{1\,\text{kg}} * \frac{1.00\,\text{kg}}{\text{L}} * \frac{\text{mol Fe}}{55{,}800\,\text{mg Fe}} = 8.96 * 10^{-6}\frac{\text{mol Fe}}{\text{L}}$$

$$c_{OH} = \sqrt[3]{\frac{3.8 * 10^{-38}\text{mol}^4 * \text{L}}{\text{L}^4 * 8.96 * 10^{-6}\text{mol}}} = \sqrt[3]{4.24 * 10^{-33}}\frac{\text{mol}}{\text{L}} = 1.62 * 10^{-11}\frac{\text{mol}}{\text{L}}$$

$$pOH = -\lg(c_{OH}) = -\lg\left(1.62 * 10^{-11}\right) = 10.8$$

$$pH + pOH = 14.0$$

$$\boldsymbol{pH} = 14.0 - pOH = 14.0 - 10.8 = 3.2$$

b) *Mass of iron(III) hydroxide precipitated per unit of time:*

$$\dot{m}_{Fe\downarrow} = \Delta C_{Fe} * \dot{m}_{Water}$$

$$\Delta C_{Fe} = 100\,\text{ppm} - 0.5\,\text{ppm} = 99.5\,\text{ppm} = \frac{99.5\,\text{mg Fe}}{1\,\text{kg Water}}$$

$$\dot{m}_{Water} = \dot{V}_{Water} * \rho_{Water} = 10\frac{\text{m}^3}{\text{h}} * 1000\frac{\text{kg}}{\text{m}^3} = 10{,}000\frac{\text{kg}}{\text{h}}$$

$$\rightarrow \dot{m}_{Fe\downarrow} = 99.5\frac{\text{mg}}{\text{kg}} * 10.000\frac{\text{kg}}{\text{h}} = 0.995\frac{\text{kg}}{\text{h}} \quad 55.8\frac{\text{g}}{\text{mol}}\text{Fe} \triangleq 106.8\frac{\text{g}}{\text{mol}}\text{Fe(OH)}_3$$

$$\dot{m}_{Fe(OH)_3\downarrow} = 0.995\frac{\text{kg}}{\text{h}} * \frac{106.8}{55.8} = 1.90\frac{\text{kg}}{\text{h}}$$

*Number of hours in a week: 7 * 24 h = 168 h*

$$\boldsymbol{m_{Fe(OH)_3\downarrow}} = 1.90\frac{\text{kg}}{\text{h}} * 168\,\text{h} = \mathbf{319\,kg}$$

→ ***Result***

a) **The pH value necessary for the precipitation of iron(III) hydroxide must be at least 3.2, but in order to avoid corrosion, one should not go below a pH value of 8.**
b) **319 kg of precipitated iron(III) hydroxide are precipitated per week, calculated as dry substance.**

3.3 Material Balances

3.3.1 Mass Balances and Stoichiometric Balances

Exercise 22
A wastewater stream of 2.2 m^3 per hour and a density of 1.050 kg/L contains 0.15 wt% sodium sulfite. What mass flow of 5 wt% aqueous hydrogen peroxide solution must be added at least to oxidize all sulfite to sulfate?

(Relative atomic masses: Na = 23 g/mol; S = 32 g/mol; O = 16 g/mol; H = 1 g/mol).

⊗ **Solution**
→ ***Strategy***
The reaction equation is set up. One mole of hydrogen peroxide is required per mole of sulfite. The molar flow of sulfite is calculated from the volume flow and the density of the wastewater. This molar flow corresponds to the minimum required molar flow of hydrogen peroxide. The necessary mass flow is calculated from the molar mass of the hydrogen peroxide and related to the concentration of the hydrogen peroxide solution.

→ ***Calculation***

$$Na_2SO_3 + H_2O_2 \rightarrow Na_2SO_4 + H_2O$$

$$\dot{m}_{Na_2SO_3} = \frac{wt\%_{Na_2SO_3}}{100wt\%} * \dot{m}_{wastewater} = \frac{wt\%_{Na_2SO_3}}{100wt\%} * \dot{V}_{wastewater} * \rho_{wastewater}$$
$$= \frac{0.15}{100} * 2.2\frac{m^3}{h} * 1050\frac{kg}{m^3} = 3.465\frac{kg}{h}$$

$$\dot{n}_{Na_2SO_3} = \frac{\dot{m}_{Na_2SO_3}}{M_{Na_2SO_3}} \text{ by } M_{Na_2SO_3} = (2 * 23 + 32 + 3 * 16)\frac{g}{mol} = 126\frac{g}{mol}$$

$$\dot{n}_{Na_2SO_3} = \frac{3.465\,kg * mol}{h * 0.126\,kg} = 27.5\frac{mol}{h} = \dot{n}_{H_2O_2}$$

$$M_{H_2O_2} = (2 * 1 + 2 * 16)\frac{g}{mol} = 34\frac{g}{mol}$$

$$\dot{m}_{H_2O_2} = \dot{n}_{H_2O_2} * M_{H_2O_2} = \frac{27.5\,\text{mol} * 0.034\,\text{kg}}{\text{h} * \text{mol}} = 0.935\frac{\text{kg}}{\text{h}}$$

5 wt% → 0.935 kg/h
100 wt% → X

$$\dot{m}_{H_2O_2-5\,\%} = \frac{100\,\text{wt\%} * 0.935\,\text{kg}}{5\,\text{wt\%} * \text{h}} = \mathbf{18.7\frac{kgH_2O_2 - Solution}{h}}$$

→ ***Result***
At least 18.7 kg of 5 wt% aqueous hydrogen peroxide solution must be added to the wastewater per hour.

Task 23
A pure sodium chloride solution of a mass flow of 200 t/h is to be prepared for the chlor-alkali electrolysis. To remove the calcium content of 400 ppm, 12 wt% aqueous sodium carbonate solution is added as 1.5 times of the stoichiometrically required amount and the formed calcium carbonate is precipitated by sedimentation (calcium carbonate is approximately completely insoluble).

a) How many kg of calcium carbonate precipitates per hour?
b) What is the feed of sodium carbonate solution in L/s?
c) How large must a maximum of 90% filled stirred tank be to cover the daily sodium carbonate solution requirement?

(Atomic masses: Ca = 40 g/mol, C = 12 g/mol, O = 16 g/mol, Na = 23 g/mol
Density 12 wt% Na_2CO_3 solution: 1120 kg/m³)

⊗ **Solution**
→ ***Strategy***

a) The reaction equation of calcium carbonate precipitation is set up. Then it is calculated how many moles of calcium must be precipitated per hour. This is also the number of moles of calcium carbonate precipitated per hour. With the molar mass of calcium carbonate, the calcium carbonate mass formed per hour is calculated.
b) For 1 mol calcium ions, 1 mol carbonate ions is required for precipitation. With the given excess of 1.5 mol carbonate ions per mol calcium and the molar mass of sodium carbonate, the necessary mass flow of sodium carbonate is calculated. It is converted to the mass flow of the required 12 wt% solution and the volume flow is determined by means of the density of the solution.
c) The daily sodium carbonate solution requirement results from the volume flow per day. This corresponds to 90% of the stirred tank volume, which is calculated by means of the proportion.

→ ***Calculation***

$Ca^{2+} + CO_3^{2-} \rightarrow CaCO_3\downarrow$

a) 1 mol Ca^{2+} *corresponds stoichiometrically* 1 mol Na_2CO_3
Since a 1.5-fold excess of Na_2CO_3 *is used:* $\dot{n}_{Na_2CO_3} = 1.5 * \dot{n}_{Ca^{2+}}$

$$1\,\text{ppm} = \frac{1\,\text{mg}}{10^6\,\text{mg}} = 1\frac{\text{mg}}{\text{kg}}$$

400 ppm Ca^{2+} = 400 mg Ca^{2+}/kg brine
In the case of a mass flow of 200 t brine/h = 200,000 kg/h →

$$\dot{m}_{Ca^{2+}} = \frac{400\,\text{mgCa} * 200{,}000\,\text{kg Brine}}{\text{kg} * \text{h}} = 8 * 10^7\frac{\text{mgCa}}{\text{h}} = 80\frac{\text{kgCa}}{\text{h}}$$

$$\dot{n}_{Ca^{2+}} = \frac{\dot{m}_{Ca}}{M_{Ca}} = \frac{80\,\text{kg} * \text{mol}}{\text{h} * 0.040\,\text{kg}} = \frac{2000\,\text{mol}}{\text{h}}$$

$$\dot{n}_{Ca^{2+}} = \dot{n}_{CaCO_3}$$

$$\dot{m}_{CaCO_3} = \dot{n}_{Ca^{2+}} * M_{CaCO_3}$$

$$M_{CaCO_3} = \frac{(40 + 12 + 3 * 16)\text{g}}{\text{mol}} = 100\,\text{g/mol}$$

$$\dot{\mathbf{m}}_{\mathbf{CaCO_3}} = \frac{2000\,\text{mol} * 0.100\,\text{kg}}{\text{h} * \text{mol}} = \mathbf{200\,kg/h}$$

b) $\dot{n}_{Na_2CO_3} = 1.5 * \frac{2000\,\text{mol}}{\text{h}} = \frac{3000\,\text{mol}}{\text{h}}$

$$\dot{m}_{Na_2CO_3} = \dot{n}_{Na_2CO_3} * M_{Na_2CO_3}$$

$$M_{Na_2CO_3} = \frac{(2 * 23 + 12 + 3 * 16)\text{g}}{\text{mol}} = 106\,\text{g/mol}$$

$$\dot{m}_{Na_2CO_3} = \frac{3000\,\text{mol} * 0.106\,\text{kg}}{\text{h} * \text{mol}} = 318\frac{\text{kg}}{\text{h}}$$

12 percent by weight Na_2CO_3 solution:

$$\dot{m}_{\text{solution}} = \frac{100\%}{12\%} * 318\frac{\text{kg}}{\text{h}} = 2650\frac{\text{kg}}{\text{h}}$$

$$\dot{V}_{\text{solution}} = \frac{\dot{m}_{\text{solution}}}{\rho_{\text{solution}}} = \frac{2650\,\text{kg} * \text{m}^3}{\text{h} * 1120\,\text{kg}} = 2.366\frac{\text{m}^3}{\text{h}} = \frac{2.366\,\text{m}^3 * 1000\,\text{L} * \text{h}}{\text{h} * \text{m}^3 * 3600\,\text{s}} = \mathbf{0.656\,L/s}$$

c) $\dot{V}_{solution} = \frac{2.35\,m^3 * 24\,h}{h * day} = 56.7\,m^3/day$

90% filling level →56.7 m³

100% → X $V_{tank} = \frac{100\% * 56.7\,m^3}{90\%} = \mathbf{63.0\,m^3}$

→ ***Result***

a) **200 kg of calcium carbonate are produced per hour.**
b) **The volumetric flow of the added sodium carbonate solution is 0.656 L/s.**
c) **The stirred tank for preparing the sodium carbonate solution must have a total volume of 63 m³.**

Task 24

An industrial-scale method for the production of diphenyl carbonate (DPC M_{DPC} = 214 g/mol) proceeds via the introduction of phosgene (M_{COCl_2}= 99 g/mol) into an aqueous, alkaline solution of phenol (Phen M_{Phen} = 94 g/mol), which is present as sodium phenolate. The solution contains an excess of sodium hydroxide (M_{NaOH} = 40 g/mol).

a) How many kg of phenol are needed for its complete conversion to 1 metric t DPC, if no side reactions of phenol occur?
b) How many kg of phosgene are required for the production of 1 metric t DPC, if 10% of the phosgene feed is lost by hydrolysis with sodium hydroxide?
c) How much kg of sodium chloride and sodium carbonate are formed in this process?

⊗ **Solution**

→ ***Strategy***

The equations of the DPC formation and the phosgene hydrolysis are formulated. The number of moles of 1 metric t DPC is calculated using its molar mass. The number of moles of phenol used is twice as high as the number of moles of DPC formed. From this, the mass of phenol used is calculated by means of the molar mass of phenol. Without hydrolysis, the number of moles of phosgene used would be equal to the number of moles of DPC formed. By hydrolysis, 10% more phosgene is required, so that only 90% of the introduced phosgene reacts to DPC. Both in the formation of DPC and in the hydrolysis, 2 moles of sodium chloride are formed per mol of phosgene. This results in the double molar amount of phosgene used, multiplied by the molar mass of sodium chloride, the mass of sodium chloride is calculated. Sodium carbonate is only formed by hydrolysis. In this process, 1 mol of carbonate is formed per mole of phosgene hydrolyzed. The mass of sodium carbonate formed thus corresponds to one tenth of the number of moles of phosgene used, multiplied by the molar mass of sodium carbonate.

→ ***Calculation***

DPC synthesis: 2 Ø-ONa + $COCl_2$ → Ø-O-CO-O-Ø + 2NaCl

Hydrolysis: $COCl_2 + 4\,NaOH \rightarrow Na_2CO_3 + 2NaCl$

$_{Phen} = n_{Phen} * M_{Phen}$ with $n_{Phen} = 2 * n_{DPC}$ and

a)
$$n_{DPC} = \frac{m_{DPC}}{M_{DPC}} = \frac{1000\,\text{kg} * \text{mol}}{0.214\,\text{kg}} = 4672.9\,\text{mol}$$

$$m_{Phen} = 2 * n_{DPC} * M_{Phen} = 2 * 4672.9\,\text{mol} * 0.094\frac{\text{kg}}{\text{mol}} = \mathbf{878.5\,kg}$$

b) *Without hydrolysis would* $n_{COCl_2-ideal} = n_{DPC}$
This corresponds to only 90% of the used phosgene. The total amount of phosgene used is therefore

$$n_{COCl_2} = \frac{100\%}{90\%} * n_{COCl_2-ideal} = \frac{100\%}{90\%} * n_{DPC} = 1.111 * 4672.9\,\text{mol} = 5192\,\text{mol}$$

$$m_{COCl_2} = n_{COCl_2} * M_{COCl_2} = 5192\,\text{mol} * 0.099\frac{\text{kg}}{\text{mol}} = \mathbf{514\,kg}$$

c) $m_{NaCl} = n_{NaCl} * M_{NaCl}$ with $n_{NaCl} = 2 * n_{COCl_2}$

$$m_{NaCl} = 2 * n_{COCl_2} * M_{NaCl} = 2 * 5192\,\text{mol} * 0.0585\frac{\text{kg}}{\text{mol}} = 607.5\,\text{kg}$$

$$m_{Na_2CO_3} = n_{Na_2CO_3} * M_{Na_2CO_3} \text{ with } n_{Na_2CO_3} = \frac{10\%}{100\%} * n_{COCl_2} = 0.1 * 5192\,\text{mol} = 519.2\,\text{mol}$$

$$m_{Na_2CO_3} = 519.2\,\text{mol} * 0.106\frac{\text{kg}}{\text{mol}} = \mathbf{55.0\,kg}$$

→ ***Result***

a) **The amount of phenol needed to produce 1** metric **t of diphenyl carbonate is 878.5 kg.**
b) **Taking into account the side reaction of phosgene hydrolysis, 514 kg of phosgene is required for this.**
c) **As by-products, 607.5 kg of sodium chloride and 55 kg of sodium carbonate are formed.**

Task 25

The concentration of chlorine gas in an exhaust stream of 20 m^3 per hour is to be completely removed in a washing column operated with 1.25 molar aqueous sodium hydroxide (M_{NaOH} = 40 g/mol).

In a laboratory test to measure the average chlorine gas concentration, 100 mL of the exhaust gas were intensively shaken with 20 mL of 0.1 normal potassium hydroxide solution, so that it can be assumed that the reaction is complete. The final concentration of KOH was 0.065 mol/L.

What volume flow of 1.25 molar sodium hydroxide must be supplied to the absorption column?

⊗ **Solution**

→ ***Strategy***

The reaction equation is set up and the HCl content in the exhaust gas is calculated. For this purpose, the molar HCl content is calculated from the laboratory result of the difference in the KOH concentration and the volume of the gas sample. From this the molar flow of HCl in the entire exhaust gas stream is calculated. The amount of NaOH required for neutralization is determined and calculated for a 1.25 molar sodium hydroxide solution.

→ ***Calculation***

$HCl + NaOH \rightarrow NaCl + H_2O$

For 100 mL exhaust gas :

$$n_{HCl} = \Delta n_{KOH} = \Delta c_{KOH} * V_{0.1NKOH} = (0.1 - 0.065)\frac{mol}{L} * 0.02\,L = 7.0 * 10^{-4}\,mol$$

For 20 m³exhaust gas/h:

$$\dot{n}_{HCl} = \frac{7.0 * 10^{-4}\,mol * 20\,m^3}{0.1\,L * h} = \frac{7.0 * 10^{-4}\,mol * 20{,}000\,L}{0.1\,L * h}$$
$$= \frac{140\,mol}{h} = \frac{140\,mol * h}{h * 3600\,s} = 0.0389\frac{mol}{s}$$

$$\dot{n}_{NaOH} = \dot{n}_{HCl} = 0.0389\frac{mol}{s}$$

$$\dot{V} = \frac{\dot{n}}{c} = \frac{0.0389\,mol * L}{s * 1.25\,mol} = \mathbf{0.0311\frac{L}{s}} = \mathbf{112\frac{L}{h}}$$

→ ***Result***

The stream of 1.25 molar sodium hydroxide to be supplied to the absorption column is 0.031 L/s corresponding to 112 L/h.

3.3.2 Conversion, Yield and Selectivity

→ ***General solution strategy***

All of the following tasks are solved using the same basic pattern:

- Setting up the reaction equation
- Determining the stoichiometric factors ν_i
- Converting from mass-based to mole-based quantities
- Inserting the data into Formula 9a, b, c (conversion), 10a, b, c (yield) and 11a, b, c (selectivity)
- If necessary, rearranging the formulas
- Calculating the results

Task 26

In a stirred tank reactor , the esterification reaction of 920 kg ethanol (M_{Et} = 46 g/mol) with 1.5 metric tons acetic acid (M_{Ac} = 60 g/mol) to ethyl acetate (M_{EE} = 88 g/mol). The final mixture contains 1.32 metric t ethyl acetate, 230 kg ethanol and 600 kg acetic acid. What is the conversion of ethanol and acetic acid. What is the yield of ester vs. ethanol?

⊗ **Solution**

→ ***Calculation***

$$C_2H_5OH + CH_3COOH \leftrightarrows CH_3COO\text{-}C_2H_5 + H_2O$$

$$n_{i_o} = \frac{m_{i_o}}{M_i}$$

$$n_{Et_o} = \frac{920\,\text{kg} * \text{mol}}{0.046\,\text{kg}} = 20{,}000\,\text{mol} \quad n_{Ac_o} = \frac{1500\,\text{kg} * \text{mol}}{0.060\,\text{kg}} = 25{,}000\,\text{mol}$$

$$n_i = \frac{m_i}{M_i}$$

$$n_{Et} = \frac{230\,\text{kg} * \text{mol}}{0.046\,\text{kg}} = 5000\,\text{mol} \quad n_{Ac} = \frac{600\,\text{kg} * \text{mol}}{0.060\,\text{kg}} = 10{,}000\,\text{mol}$$

$$n_{EE} = \frac{1320\,\text{kg} * \text{mol}}{0.088\,\text{kg}} = 15{,}000\,\text{mol}$$

	C_2H_5OH	CH_3COOH	$CH_3COO\text{-}C_2H_5$	H_2O
ν_i	−1	−1	+1	+1
n_{i_o}	20,000 mol	25,000 mol	0 mol	0 mol
n_i	5000 mol	10,000 mol	15,000 mol	

Conversions: $X_E = \frac{n_{E_o} - n_E}{n_{E_o}}$

$$X_{\text{Et}} = \frac{(20{,}000 - 5000)\text{mol}}{20{,}000\,\text{mol}} = \mathbf{0.75 \rightarrow 75.0\,\%}$$

$$X_{\text{Ac}} = \frac{(25{,}000 - 10{,}000)\text{mol}}{25{,}000\,\text{mol}} = \mathbf{0.60 \rightarrow 60.0\,\%}$$

Yield: $Y_{P/E} = \frac{\upsilon_E * (n_{P_o} - n_P)}{\upsilon_p * n_{E_o}}$ *(It is based on ethanol as a lower stoichiometric component!)*

$$Y_{\text{EE/Et}} = \frac{-1 * (0 - 15{,}000)\text{mol}}{+1 * 20{,}000\,\text{mol}} = \mathbf{0.75 \rightarrow 75.0\,\%}$$

→ ***Result***

The conversion is 0.75 or 75.0% for ethanol and 0.60 or 60.0% for acetic acid. The yield of ester based on ethanol is 0.75 or 75.0%.

Task 27

Urea $OC(NH_2)_2$ is an important basic chemical and is produced from carbon dioxide CO_2 and ammonia NH_3 at 150 °C and 40 bar. A reactor is fed per second with 4.4 kg carbon dioxide and 5.1 kg ammonia. The result is a stream of products leaving the reactor consisting of 0.044 kg carbon dioxide, 5.8 kg urea and 1.7 kg ammonia per second.

(Molar masses CO_2 = 44 g/mol; NH_3 = 17 g/mol; urea = 60 g/mol)

What is the conversion of carbon dioxide and ammonia as well as the yield of urea in relation to carbon dioxide or ammonia?

⊗ **Solution**

→ ***Calculation***

$CO_2 + 2NH_3 \rightarrow (NH_2)_2C{=}O + H_2O$

$$\dot{n}_{i_o} = \frac{\dot{m}_{i_o}}{M_i}$$

$$\dot{n}_{CO_{2_o}} = \frac{4.4\,\text{kg} * \text{mol}}{\text{s} * 0.044\,\text{kg}} = 100\frac{\text{mol}}{\text{s}} \quad \dot{n}_{NH_{3_o}} = \frac{5.1\,\text{kg} * \text{mol}}{\text{s} * 0.017\,\text{kg}} = 300\frac{\text{mol}}{\text{s}}$$

$$\dot{n}_i = \frac{m_i}{M_i}$$

$$\dot{n}_{CO_2} = \frac{0.044\,\text{kg} * \text{mol}}{\text{s} * 0.044\,\text{kg}} = 1.0\frac{\text{mol}}{\text{s}} \quad \dot{n}_{NH_3} = \frac{1.7\,\text{kg} * \text{mol}}{\text{s} * 0.017\,\text{kg}} = 100\frac{\text{mol}}{\text{s}}$$

$$\dot{n}_{urea} = \frac{5.8\,\text{kg} * \text{mol}}{\text{s} * 0.060\,\text{kg}} = 96.7\frac{\text{mol}}{\text{s}}$$

	CO_2	NH_3	$(NH_2)_2C{=}O$	H_2O
ν_i	−1	−2	+1	+1
n_{i_o}	100 mol/s	300 mol/s	0 mol/s	0 mol/s
n_i	1 mol/s	100 mol/s	96.7 mol/s	

Conversions: $X_E = \frac{\dot{n}_{E_0} - \dot{n}_E}{\dot{n}_{E_0}}$

$$X_{CO_2} = \frac{(100-1)\,\text{mol} * \text{s}}{\text{s} * 100\,\text{mol}} = \mathbf{0.99 \rightarrow 99.0\,\%}$$

$$X_{NH_3} = \frac{(300-100)\,\text{mol} * \text{s}}{\text{s} * 300\,\text{mol}} = \mathbf{0.667 \rightarrow 66.7\,\%}$$

Yields: $Y_{P/E} = \frac{v_E * (\dot{n}_{P_o} - \dot{n}_P)}{v_P * \dot{n}_{E_o}}$

$$Y_{Urea/CO_2} = \frac{-1 * (0 - 96.7)\,\text{mol} * \text{s}}{\text{s} * 1 * 100\,\text{mol}} = 0.967 \rightarrow 96.7\,\%$$

$$Y_{Urea/NH_3} = \frac{-2 * (0 - 96.7)\,\text{mol} * \text{s}}{\text{s} * 1 * 300\,\text{mol}} = 0.645 \rightarrow 64.5\,\%$$

→ Result
The conversion of carbon dioxide is 0.99 i.e. 99.0 %, that of ammonia 0.667 i.e. 66.7 %.

The yield of urea in relation to carbon dioxide is 0.967 i.e. 96.7 %. The yield of urea in relation to ammonia is 0.645 i.e. 64.5 %.

Task 28
In a test reactor, acetone (Ac) is hydrogenated to 2-propanol (Prop) in the presence of an inert solvent with hydrogen:

$$CH_3\text{-}\underset{O}{C}\text{-}CH_3 + H_2 \xrightarrow{\text{Kata}} CH_3\text{-}\underset{\underset{H}{O}}{\overset{H}{C}}\text{-}CH_3$$

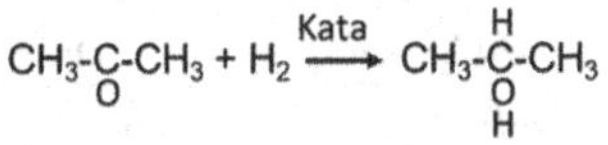

The reaction volume remains constant during the reaction. The concentrations of the feed mixture are: $c_{Ac\text{-}o} = 5.0$ mol/L and $c_{Prop\text{-}o} = 0.1$ mol/L.

After the experiment, the following concentrations are measured: $c_{Ac} = 0.9$ mol/L; $c_{Prop} = 3.2$ mol/L. What is the conversion of acetone, the yield of iso-propanol and the selectivity of iso-propanol?

⊗ Solution
Since the volume does not change during the reaction, the concentrations may be used for the calculation. Only the concentration of the reactant acetone is given. Thus, in this case, only the reference to acetone is given for the yield and selectivity of 2-propanol.

→ Calculation
The stoichiometric factor for acetone is $v_{Ac} = -1$, that for 2-propanol $v_{Prop} = +1$.

$$X_{Ac} = \frac{c_{Ac_o} - c_{Ac}}{c_{Ac_o}} = \frac{5.0 - 0.9}{5.0} = 0.82 \rightarrow 82.0\,\%$$

$$Y_{Prop/Ac} = \frac{v_{Ac} * (c_{Prop_o} - c_{Prop})}{v_{prop} * c_{Ac_o}} = \frac{-1 * (0.1 - 3.2)}{+1 * 5.0} = 0.62 \rightarrow 62.0\,\%$$

$$S_{Prop/Ac} = \frac{v_{Ac} * (c_{Prop_o} - c_{Prop})}{v_p * (c_{Ac_o} - c_{Ac})} = \frac{-1 * (0.1 - 3.2)}{+1 * (5.0 - 0.9)} = 0.756 \rightarrow 75.6\,\%$$

→ ***Result***

The conversion of acetone is 0.82 i.e. 82.0%. The yield of 2-propanol relative to acetone is 0.62 i.e. 62.0%. The selectivity of 2-propanol with respect to acetone is 0.756 i.e. 75.6%.

Task 29

Allyl chloride (ACl $M_{ACl} = 76.5$ g/mol) is reacted with dimethylamine (DMA) to allyl-dimethylamine (ADMA $M_{ADMA} = 85.0$ g/mol), whereby the resulting hydrogen chloride is neutralized with sodium hydroxide. The yield is 85% relative to allyl chloride. 1 metric ton of allyl-dimethylamine is to be produced per batch. How many kg of allyl chloride is required per batch?

⊗ **Solution**

→ ***Calculation***

$$Cl\text{-}CH_3\text{-}CH_2{=}CH_2 + (CH_3)_2NH \rightarrow CH_2{=}CH\text{-}CH_2\text{-}N(CH_3)_2 + HCl$$

$$n_{\text{ADMA}} = \frac{m_{\text{ADMA}}}{M_{\text{ADMA}}} = \frac{1000\,\text{kg} * \text{mol}}{0.085\,\text{kg}} = 11{,}765\,\text{mol}$$

	ACl	DMA	ADMA	HCl
ν_i	−1	−1	+1	+1

Definition Yield: $Y_{P/E} = \frac{\nu_E * (n_{P_0} - n_P)}{\nu_p * n_{E_0}}$ $\quad Y_{\text{ADMA/ACl}} = \frac{\nu_{\text{ACl}} * (n_{\text{ADMA}_0} - n_{\text{ADMA}})}{\nu_{\text{ADMA}} * n_{\text{ACl}_0}}$

Dissolved in $n_{\text{ACl}_0} = \frac{\nu_{\text{ACl}} * (n_{\text{ADMA}_0} - n_{\text{ADMA}})}{\nu_{\text{ADMA}} * Y_{\text{ADMA/ACl}}} = \frac{-1 * (0 - 11765)\,\text{mol}}{+1 * 0.85} = 13841\,\text{mol}$

$$m_{\text{ACl}_0} = n_{\text{ACl}_0} * M_{\text{ACl}} = 13{,}841\,\text{mol} * 0.0765\frac{\text{kg}}{\text{mol}} = \mathbf{1059\,kg} \cong \mathbf{1060\,kg}$$

→ ***Result***

1060 kg of allyl chloride is required per batch.

Task 30

The quaternary ammonium salt dimethyl-diallyl-ammonium chloride (DADMAC M = 161.6 g/mol) is prepared by the reaction of dimethylamine (DMA M = 45.1 g/mol) and allyl chloride (ACl M = 76.5g/mol):

2 ACl + DMA = DADMAC + HCl (The formed HCl is converted to NaCl by the presence of NaOH.)

For this purpose, 2000 kg of 10% DMA solution in acetone and 1200 kg of 30% sodium hydroxide solution are fed into a 15 m^3 stirred tank reactor. 5 m^3 of a 1-molar ACl solution in acetone is added within 6 h of stirring the reactor content. The reaction is completed after 7 h. The resulting solution contains 50 mol ACl, 520 mol DMA and 380 kg DADMAC.

a) What are the conversions of ACl and DMA?
b) What yields of DADMAC are achieved with respect to ACl and DMA?
c) What are the selectivities of DADMAC with respect to ACl and DMA?

⊗ **Solution**

→ ***Calculation***

$$\nu_{\text{DMA}} = -1 \quad \nu_{\text{ACl}} = -2 \quad \nu_{\text{DADMAC}} = +1 \quad \nu_{\text{HCl}} = +1$$

Starting mixture:

$$m_{\text{DMA}_0} = 2000\,\text{kg} * \frac{10\%}{100\%} = 200\,\text{kg}$$

$$n_{\text{DMA}_0} = \frac{m_{\text{DMA}_0}}{M_{\text{DMA}}} = \frac{200\,\text{kg} * \text{mol}}{0.0451\,\text{kg}} = 4434.6\,\text{mol}$$

$$n_{\text{ACl}_0} = V * c_{\text{ACl}} = 5000\,\text{L} * 1.0\frac{\text{mol}}{\text{L}} = 5000\,\text{mol}$$

$$n_{\text{DADMAC}_0} = 0\,\text{mol}$$

After the reaction:

$$n_{\text{ACl}} = 50\,\text{mol} \quad n_{\text{DMA}} = 520\,\text{mol} \quad n_{\text{DADMAC}} = \frac{m_{\text{DADMAC}}}{M_{\text{DADMAC}}} = \frac{380\,\text{kg} * \text{mol}}{0.161.6\,\text{kg}} = 2351.5\,\text{mol}$$

a) *Conversions:*

$$X_{\mathbf{DMA}} = \frac{n_{\text{DMA}_0} - n_{\text{DMA}}}{n_{\text{DMA}_0}} = \frac{(4434.6 - 520)\,\text{mol}}{4434.6\,\text{mol}} = \mathbf{0.883 \rightarrow 88.3\ \%}$$

$$X_{\mathbf{ACl}} = \frac{n_{\text{ACl}_0} - n_{\text{ACl}_0}}{n_{\text{ACl}_0}} = \frac{(5000 - 50)\,\text{mol}}{5000\,\text{mol}} = \mathbf{0.990 \rightarrow 99.0\ \%}$$

b) *Yields:*

$$Y_{\mathbf{DADMAC/DMA}} = \frac{\nu_{\text{DMA}} * \left(n_{\text{DADMAC}_0} - n_{\text{DADMAC}}\right)}{\nu_{\text{DADMAC}} * n_{\text{DMA}_0}} = \frac{-1 * (0 - 2351.5)\,\text{mol}}{+1 * 4434.6\,\text{mol}} = \mathbf{0.530 \rightarrow 53.0\ \%}$$

$$Y_{\mathbf{DADMAC/ACl}} = \frac{\nu_{\text{ACl}} * (n_{\text{DADMAC}_0} - n_{\text{DADMAC}})}{\nu_{\text{DADMAC}} * n_{\text{ACl}_0}} = \frac{-2 * (0 - 2351.5)\ \text{mol}}{+1 * 5000\ \text{mol}}$$
$$= \mathbf{0.941 \rightarrow 94.1\ \%}$$

c) *Selectivities:*

$$S_{\mathbf{DADMAC/DMA}} = \frac{\nu_{\text{DMA}} * \left(n_{\text{DADMAC}_0} - n_{\text{DADMAC}}\right)}{\nu_{\text{DADMAC}} * (n_{\text{DMA}_0} - n_{\text{DMA}})} = \frac{-1 * (0 - 2351.5)\ \text{mol}}{+1 * (4434.6 - 520)\ \text{mol}}$$
$$= \mathbf{0.601 \rightarrow 60.1\ \%}$$

$$S_{\mathbf{DADMAC/ACl}} = \frac{\nu_{\text{ACl}} * (n_{\text{DADMAC}_0} - n_{\text{DADMAC}})}{\nu_{\text{DADMAC}} * (n_{\text{ACl}_0} - n_{\text{ACl}})} = \frac{-2 * (0 - 2351.5)\text{mol}}{+1 * (5000 - 50)\text{mol}}$$
$$= \mathbf{0.950 \rightarrow 95.0\ \%}$$

→ ***Result***

a) **The conversion of dimethylamine was 0.883 i.e. 88.3%, that of allyl chloride 0.990 i.e. 99.0%.**
b) **The yield of DADMAC with respect to dimethylamine was 0.530 i.e. 53.0%.**
c) **The yield of DADMAC with respect to allyl chloride was 0.941 i.e. 94.1%.**
d) **The selectivity of DADMAC with respect to dimethylamine was 0.601 i.e. 60.1%.**

The selectivity of DADMAC with respect to allyl chloride was 0.950 i.e. 95.0%.

Task 31
The transesterification of dimethylcarbonate (DMC) with phenol (POH) to diphenylcarbonate (DPC) and methanol (MOH) is carried out on an industrial scale:

$$2\,\text{Ø-OH} + (CH_3)_2CO_3 \leftrightarrows (\text{Ø})_2CO_3 + 2CH_3OH$$

The following streams of reactants and products of the equilibrium reaction were measured in mol/s:

	Phenol	DMC	DPC	CH_3OH
Feed	50	20	4	2
Exit	28	5	15	20

a) What are the conversions of phenol and DMC?
b) What is the yield of DPC in relation to phenol and in relation to DMC?
c) What are the selectivities of DPC in relation to phenol and in relation to DMC?

⊗ **Solution**
→ ***Calculation***

$$\nu_{\text{POH}} = -2 \quad \nu_{\text{DMC}} = -1 \quad \nu_{\text{DPC}} = +1 \quad \nu_{\text{MOH}} = +2$$

a) Conversion:

$$X_E = \frac{\dot{n}_{E_o} - \dot{n}_E}{\dot{n}_{E_o}}$$

$$X_{\text{POH}} = \frac{\dot{n}_{\text{POH}_o} - \dot{n}_{\text{POH}}}{\dot{n}_{\text{POH}_o}} = \frac{(50 - 28)\ \text{mol} * \text{s}}{\text{s} * 50\ \text{mol}} = \frac{22}{50} = \mathbf{0.44 \rightarrow 44\ \%}$$

$$X_{\text{DMC}} = \frac{\dot{n}_{\text{DMC}_o} - \dot{n}_{\text{DMC}}}{\dot{n}_{\text{DMC}_o}} = \frac{(20 - 5)\ \text{mol} * \text{s}}{\text{s} * 20\ \text{mol}} = \frac{15}{20} = \mathbf{0.750 \rightarrow 75.0\ \%}$$

b) Yield:

$$Y_{P/E} = \frac{\upsilon_E * (\dot{n}_{P_o} - \dot{n}_P)}{\upsilon_p * \dot{n}_{E_o}}$$

$$Y_{\mathbf{DPC/POH}} = \frac{\upsilon_{\mathrm{POH}} * (\dot{n}_{\mathrm{DPC_o}} - \dot{n}_{\mathrm{DPC}})}{\upsilon_{\mathrm{DPC}} * \dot{n}_{\mathrm{POH_o}}} = \frac{-2 * (4 - 15)\ \mathrm{mol} * \mathrm{s}}{\mathrm{s} * 1 * 50\ \mathrm{mol}} = \frac{22}{50} = \mathbf{0.44 \to 44\ \%}$$

$$Y_{\mathbf{DPC/DMC}} = \frac{\upsilon_{\mathrm{DMC}} * (\dot{n}_{\mathrm{DPC_o}} - \dot{n}_{\mathrm{DPC}})}{\upsilon_{\mathrm{DPC}} * \dot{n}_{\mathrm{DMC_o}}} = \frac{-1 * (4 - 15)\ \mathrm{mol} * \mathrm{s}}{\mathrm{s} * 1 * 20\ \mathrm{mol}} = \frac{11}{20} = \mathbf{0.55 \to 55\ \%}$$

c) Selectivity:

$$S_{P/E} = \frac{\upsilon_E * (\dot{n}_{P_o} - \dot{n}_P)}{\upsilon_p * (\dot{n}_{E_o} - \dot{n}_E)}$$

$$S_{\mathbf{DPC/POH}} = \frac{\upsilon_{\mathrm{POH}} * (\dot{n}_{\mathrm{DPC_o}} - \dot{n}_{\mathrm{DPC}})}{\upsilon_{DPC} * (\dot{n}_{\mathrm{POH_o}} - \dot{n}_{\mathrm{POH}})} = \frac{-2 * (4 - 15)\ \mathrm{mol} * \mathrm{s}}{\mathrm{s} * 1 * (50 - 28)\ \mathrm{mol}} = \frac{22}{22} = \mathbf{1.00 \to 100\ \%}$$

$$S_{\mathbf{DPC/DMC}} = \frac{\upsilon_{\mathrm{DMC}} * (\dot{n}_{\mathrm{DPC_o}} - \dot{n}_{\mathrm{DPC}})}{\upsilon_{\mathrm{DPC}} * (\dot{n}_{\mathrm{DMC_o}} - \dot{n}_{\mathrm{DMC}})} = \frac{-1 * (4 - 15)\ \mathrm{mol} * \mathrm{s}}{\mathrm{s} * 1 * (20 - 5)\ \mathrm{mol}} = \frac{11}{15} = \mathbf{0.73 \to 73\ \%}$$

→ ***Result***

a) **The conversion of phenol is 0.44 i.e. 44%, that of dimethylcarbonate 0.75 i.e. 75%.**
b) **The yield of diphenylcarbonate in terms of phenol is 0.44 i.e. 44%. The yield of diphenylcarbonate in terms of dimethylcarbonate is 0.55 i.e. 55%.**
c) **The selectivity of diphenylcarbonate in terms of phenol is 1.00 i.e. 100%, i.e., that the phenol reacted was completely converted to diphenylcarbonate and no side reactions occurred.**

The selectivity of diphenylcarbonate in terms of dimethylcarbonate is 0.73 i.e. 73%, i.e., that 27% of the dimethylcarbonate used reacted in side reactions.

3.3.3 Reaction Rate

Exercise 32
Carbamide decomposes according to the following first-order reaction equation into an amine and carbon dioxide, so they are used, among other things, for foaming of polymers:

$$\text{R-NH}-\text{COOH} \to \text{R-NH}_2 + \text{CO}_2$$

How many mol/L or % of the initial concentration of carbamide remain after a reaction time of exactly one minute in an approach of an initial concentration of 0.15 mol/L in the polymer melt? The rate constant at the selected reaction temperature is 0.11 s^{-1}.

⊗ **Solution**

→ ***Strategy***

This is a reaction of the type A → B + C.

The concentration of carbamide (Carb) is calculated after the specified reaction time using formula 13c and the percentage residual content is determined by applying the rule of three. It should be noted that in this case the mol- or mass-related percentage is identical.

→ ***Calculation***

$c_{Carb} = c_{Carb_0} * e^{-k*t} \textit{with } t = 1\ min = 60\ s$

$$c_{\mathbf{Carb}} = 0.15\frac{\text{mol}}{\text{L}} * e^{-\frac{0.11 * 60\,\text{s}}{\text{s}}} = 0.15\frac{\text{mol}}{\text{L}} * 0.00136 = \mathbf{2.04 * 10^{-4}\frac{mol}{L}}$$

0.15 mol/L → 100 %

0.000204 mol/L → X %

$$X = \frac{0.000204\frac{\text{mol}}{\text{L}} * 100\ \%}{0.15\ \text{mol/L}} = \mathbf{0.135\ \%}$$

→ ***Result***

The polymer foam contains 2.04 * 10^{-4}mol carbamide based on the volume of the polymer matrix. This corresponds to 0.135% of the initial concentration.

Exercise 33

In a production plant for polyvinyl chloride (PVC), a solution of dilauroyl peroxide (DLPO) is used as an initiator. The storage time in the make-up tank is 30 days. Dilauroyl peroxide decomposes according to a first-order reaction.

a) To quantify the decomposition kinetics, two experiments were carried out in the plant laboratory:
   A 0.01 molar DLPO solution was held at 70 °C for 10 hours. After this time, a DLPO content of 1.54 * 10^{-3} mol/L was analyzed. A similar experiment at 50 °C and a reaction time of exactly three days resulted in a DLPO content of 4.85 * 10^{-3} mol/L. What is the value of the rate constant at both temperatures, and what is the activation energy of the decomposition reaction?
b) For the sake of uniform quality of the PVC, the DPLO content of the used solution may only decrease by a maximum of 2% relative to the content of its preparation. Is this still the case at a storage temperature of 15 °C and a storage time of 30 days?

⊗ **Solution**

→ ***Strategy***

This is a reaction of the type A → C.

a) The rate constants are calculated according to the appropriately rearranged formula 13c from the ratio of the initial to the final concentration and the reaction time. From the thus calculated rate constants and the associated temperatures, the activation energy results from the rearranged formula 17b.
b) Also from formula 17b, the rate constant at 15 °C is calculated with the determined activation energy and a known rate constant of an associated temperature (e.g. 50 °C or 70 °C), and from this the ratio of final to starting concentration is calculated with the aid of formula 13c, which indicates whether the specified condition of a DLPO degradation of <2% is fulfilled.

→ ***Calculation***

a) $k = \frac{\ln\left(c_{A_0}/c_A\right)}{t}$

50 °C → 323 K $t = \frac{3 \text{ days} * 24 \text{ h} * 3600 \text{ s}}{\text{day} * \text{h}}$

$$\mathbf{k_{50}} = \frac{\ln \frac{0.01}{0.00485}}{3 * 24 * 3600 \text{ s}} = \mathbf{2.79 * 10^{-6} s^{-1}}$$

70 °C → 343 K

$$t = \frac{10 \text{ h} * 3600 \text{ s}}{\text{h}}$$

$$\mathbf{k_{70}} = \frac{\ln \frac{0.01}{0.00154}}{36{,}000 \text{ s}} = \mathbf{5.20 * 10^{-5} s^{-1}}$$

$$\frac{k_{70}}{k_{50}} = e^{-\frac{E_A}{R} * \left(\frac{1}{323} - \frac{1}{343}\right)\text{K}^{-1}} \rightarrow E_A = \frac{0.008315 \text{ kJ} * \ln\frac{k_{70}}{k_{50}}}{\text{mol} * \text{K} * \left(\frac{1}{323 \text{ K}} - \frac{1}{343 \text{ K}}\right)} = \mathbf{134.4 \frac{kJ}{mol}}$$

b) $T = (273 + 15)\text{ K} = 288\text{ K}$

$$k_{15} = k_{50} * e^{\frac{134.4 \text{ kJ} * \text{mol} * \text{K}}{\text{mol} * 0.008315 \text{ kJ}} * \left(\frac{1}{323} - \frac{1}{288}\right)\text{K}^{-1}} = 2.79 * 10^{-6}\mathbf{s^{-1}} * \boldsymbol{e^{-6.0815}}$$

$$k_{15} = 2.79 * 10^{-6}\text{s}^{-1} * 2.29 * 10^{-3} = 6.39 * 10^{-9}\text{s}^{-1}$$

$$\frac{c_A}{c_{Ao}} = e^{-k*t} = e^{-6.39 * 10^{-9}\text{s}^{-1} * 30 * 24 * 3600 \text{ s}} = e^{-0.01656} = 0.9835$$

Relative decrease** → 100%* (1 – 0.9835) =* ***1.65%

→ ***Result***

a) **The rate constant of lauroyl peroxide at 50** °C is $2.79 * 10^{-6}\ s^{-1}$, at 70 °C it is $5.20 * 10^{-5}\ s^{-1}$. The activation energy is 134.4 kJ/mol.
b) **A 30-day storage of the lauroyl peroxide solution at 15** °C leads to a decomposition of 1.65% of the initiator. The requirements are thus given.

Exercise 34

The half-life of the decay decomposition to a first-order reaction of the commonly used polymerization initiator azobis(isobutyronitrile) (AIBN) is 22 h at 60°C. The activation energy of the reaction is 120 kJ/mol.

a) What is the rate constant of the AIBN decay at 60°C?
b) How long can AIBN be stored at 5 °C so that no more than 2% have decomposed by the time it is used?
c) What must the residence time be in a reactor operated at 120 °C of the PFR type (e.g. an extruder) so that the outlet concentration of AIBN is no more than 1% of the inlet concentration?
d) What must the residence time be in a reactor operated at 120 °C of the CSTR type (e.g. a continuous kneader) so that the outlet concentration of AIBN is no more than 1% of the inlet concentration?

⊗ **Solution**

→ ***Strategy***

This is a reaction of the type A → 2C.

a) When the half-life is reached, half of the AIBN used has decomposed: $c_{Ao}/c_A=2$. Using the rearranged equation 13c, the decay rate constant at 60 °C is calculated from this and the half-life.
b) If 2% of the AIBN has decomposed, $c_A/c_{ao} = 0.98$. Equation 13c is rearranged for time. The rate constant to be used in this relationship at 5 °C is calculated using formula 17b from the rate constant calculated at 60 °C and the activation energy.
c) Analogous to part b of the exercise, formula 17b is used to calculate the rate constant at 120 °C. With 1% residual AIBN content at the end of the reaction, $c_A/c_{A0} = 0.01$. The necessary residence time in the PFR type reactor is determined using the rearranged formula 13d.
d) In the case of the CSTR type reactor, the residence time is calculated from the rearranged formula 13e.

→ ***Calculation***

a) $\boldsymbol{k_{60}} = \frac{1}{t} * \ln\frac{c_{A_0}}{c_A} = \frac{1*\mathrm{h}}{22\ \mathrm{h}*3600\ \mathrm{s}} * \ln 2 = \mathbf{8.752 * 10^{-6}\ s^{-1}}$

b) $k_5 = k_{60} * e^{\frac{E_A}{R}*\left(\frac{1}{273+60} - \frac{1}{273+5}\right)\mathrm{K}^{-1}}$

$$k_5 = 8.752 * 10^{-6}\ \mathrm{s}^{-1} * e^{\frac{120\,\mathrm{kJ}*\mathrm{mol}*\mathrm{K}}{\mathrm{mol}*0.008315\,\mathrm{KJ}}*\left(\frac{1}{333\,\mathrm{K}} - \frac{1}{278\,\mathrm{K}}\right)} = 8.752 * 10^{-6}\ \mathrm{s}^{-1} * e^{-8.574} = 1.654 * 10^{-9}\mathrm{s}^{-1}$$

$$\frac{c_A}{c_{Ao}} = 0.98 = e^{-k*t}$$

$$t = \frac{-\ln\frac{c_A}{c_{Ao}}}{k_5} = \frac{-\ln 0.98}{1.654 * 10^{-9}\ \mathrm{s}^{-1}} = \mathbf{1.22 * 10^7\ s}$$

$$= \frac{1.22 * 10^7\ \mathrm{s} * \mathrm{h}}{3600\ \mathrm{s}} = \mathbf{3393\ h} = \frac{3393\ \mathrm{h} * \mathrm{day}}{24\ \mathrm{h}} = \mathbf{141}\ \mathrm{days}$$

c) $k_{120} = k_{60} * e^{\frac{E_A}{R} * \left(\frac{1}{[273+60]\mathrm{K}} - \frac{1}{[273+120]\mathrm{K}}\right)}$

$$\boldsymbol{k_{120}} = 8.752 * 10^{-6}\ \mathrm{s}^{-1} * e^{\frac{120\ \mathrm{kJ} * \mathrm{mol} * \mathrm{K}}{\mathrm{mol} * 0.008315\ \mathrm{kJ}} * \left(\frac{1}{333\ \mathrm{K}} - \frac{1}{393\ \mathrm{K}}\right)} = 8.752 * 10^{-6}\mathrm{s}^{-1} * e^{6.617} = \mathbf{6.544 * 10^{-3}s^{-1}}$$

$$\frac{c_A}{c_{A_o}} = 0.01$$

$$\tau = \frac{-\ln\frac{c_A}{c_{Ao}}}{k_{120}} = \frac{-\ln 0.01}{6.544 * 10^{-3}\mathrm{s}^{-1}} = \mathbf{703.7\ s = 11.7\ min}$$

d) $\tau = \frac{1}{k_{120}} * \left(\frac{c_{A_o}}{c_A} - 1\right) = \frac{1}{6.544 * 10^{-3}\ \mathrm{s}^{-1}} * (100 - 1) = \mathbf{15.128\,s = 252\,min = 4.20\,h}$

→ ***Result***

a) **The rate constant of AIBN decomposition at 60 °C is 8.75 * 10^{-6}s^{-1}.**
b) **The maximum storage time is 1.22 * 10^7s = 3393 h = 141 days under the given conditions.**
c) **The residence time in a PFR-type reactor is 703 s = 11.7 min.**
d) **The residence time in a CSTR-type reactor is 15.128 s = 252 min = 4.20 h.**

Task 35

Butadiene, dissolved in a hydrocarbon of a relatively high boiling point, is to be dimerized catalytically to octene. This is a second order reaction according to 2 * butadiene → octene. From laboratory measurements, the maximum rate constant k_o = 1.10 * 10^7 L/(mol * s) and the activation energy E_A = 75 kJ/mol are known. The reaction is to be carried out on an industrial scale at 150 °C in either a tubular reactor (PFR type) or a reactor with complete backmixing (CSTR type). The mean residence time (reaction time) is 3.5 min. The concentration of butadiene in the feed stream is 5.0 mol/L.

a) What is the rate constant at the process temperature of 150 °C?
b) How much is the butadiene conversion in the PFR type reactor? It is assumed here that the reaction volume does not change significantly.
c) How much butadiene conversion is in the CSTR type reactor? It is assumed here that the reaction volume does not change significantly.

⊗ **Solution**

→ ***Strategy***

This is a type of reaction of 2A → C.

a) The rate constant at 150 °C is calculated according to Formula 17a.
b) From the determined rate constant and the initial concentration of butadiene, the butadiene concentration in the outlet of the PFR type reactor results by means of Formula 14d. Since the reaction volume does not change, the butadiene conversion can be calculated from its initial and final concentration by means of Formula 9c.
c) The butadiene conversion is calculated analogous to the method described earlier under b), but Formula 14e is used to calculate the butadiene concentration of the stream leaving the CSTR type reactor.

→ ***Calculation***

a) *k at 150 °C:*

$$k = k_0 * e^{-E_A/R * T} = 1.10 * 10^7 \frac{\text{L}}{\text{mol} * \text{s}} * e^{-\frac{75\ \text{kJ} * \text{mol} * \text{K}}{\text{mol} * 0.008315\ \text{kJ} * (273+150)\text{K}}}$$

$$= 1.10 * 10^7 \frac{\text{L}}{\text{mol} * \text{s}} * e^{-21.324}$$

$$\mathbf{k_{150} = 6.04 * 10^{-3} \frac{L}{mol * s}}$$

b) *PFR*

$$c_{\text{But}} = \frac{c_{\text{But}_o}}{\left(1 + c_{\text{But}_o} * k * \tau\right)}$$

$$\tau = 3.5\ \text{min} * 60 \frac{\text{s}}{\text{min}} = 210\ \text{s}$$

$$c_{\text{But}} = \frac{5.0\ \text{mol}}{\text{L} * \left(1 + 5.0 \frac{\text{mol}}{\text{L}} * 6.033 * 10^{-3} \frac{\text{mol}}{\text{L} * \text{s}} * 210\ \text{s}\right)} = \frac{5.0\ \text{mol}}{\text{L} * (1 + 6.335)} = 0.682 \frac{\text{mol}}{\text{L}}$$

$$X_{\text{But}} = \frac{c_{\text{But}_o} - c_{\text{But}}}{c_{\text{But}_o}} = \frac{5.0 - 0.682}{5.0} = \mathbf{0.864 \rightarrow 86.4\ \%}$$

c) *CSTR*

$$c_{\text{But}} = \sqrt{\frac{c_{\text{But}_o}}{k * \tau} + \frac{1}{4 * k^2 * \tau^2}} - \frac{1}{2 * k * \tau}$$

$$\frac{c_{\text{But}_o}}{k * \tau} = \frac{5.0\ \text{mol} * \text{mol} * \text{s}}{\text{L} * 6.033 * 10^{-3}\ \text{L} * 210\ \text{s}} = 3.947 \frac{\text{mol}^2}{\text{L}^2}$$

$$\frac{1}{2*k*\tau}=\frac{\text{mol}*\text{s}}{2*6.033*10^{-3}\text{L}*210\text{ s}}=0.395\frac{\text{mol}}{\text{L}}$$

$$\frac{1}{4*k^2*\tau^2}=\left(\frac{1}{2*k*\tau}\right)^2=\left(2.536\frac{\text{mol}}{\text{L}}\right)^2=0.1558\frac{\text{mol}^2}{\text{L}^2}$$

$$c_{\text{But}}=\sqrt{(3.947+0.1558)\frac{\text{mol}^2}{\text{L}^2}}-0.395\frac{\text{mol}}{\text{L}}=(2.026-0.395)\frac{\text{mol}}{\text{L}}=1.63\frac{\text{mol}}{\text{L}}$$

$$X_{\text{But}}=\frac{c_{\text{But}_o}-c_{\text{But}}}{c_{\text{But}_o}}=\frac{5.0-1.63}{5.0}=\mathbf{0.674\rightarrow 67.4\ \%}$$

→ ***Result***

a) **The rate constant at 150 °C is k = 6.04 * 10^{-3}L/(mol * s).**
b) **The butadiene conversion in the PFR-type reactor is 0.864 i.e. 86.4%.**
c) **The butadiene conversion in the CSTR-type reactor is 0.674 i.e. 67.4%.**

Exercise 36
An epoxide (EP) is to be converted to a hydroxyl amine (HA) in a pilot plant using a secondary amine (AM) and the solvent hexane in a second-order reaction:

$$\text{EP}+\text{AM}\rightarrow\text{HA}$$

For this purpose, a continuous tubular reactor (PFR) with an outer diameter of 0.1 m, a wall thickness of 4 mm and a length of 15 m or a continuous stirred tank reactor with a usable volume of 100 L are available. The total volume flow rate of feedstock (epoxide, amine and hexane) is 0.85 L/s.

a) In order to determine the rate constant, a laboratory experiment was carried out: 100 mL of a 0.2 molar solution of the epoxide in hexane were reacted with 25 mL of a 0.8 molar solution of the amine in hexane in a stirred glass flask. A sample after 10 minutes of reaction time showed concentrations of 0.0683 mol/L of epoxide and amine. No side reactions occurred. What is the rate constant of this second-order reaction?
b) The epoxide and the amine are fed to the reactor at equimolar with 4.0 mol/L each in the feed stream. What are the concentrations of epoxide, amine and product in the outlet of the PFR and the CSTR under the assumption that no side reactions occur? What are the conversion and yield?
c) In another batch, a feed mixture of 3 mol epoxide/L and 5 mol amine/L is fed to the tubular reactor. The residence time in the reactor is the same as the operating mode given in point b. above. What are the concentrations of epoxide, amine and product in the reactor outlet under the assumption that no side reactions occur? What are the conversions and yields?

⊗ **Solution**

→ ***Strategy***

It is a reaction of the type A + B → C.

a) The starting molar concentrations of epoxide and amine are calculated for the laboratory experiment. If both are equal, the rearranged formula 15b can be used to calculate the rate constant.
b) Since the reactants of the reaction are equimolar in the second order, formula 15c can be used to calculate the reactor outlet concentrations necessary to determine conversion and yield for the PFR, and formula 15d can be used for the CSTR. The residence time is calculated as the ratio of reactor volume and flow rate. Since the molar feed concentrations of the epoxide and the amine are equal and no side reactions occur, the conversions and the epoxide and amine-based yields both are also equal in the case of the PFR as well as and in the case of the CSTR.
c) Formula 16d is used to calculate the epoxide concentration in the reactor effluent. (Alternatively, formula 16d can also be used to calculate the amine concentration.)

There are no side reactions, so the amine concentration in the reactor effluent must be equal to its concentration in the feed minus the difference in epoxide concentration from inlet to outlet. The product concentration in the reactor effluent is also equal to this difference in epoxide concentration. The conversions and yields are calculated as before.

→ ***Calculation***

a) *Laboratory experiment:* $V_{\text{total}} = 25\text{ mL} + 100\text{ mL} = 125\text{ mL}$

$$n = c * V \quad n_{\text{EP}_0} = \frac{0.20\text{ mol}}{\text{L}} * 0.100\text{ L} = 0.02\text{ mol}$$

$$n_{\text{AM}_0} = \frac{0.8\text{ mol}}{\text{L}} * 0.025\text{ L} = 0.02\text{ mol}$$

For the feed mixture of the laboratory experiment:

$c = \frac{n}{V} \quad c_{\text{EP}_0} = \frac{0.02\text{ mol}}{0.125\text{ L}} = 0.160\frac{\text{mol}}{\text{L}} = c_{\text{AM}_0}$

With equal starting concentrations, $c_A = c_B = \frac{c_{A_0}}{(1+c_{A_0} * k * t)} = \frac{c_{B_0}}{(1+c_{B_0} * k * t)}$

The reaction time t was 10 min = 600 s.

For the present case, rearranged to k (the equivalent data set of the amine can also be used):

$$k = \frac{c_{\text{EP}_0} - c_{\text{EP}}}{c_{\text{EP}_0} * c_{\text{EP}} * t} = \frac{(1.16 - 0.0683)\text{ mol} * \text{L}^2}{\text{L} * 0.16\text{ mol} * 0.0683\text{ mol} * 600\text{ s}} = \mathbf{1.40 * 10^{-2}\frac{L}{mol * s}}$$

b)

$$c_{\mathrm{EP}_o} = c_{\mathrm{AM}_o} = 4.0\frac{\mathrm{mol}}{\mathrm{L}} \quad \tau = \frac{V_R}{\dot{V}} \quad \dot{V} = 0.85\frac{\mathrm{L}}{\mathrm{s}} = \frac{0.00085\ \mathrm{m}^3}{\mathrm{s}}$$

PFR:

$$V = \frac{d^2 * \pi * \mathrm{L}}{4} \quad d = 0.1\ \mathrm{m} - 2 * 0.004\ \mathrm{m} = 0.092\ \mathrm{m}$$

$$V = \frac{0.092^2\ \mathrm{m}^2 * \pi * 15\ \mathrm{m}}{4} = 0.0997\ \mathrm{m}^3 \cong 0.100\ \mathrm{m}^3$$

$$\tau = \frac{0.100\ \mathrm{m}^3 * \mathrm{s}}{0.00085\ \mathrm{m}^3} = 117.6\ \mathrm{s}$$

$$c_A = c_B = \frac{c_{A_o}}{\left(1 + c_{A_o} * k * \tau\right)} = \frac{c_{B_o}}{\left(1 + c_{B_o} * k * \tau\right)}$$

$$\boldsymbol{c_{\mathrm{EP}}} = c_{\mathrm{AM}} = \frac{c_{\mathrm{EP}_o}}{\left(1 + c_{\mathrm{EP}_o} * k * \tau\right)} = \frac{c_{\mathrm{AM}_o}}{\left(1 + c_{\mathrm{AM}_o} * k * \tau\right)} = \frac{4\ \mathrm{mol}}{\mathrm{L} * \left(1 + 4\frac{\mathrm{mol}}{\mathrm{L}} * 1.40 * 10^{-2}\frac{\mathrm{L}}{\mathrm{mol} * \mathrm{s}} * 117.6\ \mathrm{s}\right)}$$

$$= \frac{4\ \mathrm{mol}}{7.590\ \mathrm{L}} = \mathbf{0.527}\frac{\mathbf{mol}}{\mathbf{L}}$$

Since no side reactions take place and with $\nu_{\mathrm{EP}} = \nu_{\mathrm{AM}} = -1 \quad \nu_{\mathrm{AH}} = +1$*, the increase in product is equal to the decrease in reactant:*

$$c_{\mathrm{AH}} = c_{\mathrm{AH}_o} + \left(c_{\mathrm{EP}_o} - c_{\mathrm{EP}}\right) = c_{\mathrm{AH}_o} + \left(c_{\mathrm{AM}_o} - c_{\mathrm{AM}}\right) \quad c_{\mathrm{AH}_o} = 0\frac{\mathrm{mol}}{\mathrm{L}}$$

$$\boldsymbol{c_{\mathrm{AH}}} = (4.0 - 0.527)\frac{\mathrm{mol}}{\mathrm{L}} = \mathbf{3.473}\frac{\mathbf{mol}}{\mathbf{L}}$$

$$X_E = \frac{\dot{n}_{E_o} - \dot{n}_E}{\dot{n}_{E_o}}$$

$$\dot{n} = c * \dot{V} * \tau$$

Since the flow rate and the residence time for product and reactant are identical, they both cancel out of the conversion and yield equation:

$$X_E = \frac{c_{E_o} - c_E}{c_{E_o}} \quad \boldsymbol{X_{\mathrm{EP}}} = \frac{c_{\mathrm{EP}_o} - c_{\mathrm{EP}}}{c_{\mathrm{EP}_o}} = \frac{c_{\mathrm{AM}_o} - c_{\mathrm{AM}}}{c_{\mathrm{AM}_o}} = \boldsymbol{X_{\mathrm{AM}}}$$

$$= \frac{(4.0 - 0.527)\mathrm{mol} * \mathrm{L}}{\mathrm{L} * 4.0\ \mathrm{mol}} = \mathbf{0.868} \rightarrow \mathbf{86.8\ \%}$$

$$Y_{P/E} = \frac{\upsilon_E * \left(n_{P_o} - n_P\right)}{\upsilon_p * n_{E_o}} = \frac{\upsilon_E * \left(c_{P_o} - c_P\right)}{\upsilon_p * c_{E_o}}$$

$$Y_{\mathrm{AH/EP}} = \frac{\upsilon_{\mathrm{EP}} * \left(c_{\mathrm{AH_o}} - c_{\mathrm{AH}}\right)}{\upsilon_{\mathrm{AH}} * c_{\mathrm{EP_o}}} = \frac{\upsilon_{\mathrm{AM}} * \left(c_{\mathrm{AH_o}} - c_{\mathrm{AH}}\right)}{\upsilon_{\mathrm{AH}} * c_{\mathrm{AM_o}}} = Y_{\mathrm{AH/AM}}$$

$$\boldsymbol{Y_{\mathrm{AH/EP}}} = \boldsymbol{Y_{\mathrm{AH/AM}}} = \frac{-1 * (0 - 3.473)\mathrm{mol} * \mathrm{L}}{+1 * \mathrm{L} * 4.0\ \mathrm{mol}} = \mathbf{0.868 \rightarrow 86.8\ \%}$$

CSTR:

$\tau = \frac{V_R}{\dot{V}} = \frac{0.100\ \mathrm{m^3} * \mathrm{s^3}}{0.00085\ \mathrm{m^3}} = 117.6\ \mathrm{s} \rightarrow$ *The residence time in the CSTR is thus equal to that of the PFR.*

$$c_A = c_B = \sqrt{\frac{c_{A_o}}{k * \tau} + \frac{1}{4 * k^2 * \tau^2}} - \frac{1}{2 * k * \tau} = \sqrt{\frac{c_{B_o}}{k * \tau} + \frac{1}{4 * k^2 * \tau^2}} - \frac{1}{2 * k * \tau}$$

$$c_{\mathrm{EP}} = c_{\mathrm{AM}} = \sqrt{\frac{c_{\mathrm{EP_o}}}{k * \tau} + \frac{1}{4 * k^2 * \tau^2}} - \frac{1}{2 * k * \tau} = \sqrt{\frac{c_{\mathrm{AM_o}}}{k * \tau} + \frac{1}{4 * k^2 * \tau^2}} - \frac{1}{2 * k * \tau}$$

$$c_{\mathrm{EP}} = \sqrt{\frac{4.0\ \mathrm{mol} * \mathrm{mol} * \mathrm{s}}{\mathrm{L} * 1.4 * 10^{-2}\mathrm{L} * 117.6\ \mathrm{s}} + \frac{1}{4 * \left(1.4 * 10^{-2}\,\mathrm{L}/(\mathrm{mol} * \mathrm{s})\right)^2 * (117.6\ \mathrm{s})^2}} - \frac{\mathrm{mol} * \mathrm{s}}{2 * 1.4 * 10^{-2}\mathrm{L} * 117.6\ \mathrm{s}}$$

$$\boldsymbol{c_{\mathrm{EP}}} = \boldsymbol{c_{\mathrm{AM}}} = \sqrt{2.522\frac{\mathrm{mol}^2}{\mathrm{L}^2}} - 0.3037\frac{\mathrm{mol}}{\mathrm{L}} = \mathbf{1.284\frac{mol}{L}}$$

With the conversion and yield relationships already mentioned for the PFR, it follows:

$$\boldsymbol{X_{\mathrm{EP}}} = \boldsymbol{X_{\mathrm{AM}}} = \frac{(4.0 - 1.284)\ \mathrm{mol} * \mathrm{L}}{\mathrm{L} * 4.0\,\mathrm{mol}} = \mathbf{0.679 \rightarrow 67.9\ \%}$$

$$c_{\mathrm{AH}} = (4.0 - 1.284)\frac{\mathrm{mol}}{\mathrm{L}} = \mathbf{2.716\frac{mol}{L}}$$

$$\boldsymbol{Y_{\mathrm{AH/EP}}} = \boldsymbol{Y_{\mathrm{AH/AM}}} = \frac{-1 * (0 - 2.716)\ \mathrm{mol} * \mathrm{L}}{+1 * \mathrm{L} * 4.0\,\mathrm{mol}} = \mathbf{0.679 \rightarrow 67.9\ \%}$$

c) $c_A = \frac{\theta * c_{A_o} * \left(c_{B_o} - c_{A_o}\right)}{c_{B_o} - \theta * c_{A_o}}$ $\quad c_{A_o} = c_{\mathrm{EP_o}} = 3.0\frac{\mathrm{mol}}{\mathrm{L}}$

$c_{B_o} = c_{\mathrm{AM_o}} = 5.0\frac{\mathrm{mol}}{\mathrm{L}}$

$$c_{\mathrm{EP}} = \frac{\theta * c_{\mathrm{EP_o}} * \left(c_{\mathrm{AM_o}} - c_{\mathrm{EP_o}}\right)}{c_{\mathrm{AM_o}} - \theta * c_{\mathrm{EP_o}}}$$

$$\text{with } \theta = e^{(c_{A_0} - c_{B_0}) * k * \tau} = e^{(c_{EP_0} - c_{AM_0}) * k * \tau}$$

$$\theta = e^{(3.0-5.0)\frac{mol}{L} * 1.4 * 10^{-2} \frac{L}{mol * s} * 117.6\,s} = e^{-3.293} = 0.03714$$

$$c_{EP} = \frac{0.03714 * 3.0\frac{mol}{L} * (5.0 - 3.0)\frac{mole}{L}}{5.0\frac{mol}{L} - (0.03714 * 3.0)\frac{mol}{L}} = \mathbf{0.0456\frac{mol}{L}}$$

Since there are no side reactions, it follows:

$$c_{AM} = c_{AM} = c_{AM_0} - \Delta c_{EP} \text{ and } c_{AH} = \Delta c_{EP}.$$

$$\Delta c_{EP} = c_{EP_0} - c_{EP} = (3.00 - 0.0456)\frac{mol}{L} = 2.955\frac{mol}{L}$$

$$c_{AM} = (5.00 - 2.955)\frac{mol}{L} = \mathbf{2.045\frac{mol}{L}}$$

$$c_{AH} = \mathbf{2.955\frac{mol}{L}}$$

$$X_{EP} = \frac{c_{EP_0} - c_{EP}}{c_{EP_0}} = \frac{(3.00 - 0.0456)\,mol * L}{L * 3.00\,mol} = \mathbf{0.985 \rightarrow 98.5\,\%}$$

$$X_{AM} = \frac{c_{AM_0} - c_{AM}}{c_{AM_0}} = \frac{(5.00 - 2.045)\,mol * L}{L * 5.00\,mol} = \mathbf{0.591 \rightarrow 59.1\,\%}$$

$$Y_{AH/EP} = \frac{\upsilon_{EP} * (c_{AH_0} - c_{AH})}{\upsilon_{AH} * c_{EP_0}} = \frac{-1 * (0 - 2.955)\,mol * L}{+1 * 3.00\,mol * L} = \mathbf{0.985 \rightarrow 98.5\,\%}$$

$$Y_{AH/AM} = \frac{\upsilon_{AM} * (c_{AH_0} - c_{AH})}{\upsilon_{AH} * c_{AM_0}} = \frac{-1 * (0 - 2.955)\,mol * L}{+1 * 5.00\,mol * L} = 0.59 \rightarrow \mathbf{59.0\,\%}$$

→ ***Result***

a) **The rate constant is 1.40 * 10^{-2}L/(mol * s).**
b) **PFR: The concentrations at the reactor outlet are as follows:**
Epoxide and amine: 0.527 mol/L
Hydroxy amine: 3.473 mol/L
The conversions of epoxide and amine were 0.868 and 86.8%, respectively. Since no side reactions occurred, the yield of hydroxy amine based on epoxide or amine is equal to the conversion.
CSTR: The concentrations at the reactor outlet are as follows:

Epoxide and amine: 1.284 mol/L
Hydroxy amine: 2.716 mol/L
The conversions of epoxide and amine were 0.679 and 67.9%, respectively. Since no side reactions occurred, the yield of hydroxy amine based on epoxide or amine is equal to the conversion.
The significantly lower conversions and yields of hydroxy amine in the CSTR are due to the dilution effect of the reactants caused by back-mixing.

c) **The concentrations in the PFR outlet are:**
Epoxide: 0.0456 mol/L
Amine: 2.045 mol/L
Hydroxy amine: 2.955 mol/L
Since no side reactions occurred, the yield of hydroxy amine based on epoxide or amine is equal to the conversion. Thus, the conversions and hydroxy amine yields with respect to epoxide and amine are:
Epoxide: 0.985 → 98.5%
Amine: 0.591 → 59.1%

Exercise 37

A diazo compound is used for the batchwise production of a dye in a stirred tank. After the reaction is complete, the reaction mixture is held for 20 min at 40 °C in order to comply with the specified maximum limit by the decomposition of the diazo compound in accordance with a first-order reaction. In the future, this limit is to be reduced by a factor of 5.

In a laboratory experiment, the reaction rates of the decomposition of the diazo compound were measured. For this purpose, the concentration of a 0.10 molar starting solution was analyzed after a reaction time of 10 min. At a reaction temperature of 20 °C, a concentration of 0.081 mol/L was obtained, at 30 °C a concentration of 0.049 mol/L.

a) What are the rate constants of the decomposition at 20 °C and at 30 °C as well as the corresponding activation energy?
b) What is the rate constant at the reaction temperature of 40 °C? What is the ratio of the concentration of the diazo compound after and before the decomposition time of 20 min?
c) How long would the reaction mixture have to be held at 40 °C in order to meet the new, fivefold lower limit?
d) To what temperature would the reaction mixture have to be brought in order to also meet the new limit at an unchanged diazo decomposition time of 20 min?

⊗ **Solution**

→ ***Strategy***

It is a reaction of type A → C.

a) The rate constants for 20 °C and 30 °C can be calculated from the data of the laboratory experiments using the appropriately rearranged formula 13b. The activation energy is obtained by using these calculated rate constants in the appropriately rearranged formula 17b.

b) Using formula 17b, the rate constant at 40 °C is calculated by using the calculated activation energy and one of the rate constants calculated under a) as well as the corresponding temperature (e.g. 20 °C or 30 °C). With formula 13d, the required ratio of the diazo concentration after the reaction to that after the decomposition phase is calculated.

c) The rearranged formula 13b is used with the rate constant at 40 °C and the ratio of the diazo concentration after to that before the decomposition phase, which is 5 times lower, and the reaction time are calculated.

d) Formula 13b is rearranged to the rate constant and the lower ratio of the diazo concentration after to that before the decomposition phase as well as the desired reaction time of 20 min are used. Using formula 17b, the activation energy and a known value pair of rate constant & temperature (e.g. 40 °C), the corresponding reaction temperature results.

→ ***Calculation***

a) $k = \frac{1}{t} * \ln\frac{c_{A_0}}{c_A}$

$$20\,°C \rightarrow \boldsymbol{k_{20}} = \frac{\text{min}}{10\,\text{min} * 60\,\text{s}} * \ln\frac{0.1}{0.081} = \mathbf{3.512 * 10^{-4} s^{-1}}$$

$$30\,°C \rightarrow \boldsymbol{k_{30}} = \frac{\text{min}}{10\,\text{min} * 60\,\text{s}} * \ln\frac{0.1}{0.049} = \mathbf{1.189 * 10^{-3} s^{-1}}$$

$$\ln\frac{k_1}{k_2} = \frac{E_A}{R} * \left(\frac{1}{T_1} - \frac{1}{T_2}\right)$$

$$\boldsymbol{E_A} = \frac{R * \ln\frac{k2}{k1}}{\frac{1}{T1} - \frac{1}{T2}} = \frac{0.008315\frac{\text{kJ}}{\text{mol} * \text{K}} * \ln\frac{0.001189}{0.0003512}}{\frac{1}{293\,\text{K}} - \frac{1}{303\,\text{K}}} = \mathbf{90.1\frac{kJ}{mol}}$$

b) $k_2 = k_1 * e^{\frac{E_A}{R} * \left(\frac{1}{T_1} - \frac{1}{T_2}\right)}$

$$\boldsymbol{k_{40}} = 1.189 * 10^{-3}\,\text{s}^{-1} * e^{\frac{90.1\,\text{kJ} * \text{mol} * \text{K}}{\text{mol} * 0.008315\,\text{kJ}} * \left(\frac{1}{303\,\text{K}} - \frac{1}{313\,\text{K}}\right)} = \mathbf{3.727 * 10^{-3} s^{-1}}$$

$$\frac{\boldsymbol{c_A}}{\boldsymbol{c_{A_0}}} = e^{-k * t} = e^{-0.003727\,\text{s}^{-1} * 20\,\text{min} * 60\frac{\text{s}}{\text{min}}} = \mathbf{0.0114}$$

c) $$\left(\frac{c_A}{c_{A_0}}\right)_{\text{future}} = \frac{1}{5} * \left(\frac{c_A}{c_{A_0}}\right)_{\text{present}} = \frac{0.0114}{5} = 0.00228$$

$$\boldsymbol{t} = \frac{\ln\frac{c_{A_0}}{c_A}}{k} = \frac{\ln 0.00228}{0.003727\,\text{s}^{-1}} = \mathbf{1632\,s = 27.2\,min}$$

d) $$k_x = \frac{1}{t} * \ln\frac{c_{A_0}}{c_A} = \frac{1}{1200\,\text{s}} * \ln\frac{1}{0.00228} = 5.07 * 10^{-3}\,\text{s}^{-1}$$

$$\frac{k_x}{k_{40}} = e^{\frac{E_A}{R} * \left(\frac{1}{[273 + 40]\,\text{K}} - \frac{1}{T_X}\right)}$$

$$\ln\frac{k_x}{k_{40}} = \frac{E_A}{R} * \left(\frac{1}{313\,\mathrm{K}} - \frac{1}{T_x}\right)$$

$$\frac{1}{T_x} = \frac{1}{313\,\mathrm{K}} - \frac{R}{E_A} * \ln\frac{k_x}{k_{40}}$$

$$\frac{1}{T_x} = \frac{1}{313\,\mathrm{K}} - \frac{0.008315\,\mathrm{kJ} * \mathrm{mol}}{\mathrm{mol} * \mathrm{K} * 90.1\,\mathrm{kJ}} * \ln\frac{0.00507\,\mathrm{s}}{0.003727\,\mathrm{s}} = \left(3.1949 * 10^{-3} - 2.84 * 10^{-5}\right)\mathrm{K}^{-1}$$
$$= 3.17 * 10^{-3}\,\mathrm{K}^{-1}$$

$$\boldsymbol{T_x = 315.5\,\mathrm{K} = 42.3\,°\mathrm{C}}$$

→ ***Result***

a) **The rate constant of diazo component degradation at 20 °C is 3.512 * $10^{-4}s^{-1}$** and at 30 °C **1.189 * $10^{-3}s^{-1}$. The activation energy of this reaction is 90.1 kJ/ mol.**
b) **The rate constant of the decomposition reaction at 40 °C is 3,727 * $10^{-3}s^{-1}$. The ratio of the concentrations of the diazo compound before to after the 20-minute decomposition phase is 0.0114 at 40 °C.**
c) **To achieve a fivefold lower value of the diazo component after the decomposition time, 27.2 min would be required at 40 °C.**
d) **To achieve the fivefold lower limit at unchanged 20 min decomposition time, a post-reaction temperature of 42.3 °C would be necessary.**

3.4 Heat

Since pipes, reactors and apparatus are not ideal, that is, not completely insulated, always a certain amount of heat or cooling capacity will be lost during operation. A complete exact capture of these losses is difficult to achieve in real operation. Such heat losses are approximately neglected in the following calculations in order to maintain the clarity of the calculation process and to focus on the basic solution paths.

3.4.1 Heating, Melting, Vaporizing, Dissolving

Exercise 38

How much condensate is formed by melting of 100 kg of naphthalene (M = 128 g/mol, $\Delta_S H$ = 18.8 kJ/mol) by saturated steam ($\Delta_V H$ = 2450 kJ/kg)? The solid naphthalene is used at its melting point. The steam condensate leaves the system at the temperature of the saturated steam.

⊗ **Solution**

→ ***Strategy***

Since the melting heat of naphthalene is given as a molar-related quantity, the mass of naphthalene used is converted to the number of moles. From this, the amount of heat required for melting can be calculated using formula 23a. This amount of heat corresponds to the condensation heat of the saturated steam. The necessary amount of saturated steam results from rearranging formula 24a and using the corresponding quantities.

→ ***Calculation***

$$n_{\mathrm{Naph}} = \frac{m_{\mathrm{Naph}}}{M_{\mathrm{Naph}}} = \frac{100\,\mathrm{kg} * \mathrm{mol}}{0.128\,\mathrm{kg}} = 781.25\,\mathrm{mol}$$

$$Q = n_{\mathrm{Naph}} * \Delta_{\mathrm{S}} H_{\mathrm{Naph}} = 781.25\,\mathrm{mol} * 18.8\frac{\mathrm{kJ}}{\mathrm{mol}} = 14{,}687.5\,\mathrm{kJ}$$

$$m_{\mathrm{Steam}} = \frac{Q}{\Delta_{V} H} = \frac{14{,}687.5\,\mathrm{kJ} * \mathrm{kg}}{2450\,\mathrm{kJ}} = \mathbf{6.00\,kg}.$$

→ ***Result***

6.0 kg of steam is required to melt the naphthalene.

Exercise 39

How much 12 bar saturated steam (T = 190 °C) is required to vaporize 2000 L of toluene (density $\rho = 0.88$ kg/L; heat capacity cp = 1.8 kJ/[kg * °C]; vaporization heat $\Delta_V H = 336$ kJ/kg; boiling point 110 °C), which is fed into the evaporator at 20 °C? The steam condensate (cp = 4.3 kJ/[kg * °C] is used to preheat the toluene and discharged at 125 °C. The condensation heat of the steam is $\Delta_V H = 2310$ kJ/kg.

⊗ **Solution**

→ ***Strategy***

The required amount of heat consists of the amount of heating the toluene from its feed temperature to its boiling point (Formula 18a) and the amount for vaporization (Formula 24a). This amount of heat is supplied by the condensation heat of the saturated vapor and the cooling of the condensate from the temperature of the vapor to its discharge temperature. Since the toluene amount is given as a volume, the heat capacity and the vaporization heat are mass-related, it must be converted to mass beforehand by multiplication with the density.

→ ***Calculation***
Toluene is denoted by the subscript Tol, steam by S and condensate by W.

$$m_{Tol} = V_{Tol} * \rho_{Tol} = 2000\,\text{L} * 0.88\frac{\text{kg}}{\text{L}} = 1760\,\text{kg}$$

$$Q_{Tol} = m_{Tol} * (cp_T * \Delta T_{Tol} + \Delta_V H_{Tol}) = 1760\,\text{kg} * \left[1.8\frac{\text{kJ}}{\text{kg} * {}^\circ\text{C}} * (110 - 20)\,{}^\circ\text{C} + 336\frac{\text{kJ}}{\text{kg}}\right]$$

$$Q_{Tol} = (285{,}120 + 591{,}360)\,\text{kJ} = 876{,}480\,\text{kJ}$$

$$Q_S = m_S * (\Delta_V H_S + cp_W * \Delta T_W)$$

$$Q_S = Q_{Tol} = 876{,}480\,\text{kJ}$$

$$m_D = \frac{Q_S}{\Delta_V H_S + cp_W * \Delta T_W} = \frac{876{,}480\,\text{kJ}}{2310\,\text{kJ} + 4.3\frac{\text{kJ}}{\text{kg}*{}^\circ\text{C}} * (190 - 125\,{}^\circ\text{C})} = \frac{876{,}480\,\text{kJ} * \text{kg}}{2310\,\text{kJ} + 279.5\,\text{kJ}}$$

$$m_S = \frac{876{,}480}{2589.5}\text{kJ} = \mathbf{338.5\,kg}$$

→ ***Result***
338.5 kg of steam is required.

Task 40
In a continuous stirred tank, the reactants A (1.2 t/h; $cp_A = 2.5$ kJ/[kg * °C)]) and B (2 t/h; $cp_B = 2.1$ kJ/[kg * °C]) and a solvent (10 t/h; $cp_S = 1.8$ kJ/[kg * °C]) are fed in. All of these feeds have a temperature of 15 °C. The amount of heat released by the reaction is 11,500 MJ/h. The reactor temperature must not exceed 80 °C. The cooling water ($cp_w = 4.2$ kJ/[kg * K]) has an inlet temperature of 20 °C, the outlet must not exceed 60 °C. How much cooling water is required?

⊗ **Solution**
→ ***Strategy***
The heat generation of the reaction must be removed by the cooling water (cw) flow minus the heat required to increase the reactor feed from 15 °C to 80 °C (feed). The heat to increase the reactor feed stream is calculated according to formula 19b. The mass flow of the cooling water results from the corresponding rearranged formula 19a.

→ ***Calculation***

$$\dot{Q}_{\text{cw}} = \dot{m}_{\text{cw}} * cp_{\text{cw}} * (T_{\text{cw-ex}} - T_{\text{cw-in}})$$

$$\dot{m}_{\text{cw}} = \frac{\dot{Q}_{\text{cw}}}{cp_{\text{cw}} * (T_{\text{cw-ex}} - T_{\text{cw-in}})}$$

$$\dot{Q}_{\text{reaction}} = \dot{Q}_{\text{feed-heating}} + \dot{Q}_{\text{cw}}$$

$$\dot{Q}_{\text{cw}} = \dot{Q}_{\text{reaction}} - \dot{Q}_{\text{feed-heating}}$$

$$\dot{Q}_{\text{feed-heating}} = \sum_i [\dot{m}_i * cp_i * (T_{\text{ex}} - T_{\text{in}})]$$

$$\dot{Q}_{\text{feed-heating}} = (\dot{m}_A * cp_A + \dot{m}_B * cp_B + \dot{m}_S * cp_S) * (T_{\text{ex}} - T_{\text{in}})$$

$$\dot{Q}_{\text{feed-heating}} = \left(1200\frac{\text{kg}}{\text{h}} * 2.5\frac{\text{kJ}}{\text{kg} * {}^\circ\text{C}} + 2000\frac{\text{kg}}{\text{h}} * 2.1\frac{\text{kJ}}{\text{kg} * {}^\circ\text{C}} + 10{,}000\frac{\text{kg}}{\text{h}} * 1.8\frac{\text{kJ}}{\text{kg} * {}^\circ\text{C}}\right) * (80 - 15)\,{}^\circ\text{C}$$

$$\dot{Q}_{\text{feed-heating}} = 25{,}200\frac{\text{kJ}}{\text{h} * {}^\circ\text{C}} * 65\,{}^\circ\text{C} = 1638{,}000\frac{\text{kJ}}{\text{h}} = 1638\frac{\text{MJ}}{\text{h}}$$

$$\dot{Q}_{\text{cw}} = (11{,}500 - 1638)\frac{\text{MJ}}{\text{h}} = 9862\frac{\text{MJ}}{\text{h}}$$

$$\dot{m}_{\text{cw}} = \frac{9862\,\text{kJ} * \text{kg} * {}^\circ\text{C}}{\text{h} * 4.2\,\text{kJ} * (60 - 20)\,{}^\circ\text{C}} = \mathbf{58,700\frac{kg}{h} = 58.7\frac{t}{h} = 16.3\frac{kg}{s}}$$

→ ***Result***
The required cooling water flow is 58.7 t/h = 16.3 kg/s.

Task 41
A liquid reaction mixture ($cp_R = 2.05$ kJ/[kg * °C]; melting point <30 °C) at a temperature of 90 °C and a mass flow of 750 kg/h is mixed with 100 kg/h of solid phenol prills at 20 °C ($cp_{\text{solid phenol}} = 1.4$ kJ/(kg * °C); $cp_{\text{liquid phenol}} = 2.25$ kJ/(kg * °C); melting point phenol = 40.85 °C; melting heat phenol = 120.6 kJ/kg).

a) What temperature does the mixture leave the mixer?
b) How high must the temperature of the reaction mixture be at least to just melt the entire amount of phenol?

⊗ **Solution**
→ ***Strategy***

a) Since the mass flow of the reaction mixture is 7.5 times greater than that of the phenol, the heat capacities of both streams are of the same order of magnitude and, in addition, the temperature of the reaction mixture is significantly above the melting temperature of the phenol at 90 °C, the following situation can be assumed: The heat released by the reaction mixture is sufficient to heat the phenol to its melting point, its complete melting and the heating of the phenol melt to the final temperature.
The heat released by the reaction mixture to the phenol is calculated according to Formula 19a. The heat absorbed by the phenol is the sum of the heat required to heat the solid phenol to the melting point (Formula 19a), the heat of melting (Formula 22b) and the heat required to heat the melt to the final temperature (Formula 19a). The thus assembled equation is solved for the final temperature. For a meaningful solution, the determined temperature must be above the melting temperature of the phenol!
b) In the case of complete melting of the phenol, but no further heating of the melt, the final temperature is equal to the melting temperature of the phenol. The procedure is the same as in a), only that the term for heating the melt is omitted. The equation is solved for the initial temperature.

→ ***Calculation***
Subscript: Reaction mixture = R; Phenol = P;
s = solid; l = liquid; S = melting;
T_o= Starttemperature; T_E= Endtemperature

a) *Heat released by the reaction mixture (R):*
$\dot{Q} = \dot{m}_R * cp_R * \left(T_E - T_{R_0}\right)$
Heat taken up by phenol (P):
$\dot{Q} = \dot{m}_P * cp_{Ps} * \left(T_{PS} - T_{P_0}\right) + \dot{m}_P * \Delta_S H + \dot{m}_P * cp_{Pl} * (T_E - T_{PS})$
Combination of both equations:

$$\dot{Q} = \dot{m}_R * cp_R * \left(T_E - T_{R_0}\right) = \dot{m}_P * cp_{Ps} * \left(T_{PS} - T_{P_0}\right) + \dot{m}_P * \Delta_S H + \dot{m}_P * cp_{Pl} * (T_E - T_{PS})$$

$$T_E = \frac{\dot{m}_R * cp_R * T_{R0} - \dot{m}_P * \left[cp_{Ps} * \left(T_{PS} - T_{P_0}\right) + \Delta_S H - cp_{Pl} * T_{PS}\right]}{\dot{m}_R * cp_R + \dot{m}_P * cp_{Pl}}$$

$$\dot{m}_R * cp_R = 750\frac{\text{kg}}{\text{h}} * 2.05\frac{\text{kg}}{\text{kJ} * {}^\circ\text{C}} = 1537.5\frac{\text{kJ}}{\text{h} * {}^\circ\text{C}}$$

$$P_R * cp_{Pl} = 100\frac{\text{kg}}{\text{h}} * 2.25\frac{\text{kg}}{\text{kJ} * {}^\circ\text{C}} = 225\frac{\text{kJ}}{\text{h} * {}^\circ\text{C}}$$

$$T_E = \frac{1537.5\frac{\text{kJ}}{\text{h}*{}^\circ\text{C}} * 90\,{}^\circ\text{C} - 100\frac{\text{kg}}{\text{h}} * \left(1.4\frac{\text{kJ}}{\text{kg}*{}^\circ\text{C}} * (40.85 - 20)\,{}^\circ\text{C} + 120.6\frac{\text{kJ}}{\text{kg}} - 2.25\frac{\text{kJ}}{\text{kg}*{}^\circ\text{C}} * 40.85\,{}^\circ\text{C}\right)}{(1537.5 + 225)\frac{\text{kJ}}{\text{h}*{}^\circ\text{C}}}$$

$$\boldsymbol{T_E = 75.2\,{}^\circ\text{C}}$$

b) *Heat removed from the reaction mixture (R):*
$\dot{Q} = \dot{m}_R * cp_R * \left(T_E - T_{R_0}\right)$
Heat tasken up by Phenol (P):
$\dot{Q} = \dot{m}_P * cp_{P_S} * \left(T_{PS} - T_{P_0}\right) + \dot{m}_P * \Delta_S H$
Combination of both equations:

$$\dot{Q} = \dot{m}_R * cp_R * \left(T_E - T_{R_0}\right) = \dot{m}_P * cp_{P_S} * \left(T_{PS} - T_{P_0}\right) + \dot{m}_P * \Delta_S H$$

$$T_{R_o} = T_{PS} + \frac{\dot{m}_P * cp_{P_S} * \left(T_{PS} - T_{P_0}\right) + \dot{m}_P * \Delta_S H}{\dot{m}_R * cp_R}$$

$$T_{R_o} = 40.85\,°C + \frac{100\frac{kg}{h} * 1.4\frac{kJ}{kg * °C} * (40.85 - 20.0)\,°C + 100\frac{kJ}{h} * 120.6\frac{kJ}{kg}}{750\frac{kg}{h} * 2.05\frac{kJ}{kg * °C}}$$

$$\boldsymbol{T_{R_o}} = 40.85\,°C + 9.74\,°C = \mathbf{50.6°C}$$

→ ***Result***

a) **The final temperature is 75.2 °. Hence all phenol is molten.**
b) **The minimum temperature of the reaction mixture to melt all phenol is 50.6 °C.**

Exercise 42
In preparation for a reaction, 1.8 metric t of a granulate solid feedstock (feed temperature = 10 °C; melting point 34 °C) is to be melted and brought to 80 °C. The heat capacity of the solid is 1.2 kJ/(kg * ° C), that of the melt 0.9 kJ/(kg * °C), the melting heat is 830 kJ/kg.

a) The heating of the feedstock is to be represented graphically: For this purpose, its temperature is to be plotted as a function of the heat supplied.
b) What amount of heat is required to heat the solid feedstock to the scheduled temperature?
c) How much saturated steam (T = 130 °C; condensation heat 2250 kJ/kg) is required for this, if the steam condensate leaves the apparatus with saturated steam temperature?

⊗ **Solution**
→ ***Strategy***

a) The heating of the feedstock takes place in three phases:
   1. Heating of the solid from 10 °C to 34 °C (melting point): Heat quantity → Formula 18a

2. The melting of the solid (34 °C): Heat quantity → Formula 22a
3. Heating the melt from 34 °C to 80 °C: Heat quantity → Formula 18a

In phases 1 & 3, the relationship between temperature and heat input is a straight line. In phase 2, the temperature does not change despite the heat input, until the entire material is melted.

b) The amount of heat required to heat the solid from 10 °C to a melt temperature of 80 °C is the sum of the heat quantities of the three phases mentioned.
c) The calculated total heat required to heat the feedstock from 10 °C to 80 °C must be provided by the condensation heat of the saturated steam, from which the amount of steam results → Formula 24a.

→ ***Calculation***

a) *Heating of the solid from 10 °C to 34 °C:*

Phase 1: $Q_1 = m * cp * (T_1 - T_0) = 1800\ kg * 1.2\ kJ/(kg * °C) * (34-10)\ °C = 51{,}840\ kJ$

Phase 2: $Q_2 = m * \Delta_S H = 1800\ kg * 830\ kJ/kg = 1{,}494{,}000\ kJ$

Phase 3: $Q_3 = m * cp * (T_1 - T_0) = 1800\ kg * 0.9\ kJ/(kg * °C) * (80-34)\ °C = 74{,}520\ kJ$

→ A T vs. Q diagram is created with the following points:

Point 1 → $Q = 0\ kJ$
$T = 10\ °C$

Point 2 → $Q = 51{,}840\ kJ$
$T = 34\ °C$

Point 3 → $Q = 51{,}840\ kJ + 1{,}494{,}000\ kJ = 1{,}545{,}840\ kJ$
$T = 34\ °C$

Point 4 → $Q = 1{,}545{,}840\ kJ + 74{,}520\ kJ = 1{,}620{,}360\ kJ$
$T = 80\ °C$

b) *Required total heat*

$$Q = Q1 + Q2 + Q3 = (51{,}840 + 1{,}494.000 + 74{,}520)\ \text{kJ} = \mathbf{1{,}620.4\ MJ}$$

c) *Required steam amount*

$$Q = m * \Delta_V H$$

$$m = \frac{Q}{\Delta_V H} = \frac{1{,}620{,}360\ \text{kJ} * \text{kg}}{2250\ \text{kJ}} = \mathbf{720\ kg}$$

→ ***Result***

a) **Diagram T vs. Q**

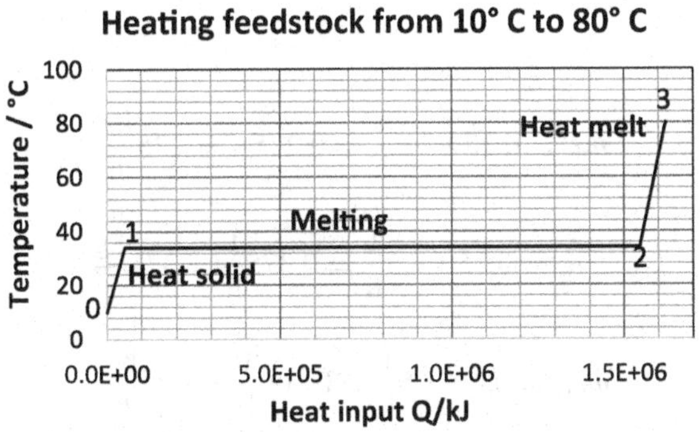

b) **The total amount of heat needed to raise the solid feedstock from 10 °C to a melt at 80 °C is Q = 1,620,360 kJ = 1,620 MJ.**
c) **This requires 720 kg of saturated steam.**

Task 43
Water gas can be produced by the partial oxidation of LPG (Liquid Petroleum Gas), possibly with the addition of water vapor. A stream of water gas produced in this way with a composition of 15 wt% hydrogen, 25 wt% carbon dioxide, 40 wt% carbon monoxide and 20 wt% water vapor, and a temperature of 400 °C and under a pressure of 10 bar is to be cooled down to 200 °C in one step by quenching with water (spraying water) in order to obtain the specified equilibrium composition. How much water at 20 °C must be injected into the gas stream per hour?

The specific heats in kJ/(kg * °C) in the temperature range under consideration are known:

Hydrogen: 14.5; CO_2: 1.10; CO: 1.10
Water liquid: 4.30; Water vapor: 10.0
Boiling point of water at 10 bar = 180 °C

The enthalpy of vaporization of water in the temperature range under consideration is 1500 kJ/kg.

⊗ **Solution**
→ ***Strategy***
The amount of heat that needs to be removed from the gas mixture is equal to the amount of heat consumed by heating and vaporizing the sprayed water. To calculate it, the mass flows of the components are calculated and inserted into Formula 19b with the corresponding heat capacities and the temperature difference of the

cooling process. The mass flow of water to be added results from the sum of the required heat flows to heat the water to its boiling point, its heat of vaporization and the heating of the steam to the final temperature. For this purpose, Formula 19a is added to Formula 24b and solved for the mass flow of water.

→ ***Calculation***

$$\dot{m}_{\mathrm{H_2}} = 0.15 * 1.5\frac{\mathrm{t}}{\mathrm{h}} = 0.225\frac{\mathrm{t}}{\mathrm{h}} = 225\frac{\mathrm{kg}}{\mathrm{h}}$$

$$\dot{m}_{\mathrm{CO_2}} = 0.25 * 1.5\frac{\mathrm{t}}{\mathrm{h}} = 0.375\frac{\mathrm{t}}{\mathrm{h}} = 375\frac{\mathrm{kg}}{\mathrm{h}}$$

$$\dot{m}_{\mathrm{CO}} = 0.4 * 1.5\frac{\mathrm{t}}{\mathrm{h}} = 0.600\frac{\mathrm{t}}{\mathrm{h}} = 600\frac{\mathrm{kg}}{\mathrm{h}}$$

$$\dot{m}_{\mathrm{H_2O}} = 0.2 * 1.5\frac{\mathrm{t}}{\mathrm{h}} = 0.300\frac{\mathrm{t}}{\mathrm{h}} = 300\frac{\mathrm{kg}}{\mathrm{h}}$$

$$\Delta\dot{Q}_{\mathrm{Gas}} = \left(\dot{m}_{\mathrm{H_2}} * cp_{\mathrm{H_2}} + \dot{m}_{\mathrm{CO_2}} * cp_{\mathrm{CO_2}} + \dot{m}_{\mathrm{CO}} * cp_{\mathrm{CO}} + \dot{m}_{\mathrm{H_2O}} * cp_{\mathrm{H_2O}}\right) * \Delta T_{\mathrm{Gas}}$$

$$\Delta\dot{Q}_{\mathrm{Gas}} = (225 * 14.5 + 375 * 1.1 + 600 * 1.1 + 300 * 10.0)\frac{\mathrm{kg} * \mathrm{kJ}}{\mathrm{h} * \mathrm{kg} * {}^\circ\mathrm{C}} * (400 - 200)\,{}^\circ\mathrm{C}$$

$$\Delta\dot{Q}_{\mathrm{Gas}} = (3262.5 + 412.5 + 660.0 + 3000.0)\frac{\mathrm{kJ}}{\mathrm{h} * {}^\circ\mathrm{C}} * 200\,{}^\circ\mathrm{C} = 1.467 * 10^6\frac{\mathrm{kJ}}{\mathrm{h}}$$

$$\Delta\dot{Q}_{\mathrm{Water}} = \dot{m}_{\mathrm{Water}} * \left[cp_{\mathrm{Water}} * (T_{\mathrm{BP}} - T_{\mathrm{in}}) + \Delta_V H_{\mathrm{Water}} + cp_{\mathrm{Steam}} * (T_{\mathrm{End}} - T_{\mathrm{BP}})\right]$$

$$\dot{m}_{\mathrm{Water}} = \frac{\dot{Q}_{\mathrm{Water}}}{cp_{\mathrm{Water}} * (T_{\mathrm{BP}} - T_{\mathrm{in}}) + \Delta_V H_{\mathrm{Water}} + cp_{\mathrm{Steam}} * (T_{\mathrm{End}} - T_{\mathrm{BP}})}$$

$$\dot{m}_{\mathrm{Water}} = \frac{1.467 * 10^6\frac{\mathrm{kJ}}{\mathrm{h}}}{4.3\frac{\mathrm{kJ}}{\mathrm{kg} * {}^\circ\mathrm{C}} * (180 - 200)\,{}^\circ\mathrm{C} + 1500\frac{\mathrm{kJ}}{\mathrm{kg}} + 10.0\frac{\mathrm{kJ}}{\mathrm{kg} * {}^\circ\mathrm{C}} * (200 - 180)\,{}^\circ\mathrm{C}}$$

$$\dot{m}_{\mathrm{Water}} = \frac{1.467 * 10^6\mathrm{kJ} * \mathrm{kg}}{\mathrm{h} * (688 + 1500 + 200)\,\mathrm{kJ}} = \mathbf{614.3\frac{kg}{h}}$$

→ ***Result***
To cool the gas stream, 614 kg of water at 20 °C are required per hour.

Exercise 44
In a dryer, 750 kg of wet cellulose granules per hour with a water content of 8.5 wt% are to be dried to a residual moisture of 0.5 wt%. The feed temperature is 70 °C.

For this purpose, nitrogen of 70 °C (water content 11 g/m³) is passed over the granules. The nitrogen stream enters at 70°C. Its saturation concentration is 220 g of water/m³. The heat of vaporization of water is ΔvH = 2100 kJ/kg.

a) What is the mass flow of pure cellulose?
b) What is the mass flow of the dried cellulose?
c) How much water is removed per hour?
d) What is the required nitrogen flow?
e) What heat power must be supplied to the dryer to keep the temperature at 70 °C?

⊗ **Solution**

→ ***Strategy***

The cellulose content and the amount of removed water can be determined by simple percentage calculation, with the reduction of water content in cellulose of 8 wt%. The amount of water to be removed results from the concentration difference of the water content of the nitrogen leaving the dryer minus that of the nitrogen feed, multiplied by the volume flow. The equation is converted to the volume flow. Since the temperature of all the streams supplied to the dryer is equal to the temperature of all the streams leaving the dryer, only the evaporation heat of the water removed from the granulate has to be supplied to the apparatus (Formula 24b).

→ ***Calculation***

Subscript: C = cellulose; W = water

a) $\dot{m}_C = \frac{100\%-8.5\%}{100\%} * \dot{m}_{C-wet} = 0.915 * 750\frac{kg}{h} = 686.25\frac{kg}{h} \cong \mathbf{686\frac{kg}{h}}$

b) $\dot{m}_{C-dry} = \frac{100\%-8\%}{100\%} * \dot{m}_{C-wet} = 0.92 * 750\frac{kg}{h} = \mathbf{690\frac{kg}{h}}$

c) $\dot{m}_{\Delta W} = \frac{8\%}{100\%} * \dot{m}_{C-wet} = 0.08 * 750\frac{kg}{h} = \mathbf{60\frac{kg}{h}}$

d) $\dot{m}_{\Delta W} = \dot{V}_{N_2} * (c_{Wex} - c_{Win})$

$$\dot{V}_{N_2} = \frac{m_{\Delta W}}{c_{Wex} - c_{Win}} = \frac{60\,\text{kg} * \dot{m}^3}{\text{h} * (0.220 - 0.011)\,\text{kg}} = 272.7\frac{\text{m}^3}{\text{h}} \cong \mathbf{273\frac{m^3}{h}}$$

e) $\dot{Q} = \dot{m}_W * \Delta_W H_W = 60\frac{\text{kg}}{\text{h}} * 2100\frac{\text{kJ}}{\text{kg}} = 216{,}000\frac{\text{kJ}}{\text{h}} = \frac{216{,}000\,\text{kJ} * \text{h}}{\text{h} * 3600\,\text{s}} = \mathbf{35\,kW}$

→ ***Result***

a) **The mass flow of pure cellulose is 686 kg per hour.**
b) **The mass flow of dried cellulose is 690 kg per hour.**
c) **60 kg of water is removed per hour.**
d) **The nitrogen flow required for drying is 273 m³/h.**
e) **The heat input required is 35 kW.**

Exercise 45

In a process for the production of diphenyl carbonate, phenol vapor at a temperature of 250°C is introduced into the reactor. For this purpose, 100 kg of solid phenol at a temperature of 20 °C is heated to the melting point of 41 °C [cp-phenol-solid = 1.9 kJ/(kg * °C)]. After complete melting [melting heat of phenol = 121 kJ/kg], the liquid phenol [cp-phenol-liquid = 2.4 kJ/(kg * °C)] is brought to the boiling point of 182 °C and vaporized there [vaporization heat = 510 kJ/kg]. The phenol vapor is heated to 250 °C [cp-phenol-vapor = 5.5 kJ/(kg * °C)] before being introduced into the reactor. How much energy is required for this part of the process?

⊗ **Solution**

→ ***Strategy***

The heating of the solid phenol stream to the superheated phenol vapor takes place in five steps. The corresponding heat flows required for this are calculated according to the formulas given below and added:

1. Heating of the solid from 20 °C to the melting point (formula 19a)
2. Melting (formula 22b)
3. Heating of the melt from the melting point to the boiling point (formula 19a)
4. Vaporization (formula 24b)
5. Heating of the vapor from the boiling point to the final temperature (formula 19a)

→ ***Calculation***

$$\dot{Q}_{\text{solid}} = \dot{m} * cp_{\text{solid}} * (T_{\text{melt}} - T_{\text{o}}) = 100\frac{\text{kg}}{\text{h}} * 1.9\frac{\text{kJ}}{\text{kg} * {}^\circ\text{C}} * (41.0 - 20.0)\,{}^\circ\text{C} = 3990\frac{\text{kJ}}{\text{h}}$$

$$\dot{Q}_{melt} = \dot{m} * \Delta_{melt}H = 100\frac{\text{kg}}{\text{h}} * 121\frac{\text{kJ}}{\text{kg}} = 12{,}100\frac{\text{kJ}}{\text{h}}$$

$$\dot{Q}_{\text{liq}} = \dot{m} * cp_{\text{liq}} * \left(T_{\text{boiling}} - T_{\text{melt}}\right) = 100\frac{\text{kJ}}{\text{h}} * 2.4\frac{\text{kJ}}{\text{kg} * {}^\circ\text{C}} * (182.0 - 41.0)\,{}^\circ\text{C}$$
$$= 33{,}840\frac{\text{kJ}}{\text{h}}$$

$$\dot{Q}_{evap} = \dot{m} * \Delta_V H = 100\frac{\text{kg}}{\text{h}} * 510\frac{\text{kJ}}{\text{kg}} = 51{,}000\frac{\text{kJ}}{\text{h}}$$

$$\dot{Q}_{vap} = \dot{m} * cp_D * \left(T_{\text{End}} - T_{\text{boiling}}\right) = 100\frac{\text{kg}}{\text{h}} * 5.5\frac{\text{kJ}}{\text{kg} * {}^\circ\text{C}} * (250 - 182.0)\,{}^\circ\text{C} = 37{,}400\frac{\text{kJ}}{\text{h}}$$

$$\dot{Q}_{\text{total}} = \dot{Q}_{\text{solid}} + \dot{Q}_{melt} + \dot{Q}_{\text{liq}} + \dot{Q}_{evap} + \dot{Q}_{vap}$$
$$= (3990 + 12{,}100 + 33{,}840 + 51{,}000 + 37{,}400)\frac{\text{kJ}}{\text{h}}$$

$$\dot{Q}_{\text{total}} = 138{,}330\frac{\text{kJ}}{\text{h}} = \frac{138{,}330\,\text{kJ} * \text{h}}{\text{h} * 3600\,\text{s}} = \mathbf{38.4\,kW}$$

→ ***Result***

The process system for the vaporization of phenol requires a heat input of 38.4 kW.

Exercise 46

What temperature is reached in the production of a 10 wt% calcium chloride solution if the water and calcium chloride are supplied at a temperature of 15 °C? The solution enthalpy of calcium chloride (M = 111.0 g/mol) is -60.8 kJ/mol. The heat capacity of the solution is 3.5 kJ/(kg * °C).

⊗ **Solution**

→ ***Strategy***

Calcium chloride has a negative solution enthalpy, so heat is released during the dissolution process (formula 26). This heat is consumed by the temperature increase of the solution (formula 18a). Based on the specification of the calcium chloride concentration in wt% 100 kg solution, i.e. 10 kg calcium chloride, is offered as a calculation basis.

→ ***Calculation***

$$Q = -n_{\text{CaCl}_2} * \Delta_L H$$

$$n_{\text{CaCl}_2} = \frac{m_{\text{Cacl}_2}}{M_{\text{CaCl}_2}} = \frac{10\,\text{kg} * \text{mol}}{0.111\,\text{kg}} = 90.1\,\text{mol}$$

$$Q = -90.1\,\text{mol} * \left(-60.8\frac{\text{kJ}}{\text{mol}}\right) = 5478\,\text{kJ}$$

$$Q = m_{\text{sol}} * cp_{\text{sol}} * (T_E - T_o)$$

$$T_E = T_o + \frac{Q}{m_{\text{sol}} * cp_{\text{sol}}} = 15\,°\text{C} + \frac{5478\,\text{kJ} * \text{kg} * °\text{C}}{100\,\text{kg} * 3.5\,\text{kJ}} = 15\,°\text{C} + 15.7\,°\text{C} = \mathbf{30.7\,°C}$$

→ ***Result***

The prepared calcium chloride solution has a temperature of 30.7 °C.

Task 47

By continuous addition of 0.3 kg/s of potassium chloride of 20 °C to an aqueous flow with a flow rate of 1.3 L/s (ρ = 1050 kg/m^3; cp_{H_2O}= 4.19 kJ/kg * °C) is produced. The resulting solution of the salt shall have a temperature of 20 °C. The solution enthalpy of potassium chloride is 13.0 kJ/mol, its molar mass 74.55 g/mol. What temperature must the used water feed have?

⊗ **Solution**

→ ***Strategy***

Since the solution enthalpy of potassium chloride is positive, the water used has to have a higher temperature than the solution produced. First, the molar flow of potassium chloride is calculated and from this the negative heat output during the dissolution process according to formula 26. The temperature of the water used is then calculated from formula 19a.

→ ***Calculation***

$$\dot{Q} = -\dot{n}_{KCl} * \Delta_L H \qquad \dot{n}_{KCl} = \frac{\dot{m}_{KCl}}{M_{KCl}} = \frac{0.3\,\text{kg} * \text{mol}}{\text{s} * 0.07455\,\text{kg}} = 4.024\frac{\text{mol}}{\text{s}}$$

$$\dot{Q} = -4.024\frac{\text{mol}}{\text{s}} * 13.0\frac{\text{kJ}}{\text{mol}} = -52.31\frac{\text{kJ}}{\text{s}}$$

$$\dot{Q} = \dot{m}_W * cp_W * (T_{ex} - T_{in})$$

$$\dot{m}_W = \rho_W * \dot{V}_W = 1050\frac{\text{kg}}{\text{m}^3} * 0.0013\frac{\text{m}^3}{\text{s}} = 1.365\frac{\text{kg}}{\text{s}}$$

$$T_{in} = T_{ex} - \frac{\dot{Q}}{\dot{m}_W * cp_W} = 20\,°\text{C} - \frac{-52.31\,\text{kJ} * \text{s} * \text{kg} * °\text{C}}{\text{s} * 1.365\,\text{kg} * 4.19\,\text{kJ}} = \mathbf{29.15\,°C}$$

→ ***Result***

The aqueous feed stream must have a temperature of 29.15 °C.

3.4.2 Calculation of the Reaction Enthalpy from Formation Enthalpies

Task 48

Is the shift reaction of the equilibrium conversion of carbon monoxide and water vapor to hydrogen and carbon dioxide exothermic or endothermic?

Molar formation enthalpies $\Delta_f H_0$ (kJ/mol):
Carbon monoxide: −110.5, water(g): −241.8, carbon dioxide: −393.5

⊗ **Solution**

→ ***Strategy***

The reaction equation is set up and the stoichiometric factors are taken from it.

$$CO + H_2O \leftrightarrow H_2 + CO_2$$

The reaction enthalpy is calculated using Formula 28 and the heat of reaction is concluded from this. (Hydrogen is an element → $\Delta_f H_o = 0$)

→ ***Calculation***

$\nu_{CO} = -1; \nu_{H_2O} = -1; \nu_{H_2} = +1; \nu_{CO_2} = +1$

$$\Delta_R H_0 = \sum_i \left(\upsilon_i * \Delta_f H_{i0}\right) = \nu_{CO} * \Delta_f H_{CO} + \nu_{H_2O} * \Delta_f H_{H_2O} + \nu_{CO_2} * \Delta_f H_{CO_2}$$

$$\mathbf{\Delta_R H_0} = -1 * \left(-110.5\frac{kJ}{mol}\right) - 1 * \left(-241.8\frac{kJ}{mol}\right) + 1 * \left(-393.5\frac{kJ}{mol}\right) = \mathbf{-41.2\frac{kJ}{mol}}$$

→ ***Result***

With a reaction enthalpy of $\Delta_R H_o$= −41.2 kJ/mol, it is an exothermic reaction.

Task 49

Sulfur trioxide is absorbed in sulfuric acid. The resulting oleum is mixed with water and the dissolved sulfur trioxide reacts to sulfuric acid. How much heat must be dissipated in the reaction of 320 kg sulfur trioxide to keep the temperature at 25 °C? The temperature of the feedstock is 25 °C.

$$H_2O + SO_3 \rightarrow H_2SO_4$$

Molar mass SO_3 = 80 g/mol
Standard enthalpies of formation:
Water (liquid): –285.9 kJ/mol; sulfur trioxide (gas): –388.8 kJ/mol;
sulfuric acid: –193.8 kcal/mol

⊗ **Solution**

→ ***Strategy***

The reaction heat to be discharged is calculated from Formula 27a. The molar amount of sulfur trioxide results from its mass used and its molar mass. The enthalpy of reaction follows from Formula 28. It should be noted here that the enthalpy of formation of sulfuric acid is not given in SI units.

→ ***Calculation***

$$n_{SO_3} = \frac{m_{SO_3}}{M_{SO_3}} = \frac{320\,\text{kg} * \text{mol}}{0.080\,\text{kg}} = 4000\,\text{mol}$$

	SO_3	H_2O	H_2SO_4
ν_i	−1	−1	+1
$\Delta_f H$	−388,8 kJ/mol	−285,9 kJ/mol	−193.8 kcal/mol = −4.19 * 193.8 kJ/mol = −798.5 kJ/mol

$$\Delta_R H = \sum_{i8} (\nu_i * \Delta_f H_i) = [-1 * (-285.9) - 1 * (-388.8) + 1 * (-812.0)] \frac{\text{kJ}}{\text{mol}} = -137.3 \frac{\text{kJ}}{\text{mol}}$$

$$Q = -n_{SO_3} * \Delta_R H = -4000\,\text{mol} * \left(-137.3 \frac{\text{kJ}}{\text{mol}}\right) = \mathbf{549{,}200\,kJ = 549\,MJ}$$

→ ***Result***

The amount of heat to be discharged is 549 MJ.

Task 50

Into a continuous stirred tank reactor are fed per hour 2 metric t 5wt% hydrochloric acid and 6 metric t 10wt% sodium carbonate solution. Both solutions have a temperature of 20 °C.

$$HCl + Na_2CO_3 \rightarrow NaHCO_3 + NaCl$$

a) What is the stoichiometric excess ratio Na_2CO3/HCl?
b) What is the reaction enthalpy?
c) How much heat is released in the reactor per unit of time?
d) What would be the reactor temperature, if no heat would be removed?

Formation enthalpies (kJ/mol):

$HCl = -92$; $Na_2CO_3 = -1131$; $NaHCO_3 = -949$; $NaCl = -412$

cp-reactor content = 4.5 kJ/(kg * °C)

⊗ **Solution**

→ ***Strategy***

First, the molar flows of HCl and sodium carbonate are calculated from the feed streams and the molar masses. The molar ratio is determined from this.

The reaction enthalpy is calculated using Formula 28. To calculate the heat generation of the reaction, the molar flow of the deficit component, i.e. HCl, is used in Formula 27b. This heat is absorbed by the two feed streams. From this, the temperature is calculated with the inverted Formula 19a, with which the solution leaves the reactor.

→ ***Calculation***

a)

$$\dot{m}_{HCl} = 2\frac{\text{t}}{\text{h}} * 0.05 = 100\frac{\text{kg}}{\text{h}} = \frac{100\,\text{kg} * \text{h}}{\text{h} * 3600\,\text{s}} = 0.02778\frac{\text{kg}}{\text{s}}$$

$$\dot{n}_{HCl} = \frac{\dot{m}_{HCl}}{M_{HCl}} = \frac{0.02778\,\text{kg} * \text{mol}}{\text{s} * 0.0365\,\text{kg}} = 0.761\frac{\text{mol}}{\text{s}}$$

$$\dot{m}_{Na_2CO_3} = 6\frac{t}{h} * 0.1 = 600\frac{kg}{h} = \frac{600\,kg * h}{h * 3600\,s} = 0.1667\frac{kg}{s}$$

$$\dot{n}_{Na_2CO_3} = \frac{\dot{m}_{Na_2CO_3}}{M_{Na_2CO_3}} = \frac{0.1667\,kg * mol}{s * 0.106\,kg} = 1.572\frac{mol}{s}$$

→ *HCl is the deficit component.*

$$\frac{\dot{\boldsymbol{n}}_{\mathbf{Na_2CO_3}}}{\dot{\boldsymbol{n}}_{\mathbf{HCl}}} = \frac{1.572}{0.761} = \mathbf{2.07}$$

b) $\Delta_R H = \sum_i (\nu_i * \Delta_f H_i = -1 * \Delta_f H_{HCl} - 1 * \Delta_f H_{Na_2CO_3} + 1 * \Delta_f H_{NaHCO_3} + 1 * \Delta_f H_{NaCl}$

$$\Delta_R \boldsymbol{H} = [-1 * (-92) - 1 * (-1131) + 1 * (-949) + 1 * (-412)]\frac{kJ}{mol} = \mathbf{-138\frac{kJ}{mol}}$$

c) $\dot{\boldsymbol{Q}} = -\dot{n}_{HCl} * \Delta_R H = -0.761\frac{mol}{s} * \left(-138\frac{kJ}{mol}\right) = \mathbf{105.0\frac{kJ}{s}} = \mathbf{105.0\,kW}$

d)

$$\dot{Q} = \dot{m}_{Reac} * cp_{Reac} * (T_{ex} - T_{in})$$

$$T_{ex} = T_{in} + \frac{\dot{Q}}{\dot{m}_{Reac} * cp_{Reac}} = 20\,°C + \frac{105.0\,kJ * h * kg * °C}{s * (2000 + 6000)\,kg * 4.5\,kJ}$$

$$= 20\,°C + \frac{105.0 * h * °C * 3600\,s}{s * 8000 * h * 4.5}$$

$$\boldsymbol{T_{ex}} = 20\,°C + 10.5\,°C = \mathbf{30.5\,°C}$$

→ ***Result***

a) **The molar ratio of the feed streams of Na_2CO_3 to HCl is 2.07. HCl is the deficit component.**
b) **The reaction enthalpy is −138 kJ/mol.**
c) **The heat generation of the reaction is 105.0 kW.**
d) **The reactor temperature is 30.5 °C.**

Exercise 51

How much heat is released in the formation of 500 kg of tetrachloromethane (M = 154 g/mol) from methane and chlorine at 150 °C?

$$CH_4 + 4Cl_2 \rightarrow CCl_4 + 4HCl$$

Standard enthalpies of formation in (kJ/mol):

$$CH_4:\ -74.9, CCl_4:\ -33.3, HCl:\ -92.3$$

Heat capacities:

CH_4: 8.536 cal/(mol * °C); Cl_2: 8.11 cal/(mol* °C); CCl_4: 0.544 kJ/(kg * °C); HCl: 29.1 J/(mol * °C)

⊗ **Solution**

→ ***Strategy***

For a reaction at a temperature different from the standard temperature of 25 °C, the reaction enthalpy is calculated according to Formula 29. For this purpose, the stoichiometric factors are first determined from the reaction equation. The heat capacities of methane and tetrachloromethane must be converted to SI or molar units. The heat released by the reaction results from the number of moles of tetrachloromethane formed and the reaction enthalpy according to Formula 27a.

→ ***Calculation***

$$Q = -\Delta n * \Delta_R H$$

$$\Delta n = \frac{m}{M} = \frac{500\,\text{kg} * \text{mol}}{0.154\,\text{kg}} = 3246.8\,\text{mol}$$

$$\Delta_R H_T = \sum_i \left[\upsilon_i * \Delta_f H_{i_0} + \upsilon_i * cp_i * (T - 298.15\,\text{K})\right] (\text{Cl}_2 \text{ is an element} \rightarrow \Delta_f H_{i_0} = 0)$$

$$\nu_{CH_4} = -1;\ \nu_{Cl_2} = -4;\ \nu_{CCl_4} = +1;\ \nu_{HCl} = +4$$

$$\sum_i \upsilon_i * \Delta_f H_{i_0} = \nu_{CH_4} * \Delta_f H_{CH_{4_0}} + \nu_{CCl_4} * \Delta_f H_{CCl_{4_0}} + \nu_{HCl} * \Delta_f H_{HCl_0}$$

$$\sum_i \upsilon_i * \Delta_f H_{i_0} = [-1 * (-74.9) + 1 * (-33.3) + 4 * (-92.3)]\frac{\text{kJ}}{\text{mol}} = -327.6\frac{\text{kJ}}{\text{mol}}$$

$$\upsilon_i * cp_i * (T - 298.15\,\text{K})] = \left(\nu_{CH_4} * cp_{CH_4} + \nu_{Cl_2} * cp_{Cl_2} + \nu_{CCl_4} * cp_{CCl_4} + \nu_{HCl} * cp_{HCl}\right) * (150 - 25)\,°\text{C}$$

$$cp_{CH_4} = \frac{8.536\,\text{cal} * 4.19\,\text{J}}{\text{mol} * °\text{C}} = 35.8\frac{\text{J}}{\text{mol} * °\text{C}}$$

$$cp_{Cl_2} = \frac{8.11\,\text{cal} * 4.19\,\text{J}}{\text{mol} * °\text{C}} = 34.0\frac{\text{J}}{\text{mol} * °\text{C}}$$

$$cp_{CCl_4} = \frac{544\,\text{J} * 0.154\,\text{kg}}{\text{kg} * °\text{C} * \text{mol}} = 83.8\frac{\text{J}}{\text{mol} * °\text{C}}$$

$$\upsilon_i * cp_i * (T - 298.15\,\text{K})] = (-1 * 35.8 - 4 * 34.0 + 1 * 83.8 + 4 * 29.1)\frac{\text{J}}{\text{mol} * °\text{C}} * (150 - 25)\,°\text{C}$$

$$\upsilon_i * cp_i * (T - 298.15\,\text{K}) = 0.0284\frac{\text{kJ}}{\text{mol} * °\text{C}} * 125\,°\text{C} = 3.55\frac{\text{kJ}}{\text{mol}}$$

$$\Delta_R H = -327.6\frac{\text{kJ}}{\text{mol}} + 3.55\frac{\text{kJ}}{\text{mol}} = -324.05\frac{\text{kJ}}{\text{mol}}$$

$$Q = -\Delta n * \Delta_R H = -3246.8\,\text{mol} * \left(-324.05\frac{\text{kJ}}{\text{mol}}\right) = 1{,}052{,}126\,\text{kJ} = \mathbf{1052\,MJ}$$

→ ***Result***
The amount of heat released is 1052 MJ.

Task 52
2 metric t of a reacted mixture after a synthesis contain 1.5 wt% remaining sulfuryl chloride. The sulfuryl chloride is destroyed by adding aqueous sodium hydroxide solution . How much heat is released at 80 °C?

$$SO_2Cl_2 + 4\ NaOH \rightarrow Na_2SO_4 + 2\ NaCl + 2\ H_2O$$

Standard enthalpies of formation:
Water (liquid): -285.9 kJ/mol; NaOH: -102 kcal/mol,
Sodium chloride: -97.8 kcal/mol; Sulfuryl chloride: -93 kcal/mol; Sodium sulfate: −331 kcal/mol

Heat capacity:
Water (liquid): 4.19 kJ/(kg * °K); NaOH: 19.2 cal/(mol * K);
Sodium chloride: 50.7 J/(mol * °C); Sulfuryl chloride: 0.65 kJ/(kg * °C); Sodium sulfate: 30.5 cal/(mol * °C)

Atomic masses:
H: 1.0 g/mol; O: 16.0 g/mol; S: 32.0 g/mol; Cl: 35.5 g/mol

⊗ **Solution**
→ ***Strategy***
The amount of heat generated is calculated according to Formula 27a from the amount of sulfuryl chloride reacted and the enthalpy of reaction. The mass of sulfuryl chloride is given by its concentration and the amount of the reaction solution. The corresponding amount of moles is calculated using the molar mass of sulfuryl chloride.

The reaction enthalpy for the reaction temperature of 80 °C, which deviates from the standard temperature of 25 °C, is calculated according to Formula 29. For this purpose, the formation enthalpies and heat capacities must be converted into molar quantities and SI units.

→ ***Calculation***

$$Q = -n * \Delta_R H$$

$$\Delta_R H = \Delta_R H_T = \sum_i \left[\nu_i * \Delta_f H_{i_o} + \nu_i * cp_i * (T - 298.15\,\mathrm{K})\right]$$

$$(T - 298.15)\,\mathrm{K} = (273.15 + 80 - 298.15)\,\mathrm{K} = 55.0\,\mathrm{K} = 55.0\,°\mathrm{C}$$

$$n_{SO_2Cl_2} = \frac{m_{SO_2Cl_2}}{M_{SO_2Cl_2}}$$

$$m_{SO_2Cl_2} = 2000\,\text{kg} * \frac{1.5\%}{100\%} = 30\,\text{kg}$$

$$M_{SO_2Cl_2} = (32.0 + 2 * 16.0 + 2 * 35.5)\frac{\text{g}}{\text{mol}} = 135.0\frac{\text{g}}{\text{mol}}$$

$$n_{SO_2Cl_2} = \frac{m_{SO_2Cl_2}}{M_{SO_2Cl_2}} = \frac{30\,\text{kg} * \text{mol}}{0.135\,\text{kg}} = 222.2\,\text{mol}$$

Stoichiometric factors ν_i:
SO_2Cl_2: −1; NaOH: −4; Na_2SO_4: +1; NaCl: +2; H_2O: +2

Standard formation enthalpies $\Delta_f H_o$→ kJ/mol

$$SO_2Cl_2 : -93\frac{\text{kcal}}{\text{mol}} = -93\frac{\text{kcal} * 4.19\,\text{kJ}}{\text{mol} * \text{kcal}} = -389.7\frac{\text{kJ}}{\text{mol}}$$

$$NaOH : -102\frac{\text{kcal}}{\text{mol}} = -102\frac{\text{kcal} * 4.19\,\text{kJ}}{\text{mol} * \text{kcal}} = -427.4\frac{\text{kJ}}{\text{mol}}$$

$$Na_2SO_4 : -331\frac{\text{kcal}}{\text{mol}} = -331\frac{\text{kcal} * 4.19\,\text{kJ}}{\text{mol} * \text{kcal}} = -1386.9\frac{\text{kJ}}{\text{mol}}$$

$$NaCl: \ -97.8\frac{\text{kcal}}{\text{mol}} = -97.8\frac{\text{kcal} * 4.19\,\text{kJ}}{\text{mol} * \text{kcal}} = -409.8\frac{\text{kJ}}{\text{mol}}$$

$$H_2O : -285.9\frac{\text{kJ}}{\text{mol}}$$

*Heat capacities cp → kJ/(mol * °C)*

$$SO_2Cl_2 : 0.65\frac{\text{kJ}}{\text{kg} * °\text{C}} = 0.65\frac{\text{kJ} * 0.135\,\text{kg}}{\text{kg} * °\text{C} * \text{mol}} = 0.0878\frac{\text{kJ}}{\text{mol} * °\text{C}}$$

$$NaOH : 19.2\frac{\text{cal}}{\text{mol} * °\text{C}} = 19.2\frac{\text{cal} * 4.19\text{J}}{\text{mol} * \text{cal}} = 0.0805\frac{\text{kJ}}{\text{mol} * °\text{C}}$$

$$Na_2SO_4 : 30.5\frac{\text{cal}}{\text{mol} * °\text{C}} = 30.5\frac{\text{cal}*4.19\,\text{J}}{\text{mol} * \text{cal} * °\text{C}} = 0.128\frac{\text{kJ}}{\text{mol} * °\text{C}}$$

$$NaCl : 50.7\frac{\text{J}}{\text{mol} * °\text{C}} = 0.051\frac{\text{kJ}}{\text{mol} * °\text{C}}$$

$$H_2O: 4.19\frac{\text{kJ}}{\text{kg} * °\text{C}} = 4.19\frac{\text{kJ} * 0.018\,\text{kg}}{\text{kg} * °\text{C} * \text{mol}} = 0.0754\frac{\text{kJ}}{\text{mol} * °\text{C}}$$

Reaction enthalpy

$$\Delta_R H_o = \sum_i \left(\nu_i * \Delta_f H_{i_o}\right)$$
$$= \nu_{SO_2Cl_2} * \Delta_f H_{SO_2Cl_{2_o}} + \nu_{NaOH} * \Delta_f H_{NaOH_o} + \nu_{Na_2SO_4} * \Delta_f H_{Na_2SO_4o} + \nu_{NaCl} * \Delta_f H_{NaClo} + \nu_{H_2O} * \Delta_f H_{H_2Oo}$$

$$\Delta_R H_o = [-1 * (-389.7) - 4 * (-427.4) + 1 * (-1386.9) + 2 * (-409.8) + 2 * (-285.9)] \frac{kJ}{mol}$$

$$\Delta_R H_o = -679.0 \frac{kJ}{mole}$$

$$\sum_i [\nu_i * cp_i * (T - 298.15)] = \sum_i (\nu_i * cp_i * 55.0\,°C)$$

$$\sum_i (\nu_i * cp_i * 55\,°C) = (\nu_{SO_2Cl_2} * cp_{SO_2Cl_{2_o}} + \nu_{NaOH} * cp_{NaOH_o} + \nu_{Na_2SO_4} * cp_{Na_2SO_4o} + \nu_{NaCl} * cp_{NaCl_o} + \nu_{H_2O} * cp_{H_2O_o}) * 55\,°C$$

$$\sum_i (\nu_i * cp_i * 55\,°C) = (-1 * 0.0878 - 4 * 0.0805 + 1 * 0.128 + 2 * 0.051 + 2 * 0.0754) \frac{kJ}{mol * °C} * 55.0\,°C$$

$$\sum_i (\nu_i * cp_i * 55\,°C) = -1.6 \frac{kJ}{mol}$$

$$\Delta_R H = -679.0 \frac{kJ}{mol} - 1.6 \frac{kJ}{mol} = -680.6 \frac{kJ}{mol}$$

$$Q = -n * \Delta_R H = -222.2\,\text{mol} * \left(-680.6 \frac{kJ}{mol}\right) = \mathbf{151{,}229\,kJ = 151.2\,MJ}$$

→ ***Result***
In the reaction, an amount of heat of 151.2 MJ is released.

3.4.3 Heating Value/Calorific Value

The definition and calculation of the high and low heating value as well as the calorific value are treated in Sect. 2.4.4.

Task 53
What are the high heating value (HHV) and the low heating value (LHV) of propane in MJ/kg or MJ/Standard-m^3?

$\Delta_f H_{propane} = -104$ kJ/mol; $\Delta_f H_{CO_2} = -391$ kJ/mol; $\Delta_f H_{H_2Ogas} = -242$ kJ/mol; $\Delta_f H_{H_2Oliq} = -286$ kJ/mol,
$\Delta_V H_{H_2O} = 44.1$ kJ/mol

Atomic masses: H = 1 g/atom; C = 12 g/atom; O = 16 g/atom

⊗ **Solution**
→ ***Strategy***
First, the reaction equation is set up and the stoichiometric signs are determined. The reaction heat is calculated from the formation enthalpies (formula 28). For the calculation of the high heating value (calorific value), the formation enthalpy of liquid water is used here, for the low heating value that of gaseous water. The number of moles corresponding to 1 kg propane or 1 Std-m^3 propane is calculated. The heat determined from this and from the formation heat according to formula 27a corresponds to the heating value.

→ ***Calculation***

$$C_3H_8 + 5O_2 \rightarrow 3CO_2 + 4H_2O$$

	C_3H_8	CO_2	H_2O
ν_i	−1	+3	+4

$$Q = -n * \Delta_R H$$

For the mass-related heating value $:n = \frac{m}{M}$

$$M_{Propane} = 44 \frac{g}{mol}$$

For 1 kg: $n = \frac{1\,kg * mol}{0.044\,kg} = 22.73\,mol$

For the standard volume-related heating value: $n = 44.63 \frac{mol}{Std-m^3}$

$$\Delta_R H = \sum_i (\nu_i * \Delta_f H_i) = \nu_P * \Delta_f H_P + \nu_{CO_2} * \Delta_f H_{CO_2} + \nu_{H_2O} * \Delta_f H_{H_2O}$$

For the high heating value, the formation enthalpy for liquid water is entered, for the low heating value that of the gaseous water.

High heating value HHV:

$$\Delta_R H = [-1 * (-104) + 3 * (-391) + 4 * (-286)] \frac{\text{kJ}}{\text{mol}} = -2213 \frac{\text{kJ}}{\text{mol}}$$

$$\boldsymbol{HHV} = Q = -22.73 \frac{\text{mol}}{\text{kg}} * (-2213) \frac{\text{kJ}}{\text{mol}} = 50{,}301 \frac{\text{kJ}}{\text{kg}} = \mathbf{50.3} \frac{\mathbf{MJ}}{\mathbf{kg}}$$

$$\boldsymbol{HHV} = Q = -44.63 \frac{\text{mol}}{\text{Std-m}^3} * (-2213) \frac{\text{kJ}}{\text{mol}} = 98{,}766 \frac{\text{kJ}}{\text{Std-m}^3} = \mathbf{98.8} \frac{\mathbf{MJ}}{\mathbf{Std-m^3}}$$

Low heating value LHV:

$$\Delta_R H = [-1 * (-104) + 3 * (-391) + 4 * (-242)] \frac{\text{kJ}}{\text{mol}} = -2037 \frac{\text{kJ}}{\text{mol}}$$

$$\boldsymbol{LHV} = Q = -22.73 \frac{\text{mol}}{\text{kg}} * (-2037) \frac{\text{kJ}}{\text{mol}} = 46{,}301 \frac{\text{kJ}}{\text{kg}} = \mathbf{46.3} \frac{\mathbf{MJ}}{\mathbf{kg}}$$

$$\boldsymbol{LHV} = Q = -44.63 \frac{\text{mole}}{\text{Std-m}^3} * (-2037) \frac{\text{kJ}}{\text{mol}} = 90{,}911 \frac{\text{kJ}}{\text{Std-m}^3} = \mathbf{90.9} \frac{\mathbf{MJ}}{\mathbf{Std-m^3}}$$

→ ***Result***

The calorific value (high heating value) of propane is 50.3 MJ/kg or 98.8 MJ/standard-m³.

The low heating value of propane is 46.3 MJ/kg or 90.9 MJ/standard–m³.

Task 54

What are the high and low heating values (MJ/standard-m³) of a mixture of 34 vol% hydrogen, 49 vol% carbon monoxide and 17 vol% methane?

Formation enthalpies $\Delta_f H$/(kJ/mol):
H_2O gaseous: −242; H_2O liquid: −286; CO: −110.5; CO_2: −393.5; CH_4: −74.9

⊗Solution

→ ***Strategy***

The heating value of gases is based on a standard-m³ = 44.63 mol. The corresponding molar numbers of hydrogen, carbon monoxide and methane in the gas mixture are calculated from the associated volume percentages. The heating value is the sum of the reaction heats of combustion of hydrogen, carbon monoxide and methane according to formula 27a. The reaction enthalpies are calculated using the formation enthalpies (formula 28). It is important to use the formation enthalpy of liquid water for the high heating value HHV and the formation enthalpy of gaseous water for the low heating value LHV.

→ ***Calculation***

$$\text{heating value} = Q_{\text{total}} = Q_{H_2} + Q_{CO} + Q_{CH_4} \text{ with } Q_{\text{total}} \text{ for one standard} - \text{m}^3$$

$$Q_i = -n_i * \Delta_R H_i \text{ und } \Delta_R H_i = \sum_i (\upsilon_i * \Delta_f H_i)$$

Hydrogen:

$$n_{H_2} = 44.63 \frac{\text{mol}}{\text{Std-m}^3} * 0.34 = 15.17 \frac{\text{mol}}{\text{Std-m}^3}$$

$$H_2 + {}^1\!/\!_2 O_2 \rightarrow H_2O$$

For the high heating value, the reaction enthalpy is calculated with the formation enthalpy of liquid water:

$$\Delta_R H_{H_2} = +1 * \left(-286 \frac{\text{kJ}}{\text{mol}}\right) = -286 \frac{\text{kJ}}{\text{mol}} \rightarrow Q_{H_2} = -15.17 \frac{\text{mol}}{\text{Std-m}^3} * \left(-286 \frac{\text{KJ}}{\text{mol}}\right) = 4339 \frac{\text{kJ}}{\text{Std-m}^3}$$

For the low heating value, the reaction enthalpy is calculated with the formation enthalpy of gaseous water:

$$\Delta_R H'_{H_2} = +1 * \left(-242 \frac{\text{kJ}}{\text{mol}}\right) = -242 \frac{\text{kJ}}{\text{mol}} \rightarrow Q'_{H_2} = -15.17 \frac{\text{mol}}{\text{Std-m}^3} * \left(-242 \frac{\text{kJ}}{\text{mol}}\right) = 3671 \frac{\text{kJ}}{\text{Std-m}^3}$$

Carbon monoxide:

$$n_{CO} = 44.63 \frac{\text{mol}}{\text{Std-m}^3} * 0.49 = 21.87 \frac{\text{mol}}{\text{Std-m}^3}$$

$$CO + {}^1\!/\!_2 O_2 \rightarrow CO_2$$

$$\Delta_R H_{CO} = [-1 * (-110.5) + 1 * (-393.5)] \frac{\text{kJ}}{\text{mol}} = -283 \frac{\text{kJ}}{\text{mol}} Q_{H_2}$$
$$= -21.87 \frac{\text{mol}}{\text{Std-m}^3} * \left(-283 \frac{\text{kJ}}{\text{mol}}\right) = 6189 \frac{\text{kJ}}{\text{Std-m}^3}$$

Methane:

$$n_{CH_4} = 44.63 \frac{\text{mol}}{\text{Std-m}^3} * 0.17 = 7.59 \frac{\text{mol}}{\text{Std-m}^3}$$

$$CH_4 + 2O_2 \rightarrow CO_2 + 2H_2O$$

For the high heating value, the reaction enthalpy is calculated with the formation enthalpy of liquid water:

$$\Delta_R H_{CH_4} = [-1 * (-74.9) + 1 * (-393.5) + 2 * (-286)] \frac{\text{kJ}}{\text{mol}} = -890.6 \frac{\text{kJ}}{\text{mol}}$$

$$\rightarrow Q_{CH_4} = -7.59\frac{\text{mol}}{\text{Std-m}^3} * \left(-890.6\frac{\text{kJ}}{\text{mol}}\right) = 6760\frac{\text{kJ}}{\text{Std-m}^3}$$

For the low heating value, the reaction enthalpy is calculated with the formation enthalpy of gaseous water:

$$\Delta_R H'_{CH_4} = [-1 * (-74.9) + 1 * (-393.5) + 2 * (-242)]\frac{\text{kJ}}{\text{mol}} = -802.6\frac{\text{kJ}}{\text{mol}} \rightarrow$$

$$\rightarrow Q'_{CH_4} = -7.59\frac{\text{mol}}{\text{Std-m}^3} * \left(-802.6\frac{\text{kJ}}{\text{mol}}\right) = 6092\frac{\text{kJ}}{\text{Std-m}^3}$$

$$\boldsymbol{HHV} = (Q_{H_2} + Q_{CO} + Q_{CH_4}) = (4339 + 6189 + 6760)\frac{\text{kJ}}{\text{Std-m}^3} = 17{,}288\frac{\text{kJ}}{\text{Std-m}^3} = \mathbf{17.3}\frac{\mathbf{MJ}}{\mathbf{Std-m^3}}$$

$$\boldsymbol{LHV} = (Q'_{H_2} + Q_{CO} + Q'_{CH_4}) = (3671 + 6189 + 6092)\frac{\text{kJ}}{\text{Std-m}^3} = 15{,}952\frac{\text{kJ}}{\text{Std-m}^3} = \mathbf{15.95}\frac{\mathbf{MJ}}{\mathbf{Std-m^3}}$$

→ ***Result***

The high heating value of the gas mixture is 17.3 MJ per standard-m^3.
The low heating value is 15.95 MJ per standard-m^3.

Task 55

In a production process, 50 metric t of by-product have been produced, which consists practically completely of stearic acid ($C_{17}H_{35}COOH$; M = 285 g/mol; $\Delta_f H_0 = -950$ kJ/mol). It cannot be further processed into product because of its intense smell and is to be led to the thermal utilization in a boiler house.

Standard enthalpies of formation:

$\Delta_f Ho_{H_2O-gas} = -241.8$ kJ/mol; $\Delta_f Ho_{H_2O-liq} = -285.8$ kJ/mol; $\Delta_f Ho_{CO_2} = -393.5$ kJ/mol,

$M_{H_2O} = 18.0$ g/mol; heat of vaporization of water $\Delta_v H = 2450$ kJ/kg

a) What is the high heating value (HHV) of stearic acid?
b) What is the low heating value (LHV) of stearic acid? For the calculation, use the solution path via the enthalpy of formation of gaseous water and the other via the heat of vaporization of water.
c) How much electrical energy in MWh can be generated from the by-product stream if the overall electrical efficiency (boiler plant and power generator) is 45%?

⊗ **Solution**

→ ***Strategy***

The reaction equation is set up. The heating value of a liquid or a solid is the reaction heat that is generated when 1 kg of this substance is burned. Thus, the reaction enthalpy is first calculated from the enthalpies of formation according to Formula 28. This, multiplied by the corresponding molar amount that corresponds to 1 kg of fuel, gives the heating value (Formula 27a). If the enthalpy of formation of water is set to the value for its liquid state, the high heating value HHV is obtained. If the enthalpy of formation of gaseous water is used, the low heating value LHV results. The low heating value can be calculated by subtracting the heat of vaporization of the water formed during combustion from the high heating value (Formula 24a).

The amount of heat released by the combustion of stearic acid is calculated by multiplying the lower heating value by 50 t. Since only 45% of the heat generated can be converted into electrical energy, this amount is multiplied by 0.45 and converted into MWh.

→ ***Calculation***

$C_{17}H_{35}COOH + 26O_2 \rightarrow 18CO_2 + 18H_2O$

	$C_{17}H_{35}COOH$	CO_2	H_2O
ν_i	−1	+18	+18

a) $Q = -n_{\text{Stea}} * \Delta_R H$

$$n_{\text{Stea}} = \frac{m_{\text{Stea}}}{M_{\text{Stea}}} = \frac{1\,\text{kg} * \text{mol}}{0.285\,\text{kg}} = 3.51 \text{ mol per 1 kg of stearic acid}$$

$$\Delta_R H = \sum_i \left(\nu_i * \Delta_f H_i\right) = \nu_{\text{Stea}} * \Delta_f H_{\text{Stea}} + \nu_{\text{CO}_2} * \Delta_f H_{\text{CO}_2} + \nu_{\text{H}_2\text{O}} * \Delta_f H_{\text{H}_2\text{Oliq}}$$

$$\Delta_f H_{\text{O}_2} = 0$$

$$\Delta_R H = [-1 * (-950) + 18 * (-393.5) + 18 * (-285.8)]\frac{\text{kJ}}{\text{mol}} = -11{,}277.4\frac{\text{kJ}}{\text{mol}}$$

$$HHV = -3.51\frac{\text{mol}}{\text{kg}} * (-11{,}277.4)\frac{\text{kJ}}{\text{mol}} = \mathbf{39,584\frac{kJ}{kg}}$$

b)

$$\Delta_R H = \sum_i \left(\nu_i * \Delta_f H_i\right) = \nu_{\text{Stea}} * \Delta_f H_{\text{Stea}} + \nu_{\text{CO}_2} * \Delta_f H_{\text{CO}_2} + \nu_{\text{H}_2\text{O}} * \Delta_f H_{\text{H}_2\text{Ogas}}$$

$$\Delta_R H = [-1 * (-950) + 18 * (-393.5) + 18 * (-241.8)]\frac{\text{kJ}}{\text{mol}} = 10{,}485\frac{\text{kJ}}{\text{mol}}$$

$$LHV = -3.51\frac{\text{mol}}{\text{kg}} * (-10{,}485)\frac{\text{kJ}}{\text{mol}} = 36{,}804\frac{\text{kJ}}{\text{kg}} \cong \mathbf{36{,}800\frac{kJ}{kg}}$$

Alternatively, from the calorific value and the heat of vaporization of the resulting water:

$$LHV = HHV - \Delta Q$$

$$\Delta Q = m_{H_2O} * \Delta_V H_{H_2O} = n_{H_2O} * M_{H_2O} * \Delta_V H_{H_2O}$$

18 moles of water are formed from 1 mol of stearic acid.

$$\rightarrow n_{H_2O} = 18 * 3.51 \frac{\text{mol}}{\text{kgStea}} = 63.18 \frac{\text{mol}}{\text{kgStea}}$$

$$\rightarrow \Delta Q = 63.18 \frac{\text{mol}}{\text{kg}} * 0.018 \frac{\text{kg}}{\text{mol}} * 2450 \frac{\text{kJ}}{\text{kg}} = 2786 \frac{\text{kJ}}{\text{kg}}$$

$$\boldsymbol{LHV} = (39{,}584 - 2786) \frac{\text{kJ}}{\text{kg}} = 36{,}798 \frac{\text{kJ}}{\text{kg}} \cong \mathbf{36{,}800 \frac{kJ}{kg}}$$

c) $Q = LHV * m_{\text{Stea}} = 36{,}800 \frac{\text{kJ}}{\text{kg}} * 50.000\,\text{kg} = 1.84 * 10^9\,\text{kJ}$

$$Q_{\textbf{Electr}} = 0.45 * 1.84 * 10^8\,\text{kJ} = 8.28 * 10^8\,\text{kJ} = 8.28 * 10^5\text{MW} * \text{s} = \frac{8.28 * 10^5\,\text{MW} * \text{s} * \text{h}}{3600\,\text{s}} = \mathbf{230\,MWh}$$

→ ***Result***

a) **The high heating value is HHV= 39,584 kJ/kg = 39,6 MJ/kg.**
b) **For the low heating value, LHV= 36,800 kJ/kg = 36,8 MJ/kg was calculated using both methods.**
c) **From the 50 metric t stearic acid, 230 MWh of electrical energy can be generated under the given conditions.**

3.4.4 Heat Transfer

Task 56

An exothermic reaction of a heat output of 35.5 kJ/s is to be carried out at 110 °C in a continuous stirred tank with a cooling area of 2.9 m^2. The associated heat transfer coefficient is Kw = 200 W/(m^2 * °C). The inlet temperature of the cooling water is 25 °C. The mean cooling water temperature is approximately the arithmetic mean of its inlet and outlet temperature.

a) What maximum mean temperature must the cooling water not exceed in order to keep the reaction temperature at a maximum of 110 °C?
b) What outlet temperature of the cooling water is to be expected?
c) Which cooling water flow is required (cp_W = 4.2 kJ)/(kg * °C)?

⊗ **Solution**

→ ***Strategy***

a) Formula 30 is converted to the temperature difference reactor content to cooling water and the heat output, the heat transfer coefficient and the exchange surface is used. From the temperature difference results with the reactor temperature, the minimum necessary mean cooling water temperature.
b) The equation of the arithmetic mean is converted to the cooling water outlet temperature and the mean and the inlet temperature of the cooling water is used.
c) Equation 19a is converted to the mass flow.

→ ***Calculation***

Subscript R → Reactor content
Subscript W → Cooling water

$$\overline{T}_W = \text{Mean Cooling Water Temperature}$$

a) $\Delta T = \frac{\dot{Q}}{K_W * A} = \frac{35.5\,\text{kJ} * \text{m}^2 * {}^\circ\text{C}}{\text{s} * 200\,\text{W} * 2.9\,\text{m}^2} = \frac{35{,}500\,\text{W} * \text{m}^2 * {}^\circ\text{C}}{200\,\text{W} * 2.9\,\text{m}^2} = 61.2\,{}^\circ\text{C}$

$$\Delta T = T_R - \overline{T}_W$$

$$\overline{T}_W = T_R - \Delta T = (110.0 - 61.2)\,{}^\circ\text{C} = \mathbf{48.8\,{}^\circ C}$$

b) $\overline{T}_W = \frac{T_{W-\text{in}} + T_{W-\text{ex}}}{2}$

$$T_{W-\text{ex}} = 2 * \overline{T}_W - T_{W-\text{in}} = (2 * 48.8 - 25.0)\,{}^\circ\text{C} = 72.6\,{}^\circ\text{C}$$

c) $\dot{m}_W = \frac{\dot{Q}}{cp_W * (T_{W-\text{ex}} - T_{W-\text{in}})} = \frac{35.5\,\text{kJ} * \text{kg} * {}^\circ\text{C}}{\text{s} * 4.2\,\text{kJ} * (72.6 - 25.0)\,{}^\circ\text{C}} = \mathbf{0.178\,\frac{kg}{s}} = \mathbf{639.3\,\frac{kg}{h}}$

→ ***Result***

a) **The mean cooling water temperature must not exceed 48.8 °C.**
b) **For the cooling water outlet temperature of 72.6 °C are to be expected.**
c) **A cooling water flow of 0.178 kg/s = 639.3 kg/h is required.**

Task 57

A stirred vessel with a cooling surface of 5.5 m^2 is cooled by water of an average temperature of 40 °C. The reacting mixture is kept at 85 °C. The heat transfer coefficient on the reaction side is 250 W/(m^2 * °C), the one on the cooling water side is 400 W/(m^2 * °C), the reactor wall is 4 mm thick. The thermal conductivity of the wall is $\lambda = 60$ W/(m * °C).

a) How much heat is transferred per unit of time?
b) Since the heat transfer of this reactor represents the limitation of the capacity of the entire plant and the cooling water temperature is not to be changed,

the water flow is increased accordingly and, in addition, by means of installations, the heat transfer coefficient on the water side is increased by 20%. Due to the higher cooling water flow, the average cooling water temperature has only changed slightly, despite the higher heat flow through the reactor wall. As an approximation this temperature increase may be neglected. By what percentage is the heat transfer from the reactor increased by this change of cooling water operation?

⊗ **Solution**

→ ***Strategy***

a) The heat flow through the reactor wall is calculated using Formula 30. The heat transfer area and the temperature difference are known. The heat transfer coefficient Kw must be calculated according to Formula 31a from α_1, α_2 and s/λ.
b) The heat transfer coefficient is calculated with the 20% higher water-side heat transfer coefficient, the relevant heat flow is determined and compared with that of case a).

→ ***Calculation***

a) $\dot{Q} = Kw * A * \Delta T$

$$A = 5.5\,\mathrm{m}^2 \quad \Delta T = (85 - 40)\,°\mathrm{C}$$

$$\frac{1}{Kw} = \frac{1\,\mathrm{m}^2 * °\mathrm{C}}{400\,\mathrm{W}} + \frac{0.004\,\mathrm{m} * \mathrm{m} * °\mathrm{C}}{60\,\mathrm{W}} + \frac{1\,\mathrm{m}^2 * °\mathrm{C}}{250\,\mathrm{W}}$$

$$= (0.0025 + 0.0000667 + 0.004)\frac{\mathrm{m}^2 * °\mathrm{C}}{\mathrm{W}} = 0.00657\frac{\mathrm{m}^2 * °\mathrm{C}}{\mathrm{W}}$$

$$Kw = 152.3\frac{\mathrm{W}}{\mathrm{m}^2 * °\mathrm{C}}$$

$$\dot{Q} = 152.3\frac{\mathrm{W}}{\mathrm{m}^2 * °\mathrm{C}} * 5.5\,\mathrm{m}^2 * 45\,°\mathrm{C} = \mathbf{37{,}690\,W = 37.69\,kW}$$

b) $\alpha_{\text{water}-b} = 1.2 * \alpha_{\text{water}-a} = 1.2 * 400\frac{\mathrm{W}}{\mathrm{m}^2 * °\mathrm{C}} = 480\frac{\mathrm{W}}{\mathrm{m}^2 * °\mathrm{C}}$

$$\frac{1}{Kw} = \frac{1\,\mathrm{m}^2 * °\mathrm{C}}{480\,\mathrm{W}} + \frac{0.004\,\mathrm{m} * \mathrm{m} * °\mathrm{C}}{60\,\mathrm{W}} + \frac{1\,\mathrm{m}^2 * °\mathrm{C}}{250\,\mathrm{W}} = (0.002083 + 0.0000667 + 0.004)\frac{\mathrm{m}^2 * °\mathrm{C}}{\mathrm{W}}$$

$$\frac{1}{Kw} = 0.00615\frac{\mathrm{m}^2 * °\mathrm{C}}{\mathrm{W}}$$

$$Kw = 162.6\frac{\mathrm{W}}{\mathrm{m}^2 * °\mathrm{C}}$$

$$\dot{Q} = 162.6 \frac{\text{W}}{\text{m}^2 * {}^\circ\text{C}} * 5.5\,\text{m}^2 * 45\,{}^\circ\text{C} = \mathbf{40{,}244\,W} = \mathbf{40.24\,kW}$$

*The heat flow increases by (40.24 - 37.69) kW = 2.55 kW due to the higher water-side heat transfer coefficient in case b). Based on the original heat output, this corresponds to an increase of 100 % * 2.55 kW/37.69 kW =* ***6.77 %.***

→ ***Result***

The cooling heat flow is 37.69 kW in case a). It increases by 6.77% to 40.24 kW due to the 20% higher water-side heat transfer coefficient in case b).

Task 58

A CSTR [wall thickness 6 mm, thermal conductivity λ = 60 W/(m * °C)] is operated with a reaction mixture at a temperature of 60 °C and cooled with an area of 5 m^2 with water (mean cooling water temperature = 50 °C). The inner heat transfer coefficient (reaction mixture) is α_1 = 300 W/(m^2 * °C), the water-side coefficient α_2 = 1200 W/(m^2 * °C).

a) How much heat is transferred per unit of time from the reactor contents to the cooling water?
b) For corrosion protection, the reactor is given an inner coating of 2mm thick epoxy resin [thermal conductivity λ = 0.25 W/(m * °C)]. By how many °C must the mean cooling water temperature be lowered, to achieve the same heat removal as before without the epoxy resin layer?

⊗ **Solution**

→ ***Strategy***

The heat flow is calculated using Formula 30. The heat transfer area is given, as is the driving temperature difference as the difference in temperature of the reactor contents to the average cooling water temperature. The heat transfer coefficient Kw results according to Formula 31a. In the second part of the task, the additional thermal resistance of the epoxy resin layer is taken into account by calculating Kw using Formula 31b. The same amount of heat is to be dissipated as before without the epoxy resin layer. Formula 30 is rearranged to calculate the temperature difference and the newly calculated Kw, and the given heat exchange area and the heat flow are used. This results in the now necessary average cooling water temperature.

→ ***Calculation***

a) $\dot{Q} = Kw * A * \Delta T = \left(T_{\text{reaktorcontent}} - T_{\text{coolingwater}}\right)$

$$\frac{1}{Kw} = \frac{1\,\text{m}^2 * {}^\circ\text{C}}{300\,\text{W}} + \frac{0.006\,\text{m} * \text{m} * {}^\circ\text{C}}{60\,\text{W}} + \frac{1\,\text{m}^2 * {}^\circ\text{C}}{1200\,\text{W}} = (0.00333 + 0.00010 + 0.000833)\frac{\text{m}^2 * {}^\circ\text{C}}{\text{W}}$$

$$= 0.004263 \frac{\text{m}^2 * {}^\circ\text{C}}{\text{W}}$$

$$Kw = 234.6 \frac{\text{W}}{\text{m}^2 * °\text{C}}$$

$$\dot{Q} = 234.6 \frac{\text{W}}{\text{m}^2 * °\text{C}} * 5\,\text{m}^2 * (60 - 50)\,°\text{C} = 11{,}730\,\text{W} = \mathbf{11.73\,kW}$$

b $\Delta T = (T_{\text{reactorcontent}} - T_{\text{coolingwater}}) = \frac{\dot{Q}}{Kw*A}$

$$\frac{1}{Kw} = \frac{1}{\alpha_1} + \sum_i \frac{s_i}{\lambda_i} + \frac{1}{\alpha_2} = \frac{1}{\alpha_1} + \frac{s_{\text{wall}}}{\lambda_{\text{wall}}} + \frac{s_{\text{epoxy}}}{\lambda_{\text{epoxy}}} + \frac{1}{\alpha_2}$$

$$\frac{s_{\text{Epoxy}}}{\lambda_{\text{epoxy}}} = \frac{0.002\,\text{m} * \text{m} * °\text{C}}{0.25\,\text{W}} = 0.008 \frac{\text{m}^2 * °\text{C}}{\text{W}}$$

$$\frac{1}{Kw} = (0.00333 + 0.00010 + 0.00800 + 0.000833) \frac{\text{m}^2 * °\text{C}}{\text{W}} = 0.012263 \frac{\text{m}^2 * °\text{C}}{\text{W}}$$

$$Kw = 81.55 \frac{\text{W}}{\text{m}'^2 * °\text{C}}$$

$$\Delta T = (T_{\text{reactorcontent}} - T_{coolingwater}) = \frac{11{,}730\,\text{W*m}^2 * °\text{C}}{81.55\,\text{W*5}\,\text{m}^2} = 28.8\,°\text{C}$$

$$T_{\textbf{coolingwater}} = (60 - 28.8)\,°\text{C} = \mathbf{31.2\,°C}$$

→ ***Result***

a) **The heat flow transferred from the reactor to the cooling water is 11.73 kW.**
b) **After applying the epoxy resin protective layer, this heat flow to be removed from the reactor is ensured by lowering the average cooling water temperature from 50 °C to 31 °C.**

Exercise 59

115 °C hot saturated process steam flows through a 100 m long 4-inch steel pipe [wall thickness = 2 mm, thermal conductivity λ = 50 W/(m * °C)]. The heat transfer coefficient at the steam side is α_1 = 1000 W(m^2 * °C), that of the surrounding air in average is α_2 = 70 W/(m^2 * °C). The surrounding air has an average temperature of 15 °C.

a) What is the heat loss (heat output) under these conditions?
b) What would this heat loss per unit of time be if, under otherwise unchanged conditions, the pipe would be insulated by a 5 cm thick rock wool layer [thermal conductivity of rock wool λ = 0.05 W/(m * °C)]?
c) How much steam condensate is formed in case a) or case b) per hour ($\Delta_V H_{water}$ = 2215 kJ/kg)?

⊗ **Solution**

→ ***Strategy***

a) The heat flow through the pipe wall is calculated using Formula 30. For this, the heat transfer coefficient Kw must be calculated from α_1, α_2 and s/λ according to Formula 31a. The heat transfer area is the pipe wall. Since the pipe wall thickness is significantly smaller than the inner pipe diameter ($d_i = d_a - 2s$), the arithmetic mean of the diameter can be used for this. The temperature difference is ΔT = 100 °C.

b) The addition of the rock wool layer changes the heat transfer coefficient Kw. It is calculated using Formula 31b. With a rock wool layer of 5 cm, the total thickness of the pipe wall (steel wall + insulation layer) is in the order of magnitude of the pipe diameter. Therefore, the logarithmic mean of the pipe diameter (see Sect. 1.2.5) must be used to calculate the heat transfer area. This results in the heat flow for the insulated pipe according to Formula 30.

c) The amount of steam condensate is calculated from the heat flows calculated for case a) and case b) using Formula 24b.

→ ***Calculation***

a) $\dot{Q} = Kw * A * \Delta T$

$$\frac{1}{Kw} = \frac{1\,\mathrm{m}^2 * °\mathrm{C}}{1000\,\mathrm{W}} + \frac{0.002\,\mathrm{m} * \mathrm{m} * °\mathrm{C}}{50\,\mathrm{W}} + \frac{1\,\mathrm{m}^2 * °\mathrm{C}}{70\,\mathrm{W}}$$

$$= (0.001 + 0.00004 + 0.01429)\frac{\mathrm{m}^2 * °\mathrm{C}}{\mathrm{W}} = 0.01533\frac{\mathrm{m}^2 * °\mathrm{C}}{\mathrm{W}}$$

$$Kw = 65.23\frac{\mathrm{W}}{\mathrm{m}^2 * °\mathrm{C}}$$

$$A = \overline{d}_m * \pi * L$$

$$\overline{d}_m = \frac{d_a + d_i}{2} \text{ with } d_a = \frac{25.4\,\mathrm{mm}}{\mathrm{inch}} * 4'' = 0.1016\,\mathrm{m}$$

$$d_i = (0.1016 - 2 * 0.002)\,\mathrm{m} = 0.0976\,\mathrm{m}$$

$$\overline{d}_m = \frac{0.1016 + 0.0976}{2}\mathrm{m} = 0.0996\,\mathrm{m}$$

$$A = 0.0996\,\mathrm{m} * \pi * 100\,\mathrm{m} = 31.27\,\mathrm{m}^2$$

$$\dot{Q} = 65.23\frac{\mathrm{W}}{\mathrm{m}^2 * °\mathrm{C}} * 31.274\,\mathrm{m}^2 * 100\,°\mathrm{C} = 201{,}404\,\mathrm{W} = \mathbf{204.0\,kW} = \mathbf{204.0\,kJ/s}$$

b) $\dot{Q} = Kw * A * \Delta T$

$$\frac{1}{Kw} = \frac{1\,\mathrm{m}^2 * °\mathrm{C}}{1000\,\mathrm{W}} + \frac{0.002\,\mathrm{m} * \mathrm{m} * °\mathrm{C}}{50\,\mathrm{W}} + \frac{0.05\,\mathrm{m} * \mathrm{m} * °\mathrm{C}}{0.05\,\mathrm{W}} + \frac{1\,\mathrm{m}^2 * °\mathrm{C}}{70\,\mathrm{W}}$$
$$= (0.001 + 0.00004 + 1 + 0.01429)\frac{\mathrm{m}^2 * °\mathrm{C}}{\mathrm{W}} = 1.01533\frac{\mathrm{m}^2 * °\mathrm{C}}{\mathrm{W}}$$

$$Kw = 0.985\frac{\mathrm{W}}{\mathrm{m}^2 * °\mathrm{C}}$$

$$A = \overline{d}_m * \pi * L$$

With the logarithmic pipe diameter according to:
$$\overline{d}_m = \frac{\mathrm{d_a} - \mathrm{d_i}}{\ln\frac{d_a}{d_i}}$$

$d_i = 0.0976\ m$
$d_a = 0.1016\ m + 2 * 0.05\ m = 0.2016\ m$

$$\overline{d}_m = \frac{\mathrm{d_a} - \mathrm{d_i}}{\ln\frac{d_a}{d_i}} = \frac{0.2016 - 0.0976}{\ln\frac{0.2016}{0.0976}}\mathrm{m} = 0.143\,\mathrm{m}$$

$$A = 0.143\,\mathrm{m} * \pi * 100\,\mathrm{m} = 44.90\,\mathrm{m}^2$$

$$\dot{Q} = 0.985\frac{\mathrm{W}}{\mathrm{m}^2 * °\mathrm{C}} * 44.90\,\mathrm{m}^2 * 100\,°\mathrm{C} = 4423\,\mathrm{W} = \mathbf{4.42\,kW} = \mathbf{4.42\,kJ/s}$$

c) *Formula 24b is rearranged:* $\dot{m} = \dot{Q}/\Delta_V H$
For the situation a → Condensate amount $\dot{m}_a = \frac{204.0\,\mathrm{kJ} * \mathrm{kg}}{2215\,\mathrm{kJ} * \mathrm{s}} = \mathbf{0.0921\,\frac{kg}{s}} = \mathbf{331.6\,\frac{kg}{h}}$
For the situation b → Condensate amount $\dot{m}_b = \frac{4.42\,\mathrm{kJ} * \mathrm{kg}}{2215\,\mathrm{kJ} * \mathrm{s}} = \mathbf{0.0020\,\frac{kg}{s}} = \mathbf{7.2\,\frac{kg}{h}}$

→ ***Result***

a) **The heat loss through the wall of the non-insulated line is 204 kW.**
b) **The heat loss through the wall of the rock-wool-insulated line is 4.42 kW.**
c) **In operation of the non-insulated line, 0.0921 kg of steam condensate per second is produced, which corresponds to 331.6 kg per hour. In operation of the insulated line, 0.0020 kg of steam condensate per second is produced, which corresponds to 7.2 kg per hour.**

Task 60
In a brainstorming session with the goal of increasing the capacity of an ethyl-propyl-ketone plant, it was found out that the heat transfer in the stainless steel condenser of the finishing column is the main "bottleneck". The heat transfer

coefficient Kw of the condenser was determined at operating conditions of 700 W/(m^2 * K). The thickness of the stainless steel exchange wall is 4 mm, the thermal conductivity of the steel is 21 W/(m * °C). In the discussion it was proposed to replace the stainless steel condenser by a copper one (copper thermal conductivity = 393 W/(m * K). The copper exchange all should also be 4 mm thick. By what percentage would the heat transfer in the condenser and thus the production capacity of the plant increase by such a change?

⊗ **Solution**

→ ***Strategy***

The increase of the heat flow of the condensator is achieved by an increased heat transfer coefficient, due to the substitution of stainless steel by copper with a comparatively higher thermal conductivity. All other sizes, such as the heat transfer coefficients α_1 and α_2, the heat transfer area and the temperature difference, remain unchanged. The increase in heat transmission is directly proportional to the increase of the heat transfer coefficient. The heat transfer coefficient is calculated according to Formula 31a. Since for the exchanger made of copper, only the values of the wall thickness and the thermal conductivity are available, the heat transfer coefficients α_1 and α_2 must be determined. This is possible from the data given for the stainless steel exchanger by Formula 31a, since no change in the flow conditions takes place and thus also α_1 and α_2 do not change.

→ ***Calculation***

Stainless steel condenser:

$$\frac{1}{Kw_{\text{Steel}}} = \frac{1}{\alpha_1} + \frac{s_{\text{Steel}}}{\lambda_{\text{Steel}}} + \frac{1}{\alpha_2} \rightarrow \frac{1}{\alpha_1} + \frac{1}{\alpha_2} = \frac{1}{Kw_{\text{Steel}}} - \frac{s_{\text{Steel}}}{\lambda_{\text{Steel}}} = \frac{\text{m}^2 * \text{K}}{700\,\text{W}} - \frac{0.004\,\text{m} * \text{m} * {}^\circ\text{C}}{21\,\text{W}}$$

Since temperature differences are used in the calculation of heat flows and the expansion of the Kelvin and Celsius scales is identical, in this case the units Kelvin and °C are identical.

$$\frac{1}{\alpha_1} + \frac{1}{\alpha_2} = (0.001429 - 0.00019)\frac{\text{m}^2 * {}^\circ\text{C}}{\text{W}} = 0.001239\frac{\text{m}^2 * {}^\circ\text{C}}{\text{W}}$$

Copper condenser:

$$\frac{1}{Kw_{\text{Copper}}} = \frac{1}{\alpha_1} + \frac{1}{\alpha_2} + \frac{s_{\text{Copper}}}{\lambda_{\text{Copper}}}$$

Insert the calculated term $\frac{1}{\alpha_1} + \frac{1}{\alpha_2} = 0.001239\frac{\text{m}^2 * {}^\circ\text{C}}{\text{W}}$

and the wall thickness and thermal conductivity of the copper wall:

$$\frac{1}{Kw_{\text{Copper}}} = 0.001239\frac{\text{m}^2 * {}^\circ\text{C}}{\text{W}} + \frac{0.004\,\text{m} * \text{m} * {}^\circ\text{C}}{393\,\text{W}} = (0.001239 + 0.00010)\frac{\text{m}^2 * {}^\circ\text{C}}{\text{W}}$$
$$= 0.001339\frac{\text{m}^2 * {}^\circ\text{C}}{\text{W}}$$

$$Kw_{Copper} = 747 \frac{W}{m^2 * °C}$$

The percentage increase in heat flow and thus production capacity is calculated from the difference in heat transfer coefficients copper to steel capacitor based on 100 %:

$$700 \frac{W}{m^2 * °C} \triangleq 100\% \quad 747 \frac{W}{m^2 * °C} \triangleq x \quad x = \frac{747}{700} * 100\% = \mathbf{106.7\ \%}$$

→ ***Result***

The production capacity would be increased by the exchange of stainless steel for a copper capacitor by only 6.7 %.

→ Note: The result shows that in such a case the main heat resistance lies in the laminar boundary layers and not in the wall material. Compare the heat conduction through the wall, i.e. the size of λ/s = (5250 W/[m^2 * °C] for stainless steel or 98,250 W/[m^2 * °C] for copper) with the reciprocal of the sum of the reciprocal heat transfer coefficients (~800W/[m^2 * °C]).

Task 61

A reaction mixture (density = 0.85 g/cm^3; cp = 2.4 kJ/[kg * °C]) leaves the reactor at a temperature of 90 °C with a flow rate of 1.5 L/s. It is to be cooled in a tube bundle heat exchanger with 25 steel tubes (outer diameter = 2"; wall thickness = 3 mm; thermal conductivity of steel = 45 W/[m * °C]) in countercurrent with cooling water to 35 °C. The cooling water (cp = 4.2 kJ/(kg * °C) is supplied to the exchanger at 20 °C. The outlet temperature must not exceed 40 °C. The heat transfer coefficients were estimated as α_{inside} = 350 W/(m^2 * °C) and $\alpha_{outside}$ = 750 W/(m^2 * °C).

a) What cooling water flow is required for this task?
b) What is the average logarithmic temperature difference?
c) What exchange area and length must the heat exchanger have at least?

⊗ **Solution**

→ ***Strategy***

a) The heat flow from the reaction mixture to the cooling water is calculated according to formula 19a with the data of the reaction mixture. To calculate the cooling water flow required for this, formula 19a is rearranged and the heat flow just calculated and the material data of the water are used.
b) The average logarithmic temperature difference results from the *corresponding* section given in 1.2.5.
c) The total exchange area is calculated by the rearranged formula 30 from the calculated heat flow, the mean temperature difference and the heat transmission coefficient, which can be calculated from the heat transfer coefficients and the

ratio of the thermal conductivity of the steel and the wall thickness of the pipe according to formula 31a. This total area is the sum of the wall areas of the 25 individual pipes, from which the necessary pipe length is calculated according to the formula of a cylinder area.

→ ***Calculation***

Subscript R → Reaction mixture

Subscript W → Cooling water

a) $\dot{Q} = \dot{m}_R * cp_R * (T_{R-\text{in}} - T_{R-\text{ex}})$

$$\dot{m}_R = \dot{V}_R * \rho_R$$

$$\dot{Q} = \dot{V}_R * \rho_R * cp_R * (T_{R-\text{in}} - T_{R-\text{ex}})$$
$$= 0.0015\frac{\text{m}^3}{\text{s}} * 850\frac{\text{kg}}{\text{m}^3} * 2.4\frac{\text{kJ}}{\text{kg} * {}^\circ\text{C}} * (90 - 35)\,{}^\circ\text{C}$$

$$\dot{Q} = 168.3\frac{\text{kJ}}{\text{s}} = 168.3\,\text{kW}$$

$$\dot{Q} = \dot{m}_W * cp_W * (T_{W-\text{ex}} - T_{W-\text{in}})$$

$$\dot{m}_W = \frac{\dot{Q}}{cp_W * (T_{W-\text{ex}} - T_{W-\text{in}})} = \frac{168\,\text{kJ} * \text{kg} * {}^\circ\text{C}}{\text{s} * 4.2\,\text{kJ} * (40 - 20){}^\circ\text{C}} = \mathbf{2.00\frac{kg}{s}} = \mathbf{7.2\frac{t}{h}}$$

b) $\overline{T}_m = \frac{\Delta T_1 - \Delta T_2}{\ln\frac{\Delta T_1}{\Delta T_2}}$.

$$\Delta T_1 = T_{R-\text{in}} - T_{W-\text{ex}} = (90 - 40)\,{}^\circ\text{C} = 50\,{}^\circ\text{C}$$

$$\Delta T_2 = T_{R-\text{ex}} - T_{W-\text{in}} = (35 - 20)\,{}^\circ\text{C} = 15\,{}^\circ\text{C}$$

$$\overline{T}_m = \frac{50 - 15}{\ln\frac{50}{15}}{}^\circ\text{C} = \mathbf{29.1^\circ C}$$

c) $\dot{Q} = Kw * A * \Delta\dot{T}_M \rightarrow A = \frac{\dot{Q}}{Kw * \Delta\overline{T}_M}$

$$\dot{Q} = 168.3\frac{\text{kJ}}{\text{s}} = 168{,}300\,\text{W}$$

$$\frac{1}{Kw} = \frac{1}{\alpha_i} + \frac{s_{\text{Steel}}}{\lambda_{\text{Steel}}} + \frac{1}{\alpha_a} = \frac{\text{m}^2 * {}^\circ\text{C}}{350\,\text{W}} + \frac{0.003\,\text{m} * \text{m} * {}^\circ\text{C}}{45\,\text{W}} + \frac{\text{m}^2 * {}^\circ\text{C}}{750\,\text{W}}$$

$$\frac{1}{Kw} = (0.00285 + 0.000067 + 0.00133)\frac{\text{m}^2 * {}^\circ\text{C}}{\text{W}} = 0.004247\frac{\text{m}^2 * {}^\circ\text{C}}{\text{W}}$$

$$Kw = 235.5 \frac{\mathrm{W}}{\mathrm{m}^2 * {}^\circ\mathrm{C}}$$

$$A_{\text{total}} = \frac{168{,}300\,\mathrm{W} * \mathrm{m}^2 * {}^\circ\mathrm{C}}{235.5\,\mathrm{W} * 29.1\,{}^\circ\mathrm{C}} = 24.56\,\mathrm{m}^2$$

$$A_{\text{total}} = n_{\text{Pipes}} * A_{\text{Pipe}}$$

$$A_{\text{Pipe}} = \overline{d}_m * \pi * L$$

*The outer diameter of the pipes with 2″ = 2 * 0.0254 m is significantly larger than the pipe wall thickness. Thus, the accuracy of the arithmetic mean of the pipe diameter is sufficient.*

$$\overline{d}_m = \frac{d_a + d_i}{2} = \frac{2 * 0.0254 + 2 * 0.0254 - 2 * 0.003}{2}\,\mathrm{m} = 0.0478\,\mathrm{m}$$

$$L = \frac{A_{\text{Pipe}}}{\overline{d}_m * \pi} = \frac{A_{\text{total}}}{n_{\text{Pipes}} * \overline{d}_m * \pi} = \frac{24.56\,\mathrm{m}^2}{25 * 0.0478\,\mathrm{m} * 3.14} = \mathbf{6.55\,m}$$

→ ***Result***

a) **The required amount of cooling water is 2.00 kg/s = 7.20 t/h.**
b) **The average temperature difference is 29.1 °C.**
c) **The length of the heat exchanger is to be designed at 6.55 m.**

Exercise 62

1500 kg of reaction mixture (cp = 1.36 kJ/[kg * °C]) per hour is to be cooled from 90 °C to 40 °C. For this purpose, a tube bundle heat exchanger is to be designed. The reaction mixture is fed to a bundle of steel pipes (outer pipe diameter: 2″; wall thickness: 3 mm; thermal conductivity of steel: 45 W/[m * °C]). The length of the pipes must not exceed 3.2 m for reasons of the available installation space. Cooling water (cp = 4.2 kJ/[kg * °C]) at 10 °C is fed countercurrent to the shell of the exchanger. The maximum temperature of the heated cooling water is given as 30 °C. The heat transfer coefficient on the side of the reaction mixture is $\alpha_i = 90$ W/(m² * °C), the cooling water side $\alpha_a = 800$ J/(s * m² * °C).

a) Which cooling water flow is required for this?
b) What is the average logarithmic temperature difference in such a countercurrent operation?
c) What is the heat transfer coefficient?
d) How many pipes does the heat exchanger bundle have to consist of in such a countercurrent operation?
e) What would be the average logarithmic temperature difference in a cocurrent operation of reaction mixture and cooling water?
f) How many pipes would be required in a cocurrent operation under the assumption of unchanged heat transfer coefficients?

⊗ **Solution**

→ ***Strategy***

a) The heat flow is calculated from the mass flow, the heat capacity and the temperature difference of the reaction mixture using Formula 19a. To determine the required cooling water flow, Formula 19a is switched to the mass flow and the heat flow just calculated, the heat capacity of the water and its temperature difference $T_{in} - T_{ex}$ are used.
b) For this purpose, the formula of the logarithmic mean of the temperature given in Sect. 1.2.5 is used.
c) The heat transfer coefficient is calculated according to Formula 31a.
d) Formula 30 is switched to the total area A required for the heat flow calculated under a) and additionally the heat transfer coefficient and the mean logarithmic temperature difference calculated under b) are used. The necessary number of pipes results from the total heat transfer area, divided by the heat transfer area (pipe wall) of one pipe. Since the pipe diameter is significantly larger than the thickness of the wall, the arithmetic mean of the pipe diameter may be used for this purpose.
e) & f) The calculation for cocurrent operation is carried out analogously to the procedure described under b), d) and c).

→ ***Calculation***

Subscript R → Reaction mixture

Subscript W → Coolant (water)

a) $\dot{Q} = \dot{m}_R * cp_R * (T_{R\text{-in}} - T_{R\text{-ex}}) = \frac{1500\,\text{kg}}{3600\,\text{s}} * 1.36 \frac{\text{kJ}}{\text{kg} * °\text{C}} * (90 - 40)\,°\text{C} = 28.33 \frac{\text{kJ}}{\text{s}} = 28.33\,\text{kW}$

$$\dot{m}_W = \frac{\dot{Q}}{cp_W * (T_{W-\text{ex}} - T_{W-\text{in}})} = \frac{28.33\,\text{kJ} * \text{kg} * °\text{C}}{\text{s} * 4.2\,\text{kJ} * (30 - 10)°\text{C}}$$

$$= \mathbf{0.337 \frac{kg}{s}} = \mathbf{1214 \frac{kg}{h}}$$

b) $\Delta \overline{T}_m = \frac{\Delta T_1 - \Delta T_2}{\ln \frac{\Delta T_1}{\Delta T_2}}$ for countercurrent → $\Delta T_1 = T_{R\text{-in}} - T_{W\text{-ex}} = (90 - 30)\,°\text{C} = 60\,°\text{C}$

$$\Delta T_2 = T_{R-\text{ex}} - T_{W-\text{in}} = (40 - 10)\,°\text{C} = 30\,°\text{C}$$

$$\overline{T}_m = \frac{60 - 30}{\ln \frac{60\,°\text{C}}{30\,°\text{C}}}\,°\text{C} = \mathbf{43.2\,°C}$$

c)
$$\frac{1}{Kw} = \frac{1}{\alpha_i} + \frac{s_{\text{pipe}}}{\lambda_{\text{steel}}} + \frac{1}{\alpha_a} = \frac{\text{m}^2 * °\text{C}}{90\,\text{W}} + \frac{0.003\,\text{m*m}}{45\,\text{W} * °\text{C}} + \frac{\text{m}^2 * °\text{C}}{800\,\text{W}}$$

$$= 0.01243 \frac{\text{m}^2 * °\text{C}}{\text{W}}$$

$$\mathbf{Kw = 80.5\frac{W}{m^2*°C}}$$

d) $A_{\text{total}} = \frac{\dot{Q}}{Kw*\Delta T_M} = \frac{28{,}330\ \text{W}*\text{m}^2*°\text{C}}{80.5\ \text{W}*43.2\ °\text{C}} = 8.146\ \text{m}^2$

$$A_{\text{pipe}} = \overline{d}_m * \pi * L$$

$$1'' = 0.0254\ \text{m} \rightarrow d_a = 2 * 0.0254\ \text{m} = 0.0508\ \text{m}$$
$$d_i = 2 * 0.254\ \text{m} - 2 * 0.003\ \text{m} = 0.0448\ \text{m}$$

$$\overline{d}_m = \frac{\text{d}_\text{a} + \text{d}_\text{i}}{2} = \frac{0.0508 + 0.0448}{2}\text{m} = 0.0478\ \text{m}$$

$$A_{\text{pipe}} = 0.0478\ \text{m} * \pi * 3.2\ \text{m} = 0.481\ \text{m}^2$$

$$\textbf{Number of Pipes CounterCurrent Flow} = \frac{A_{\text{total}}}{A_{\text{pipe}}} = \frac{8.146\ \text{m}^2}{0.481\ \text{m}^2} = 16.9 \sim \textbf{17 Pipes}$$

e) Cocurrent operation

$$\Delta T_1 = T_{R-\text{in}} - T_{W-\text{in}} = (90 - 10)\ °\text{C} = 80\ °\text{C}$$

$$\Delta T_2 = T_{R-\text{ex}} - T_{W-\text{ex}} = (40 - 30)\ °\text{C} = 10\ °\text{C}$$

$$\overline{\boldsymbol{T}}_{\boldsymbol{m}} = \frac{80 - 10\ °\text{C}}{\ln\frac{80\ °\text{C}}{10\ °\text{C}}} = \mathbf{33.7\ °C}$$

f) Cocurrent operation

$$A_{\text{total}} = \frac{\dot{Q}}{Kw * \Delta T_M} = \frac{28{,}330\ \text{W} * \text{m}^2 * °\text{C}}{80.5\ \text{W} * 33.7\ °\text{C}} = 10.44\ \text{m}^2$$

$$\textbf{Number of Pipes Cocurrent–Operation} = \frac{A_{\text{total}}}{A_{\text{pipe}}} = \frac{10.44\ \text{m}^2}{0.481\ \text{m}^2} = 21.7 \sim \textbf{22 Pipes}$$

→ ***Result***

a) **The necessary cooling water flow is 0.337 kg/s = 1,214 kg/h.**
b) **The mean logarithmic temperature difference in countercurrent operation is 43.2 °C.**
c) **The heat transfer coefficient is 80.5 W/(m²*°C).**
d) **For the given conditions in countercurrent operation, the tube bundle of the heat exchanger must be equipped with at least 17 tubes.**
e) **The mean logarithmic temperature difference in cocurrent operation is 33.7 °C.**
f) **For the given conditions in cocurrent operation, the tube bundle of the heat exchanger must be equipped with at least 22 tubes.**

3.4.5 Heat Balances

Exercise 63

In a production plant, 2.4 t/h of a 10% aqueous solution of magnesium chloride is to be produced continuously. For this purpose, the water stream at 10 °C and the solid dry magnesium chloride (20 °C) are fed into a cascade of two CSTRs. The solution enthalpy of magnesium chloride in water is -159 kJ/mol. The heat capacity of water is 4.2 kJ/(kg * °C), that of magnesium chloride is 71.1 J/(mol * °C). The molar mass of magnesium chloride is 95.2 g/mol.

a) What mass flows of water and magnesium chloride must be fed to the cascade?
b) What heat output in kW results from the complete dissolution of magnesium chloride?
c) What is the temperature of the solution leaving the cascade, assuming complete dissolution of the salt?
d) How large must a cooling water stream be to cool down the magnesium chloride solution to 20 °C? Cooling water at 5 °C is available, but it must not exceed a maximum temperature of 30 °C at the outlet of the heat exchanger.

⊗ **Solution**

→ ***Strategy***

a) The mass flows of water and magnesium chloride result from the percentage of the solution and its mass flow.
b) With formula 26b, the heat formation can be calculated. For this purpose, the mass flow of magnesium chloride must be converted into the molar flow.
c) By the heat generation, both the water content and the magnesium chloride are heated according to formula 19a (mass-related cp_{Water}) or 21a (molar-related cp_{MgCl_2}). The thus composed formula is solved for the final temperature.
d) The formula for calculating the heat necessary for cooling the solution is again composed of the proportion of water and the magnesium chloride. If you set this value of the heat formation together with the temperature difference of the outgoing to the incoming cooling water and the heat capacity of the water in formula 19b, you will get the required cooling water stream after appropriate transformation.

→ ***Calculation***

Subscripts:

S = $MgCl_2$*-Solution, W = Water, Mg = Magnesiumchloride, CW = Cooling Water,*

T_{End}*= Endtemperature Solution,* T_{SEx}*= Temperature Solution after Cooling*

T_{Wo}*= Feed Temperature Water,* T_{Mgo}*= Feed Temperature* $MgCl_2$

a) $\dot{m}_{Mg} = 0.1 * \dot{m}_S = 0.1 * 2400 \frac{kg}{h} = \mathbf{240 \frac{kg}{h}}$

$$\dot{m}_W = 0.9 * \dot{m}_S = 0.9 * 2400\frac{\text{kg}}{\text{h}} = \mathbf{2160\frac{kg}{h}}$$

b) $\dot{Q} = -\dot{n}_{\text{Mg}} * \Delta_L H$

$$\dot{n}_{\text{Mg}} = \frac{\dot{m}_{\text{Mg}}}{M_{\text{Mg}}} = \frac{240\,\text{kg} * \text{mol}}{\text{h} * 0.0952\,\text{kg}} = 2521\frac{\text{mol}}{\text{h}}$$

$$\dot{Q} = -2521\frac{\text{mol}}{\text{h}} * \left(-159\frac{\text{kJ}}{\text{mol}}\right) = 400{,}839\frac{\text{kJ}}{\text{h}} = \frac{400{,}839\,\text{kJ*h}}{\text{h} * 3600\,\text{s}} = \mathbf{111.3\,kW}$$

c) $\dot{Q} = \dot{Q}_W + \dot{Q}_{\text{Mg}}$

$$\dot{Q}_W = \dot{m}_W * cp_W * \left(T_{\text{End}} - T_{W_o}\right)$$

$$\dot{Q}_{\text{Mg}} = \dot{n}_{\text{Mg}} * cp_{\text{Mg}} * \left(T_{\text{End}} - T_{\text{Mg}_o}\right) \text{ da } \text{cp}_{\text{Mg}} \text{ is present as molar size}$$

$$\dot{Q} = \dot{m}_W * cp_W * T_{\text{End}} - \dot{m}_W * cp_W * T_{W_o} + \dot{n}_{\text{Mg}} * cp_{\text{Mg}} * T_{\text{End}} - \dot{n}_{\text{Mg}} * cp_{\text{Mg}} * T_{\text{Mg}_o}$$

$$T_{\text{End}} = \frac{\dot{Q} + \dot{m}_W * cp_W * T_{W_o} + \dot{n}_{\text{Mg}} * cp_{\text{Mg}} * T_{\text{Mg}_o}}{\dot{m}_W * cp_W + \dot{n}_{\text{Mg}} * cp_{\text{Mg}}}$$

$$\dot{m}_W * cp_W = 2160\frac{\text{kg}}{\text{h}} * 4.2\frac{\text{kJ}}{\text{kg} * {}^\circ\text{C}} = 9072\frac{\text{kJ}}{\text{h} * {}^\circ\text{C}}$$

$$\dot{n}_{Mg} * cp_{Mg} = 2521\frac{\text{mol}}{\text{h}} * 0.0711\frac{\text{kJ}}{\text{h} * {}^\circ\text{C}} = 179\frac{\text{kJ}}{\text{h} * {}^\circ\text{C}}$$

$$\boldsymbol{T}_{\textbf{End}} = \frac{400{,}839\frac{\text{kJ}}{\text{h}} + 9072\frac{\text{kJ}}{\text{h}*{}^\circ\text{C}} * 10\,{}^\circ\text{C} + 179\frac{\text{kJ}}{\text{h}*{}^\circ\text{C}} * 20\,{}^\circ\text{C}}{(9072+179)\frac{\text{kJ}}{\text{h}*{}^\circ\text{C}}} = \frac{495{,}139}{9251}\,{}^\circ\text{C} = \mathbf{53.5\,{}^\circ C}$$

d) *Heat Take-Up Cooling Water:*

$$\dot{Q}_{CW} = \dot{m}_{CW} * cp_W * (T_{\text{CWex}} - T_{\text{CWin}})$$

$$\dot{m}_{CW} = \frac{\dot{Q}_{CW}}{cp_W * (T_{\text{CWex}} - T_{\text{CWin}})} = \frac{\dot{Q}_{CW}}{4.2\frac{\text{kJ}}{\text{kg}*{}^\circ\text{C}} * (30 - 5)\,{}^\circ\text{C}} = \frac{\dot{Q}_{CW} * \text{kg}}{105\,\text{kJ}}$$

Heat Take-Up of Cooling Water = Heat removed from Solution:

$$\dot{Q}_{CW} = (\dot{m}_W * cp_W + \dot{n}_{\text{Mg}} * cp_{\text{Mg}}) * (T_{\text{End}} - T_{\text{Sex}}) = 9251\frac{\text{kJ}}{\text{h} * {}^\circ\text{C}} * (53.5 - 20)\,{}^\circ\text{C} = 309{,}909\frac{\text{kJ}}{\text{h}}$$

$$\dot{\boldsymbol{m}}_{\boldsymbol{K}} = \frac{309{,}909\,\text{kJ} * \text{kg}}{\text{h} * 105\,\text{kJ}} = \mathbf{2950\frac{kg}{h}} \approx \mathbf{3\frac{t}{h}}$$

→ ***Result***

a) **The cascade must be supplied with 240 kg magnesium chloride and 2160 kg of water per hour.**
b) **The heat formation of the solution process is 111.3 kW.**
c) **Without cooling, the final temperature of the magnesium chloride solution is 53.5 °C.**
d) **3 metric t of cooling water per hour is required to cool the solution.**

Task 64
In a production process, 5 metric t of dichloromethane (vaporization enthalpy 330 kJ/kg) are to be evaporated from a boiling reaction mixture per hour.

a) How much heat is required for this per hour?
b) How much saturated steam is required for this per hour if the condensate leaves the evaporator with saturated steam temperature (vaporization enthalpy of water: 2300 kJ/kg)?
c) The boiling point of the reaction mixture is 40 °C, the steam temperature is 120 °C. The heat transfer coefficient is Kw = 600 W/(m ^{2} * °C). Calculate the necessary exchange area of the evaporator.

⊗ **Solution**
→ ***Strategy***
The heat to be used for the evaporation of dichloromethane ($MeCl_2$) results from Formula 24b, likewise the amount of steam required for this. The heat transfer area of the evaporator is calculated from Formula 30 with the previously calculated heat flow required for the evaporation.

→ ***Calculation***

a) $\dot{Q} = \dot{m}_{\text{MeCl}_2} * \Delta_V H_{\text{MeCl}_2} = 5000\frac{\text{kg}}{\text{h}} * \frac{330\,\text{kJ}}{\text{kg}} = 1{,}650{,}000\frac{\text{kJ}}{\text{h}} = 458\frac{\text{kJ}}{\text{s}} = \mathbf{458\,kW}$

b) $\dot{m}_{\text{Steam}} = \frac{\dot{Q}}{\Delta_v H_{\text{Steam}}} = \frac{\mathbf{458\,kJ*kg}}{\mathbf{s*2300\,kJ}} = \mathbf{0.199\frac{kg}{s}} = \mathbf{717\frac{kg}{h}}$

c) $A = \frac{\dot{Q}}{Kw*\Delta T} = \frac{458\,\text{kW}*\text{m}^2*°\text{C}}{600\,\text{W}*(120-40)\,°\text{C}} = \mathbf{9.54\,m^2}$

→ ***Result***

a) **The heat flow required for the evaporation of dichloromethane is 458 kW.**
b) **0.199 kg of steam per second or 717 kg of steam per hour is required for this.**
c) **The heat transfer area of the evaporator must be at least 9.54 m^2.**

Exercise 65
In order to continuously produce an industrial degreaser that is to remove both polar and non-polar impurities, ethanol, isopropanol and toluene are fed to a

continuous double-walled stirred tank for the purpose of mixing. No chemical reaction takes place. The mixture is to leave the stirred tank at a temperature of 15 °C. The heat transfer coefficient is 600 W/(m² * °C).

	Ethanol	Isopropanol	Toluene
Mass flow/(kg/h)	450	215	550
Feed temperature/°C	55	5	80
cp/[kJ/(kg * °C)]	2.5	2.9	1.9

a) What mass flow of cooling liquid with a heat capacity of cp = 4.35 kJ/(kg * °C) and a temperature of 0 °C must be fed to the cooling area of the tank, if it is to be heated to a maximum of 20 °C?
b) How large must the heat transfer area of the double-walled stirred tank be at least, if the mean cooling stream temperature can be assumed to be 10 °C?

⊗ **Solution**

→ ***Strategy***

a) First, the amount of heat given off or absorbed by the streams of ethanol, isopropanol, and toluene is calculated using Formula 19a. The sum of these heat flows must be carried away by the cooling fluid. The mass flow of the cooling fluid is calculated from the formula to be used for this purpose.
b) The necessary exchange surface for the heat flow to be discharged results from the rearranged Formula 30.

→ ***Calculation***

a) $\dot{Q}_{\text{total}} = \Delta\dot{Q}_{\text{Et}} + \Delta\dot{Q}_{\text{IP}} + \Delta\dot{Q}_{T}$

$$\Delta\dot{Q}_{\text{Et}} = \dot{m}_{\text{Et}} * cp_{\text{Et}} * (T_{\text{Et-in}} - T_{\text{Ex}}) = 450\frac{\text{kg}}{\text{h}} * 2.5\frac{\text{kJ}}{\text{kg} * {}^{\circ}\text{C}} * (55 - 15)\,{}^{\circ}\text{C} = 45{,}000\frac{\text{kJ}}{\text{h}}$$

$$\Delta\dot{Q}_{\text{IP}} = \dot{m}_{\text{IP}} * cp_{\text{IP}} * (T_{\text{IP-in}} - T_{\text{Ex}}) = 215\frac{\text{kg}}{\text{h}} * 2.9\frac{\text{kJ}}{\text{kg} * {}^{\circ}\text{C}} * (5 - 15)\,{}^{\circ}\text{C} = -6235\frac{\text{kJ}}{\text{h}}$$

$$\Delta\dot{Q}_{T} = \dot{m}_{T} * cp_{T} * (T_{T\text{-in}} - T_{\text{Ex}}) = 550\frac{\text{kg}}{\text{h}} * 1.9\frac{\text{kJ}}{\text{kg} * {}^{\circ}\text{C}} * (80 - 15)\,{}^{\circ}\text{C} = 67{,}925\frac{\text{kJ}}{\text{h}}$$

$$\dot{Q}_{\text{total}} = (45{,}000 - 6235 + 67{,}925)\frac{\text{kJ}}{\text{h}} = 106{,}690\frac{\text{kJ}}{\text{h}} = \frac{106{,}690\,\text{kJ} * \text{h}}{\text{h} * 3600\,\text{s}} = 29.64\,\text{kW}$$

$$\dot{Q}_{\text{total}} = \dot{m}_{\text{CoolFld}} * cp_{\text{CoolFld}} * (T_{\text{CoolFld-ex}} - T_{\text{CoolFld-in}})$$

$$\dot{m}_{\textbf{CoolFld}} = \frac{\dot{Q}_{\text{total}}}{cp_{\text{CoolFld}} * (T_{\text{CoolFld-ex}} - T_{\text{CoolFld-in}})} = \frac{106{,}690\,\text{kJ} * \text{kg} * {}^{\circ}\text{C}}{\text{h} * 4.35\,\text{kJ} * (20 - 0)\,{}^{\circ}\text{C}} = \mathbf{1226\frac{kg}{h}}$$

b) $\dot{Q}_{\text{total}} = Kw * A * \Delta T$

$$A = \frac{\dot{Q}_{\text{total}}}{Kw * \Delta T} = \frac{29{,}640\,\text{W} * \text{m}^2 * {}^\circ\text{C}}{600\,\text{W} * (15 - 10)\,{}^\circ\text{C}} = 9.88\,\text{m}^2 \cong \mathbf{10\,m^2}$$

→ ***Result***

a) **1226 kg of cooling fluid per hour are required.**
b) **The required cooling surface is 9.88 m², i.e. 10 m².**

Task 66
In a double-pipe heat exchanger, a product mixture flowing through its inner pipe (da = 10 cm; s = 2.5 mm) is to be cooled from 70 °C to 30 °C. The mixture has a heat capacity of cp_p = 1.9 kJ/(°C * kg). The mass flow is 1500 kg per hour. The cooling water flowing on the outside has a temperature of 10 °C and may be heated to a maximum of 30 °C. The heat capacity of the cooling water is cp_w = 4.2 kJ/(kg * °C).

The water-side heat transfer coefficient is α_1 = 1000 W/(m² * °C), that of the product side is α_2 = 300 W/(m² * °C). The thermal conductivity of the steel of the pipe wall is λ = 40 W/(m * °C).

a) What heat flow goes from the product to the cooling water?
b) How large is the cooling water flow?
c) How long must the pipe be in countercurrent operation to achieve the goal?
d) How long would it have to be in cocurrent operation?

⊗ **Solution**
→ ***Strategy***

a) The heat flow results from formula 19a from the mass flow of the product mixture, its heat capacity in connection with the difference between its inlet and outlet temperature.
b) The necessary cooling water flow is also calculated according to formula 19a from the heat flow determined under a), the heat capacity of the cooling water and the difference between the water inlet and outlet temperature.
c. & d. The necessary cooling area results from the formula 30 rearranged accordingly. The heat flow is known from solution a). The heat transfer coefficient is calculated according to formula 31a. The mean logarithmic value is used as the temperature difference. The length of the heat exchanger pipe can be determined from the thus calculated area by the mean pipe diameter (arithmetic or logarithmic mean).

→ ***Calculation***

a) $\dot{Q} = \dot{m}_P * cp_P * \Delta T_P = 1500\frac{\text{kg}}{\text{h}} * 1.9\frac{\text{kJ}}{\text{kg}*{}^\circ\text{C}} * (70 - 30)\,{}^\circ\text{C}$

$$\dot{Q} = \mathbf{114,000}\frac{\mathbf{kJ}}{\mathbf{h}} = \frac{114{,}000\,\mathrm{kJ} * \mathrm{h}}{\mathrm{h} * 3600\,\mathrm{s}} = \mathbf{31.7\,kW}$$

b) $\dot{Q} = \dot{m}_W * cp_W * \Delta T_W$

$$\dot{m}_W = \frac{\dot{Q}}{cp_W * \Delta T_W} = \frac{114{,}000\,\mathrm{kJ} * \mathrm{kg} * {}^\circ\mathrm{C}}{\mathrm{h} * 4.2\,\mathrm{kJ} * (30 - 10){}^\circ\mathrm{C}} = \mathbf{1,357}\frac{\mathbf{t}}{\mathbf{h}} = \frac{1{,}357\,\mathrm{kg} * \mathrm{h}}{\mathrm{h} * 3600\,\mathrm{s}} = \mathbf{0.377}\frac{\mathbf{kg}}{\mathbf{s}}$$

c) $\dot{Q} = Kw * A * \Delta T_M$

$$A = d_M * \pi * L$$

$$\dot{Q} = Kw * d_M * \pi * L * \Delta T_M$$

$$L = \frac{\dot{Q}}{Kw * d_M * \pi * \Delta T_M}$$

$$d_M = \frac{d_a + d_i}{2} = \frac{d_a + (d_a - 2 * s)}{2} = \frac{0.1\,\mathrm{m} - (0.1\,\mathrm{m} - 2 * 0.0025)\,\mathrm{m}}{2} = 0.0975\,\mathrm{m}$$

$$\frac{1}{Kw} = \frac{1}{\alpha_1} + \frac{s}{\lambda} + \frac{1}{\alpha_2} = \frac{\mathrm{m}^2 * {}^\circ\mathrm{C}}{1000\,\mathrm{W}} + \frac{0.0025\,\mathrm{m} * \mathrm{m} * {}^\circ\mathrm{C}}{40\,\mathrm{W}} + \frac{\mathrm{m}^2 * {}^\circ\mathrm{C}}{300\,\mathrm{W}} = 0.0044\frac{\mathrm{m}^2 * {}^\circ\mathrm{C}}{\mathrm{W}}$$

$$\mathrm{Kw} = 227.5\frac{\mathrm{W}}{\mathrm{m}^2 * {}^\circ\mathrm{C}}$$

$$\Delta T_{M-\,\mathrm{countercurrent}} = \frac{\Delta T_1 - \Delta T_2}{\ln\frac{\Delta T_1}{\Delta T_2}} \text{ with } \Delta T_1 = (70 - 30)\,{}^\circ\mathrm{C} = 40\,{}^\circ\mathrm{C} \text{ and}$$

$\Delta T_2 = (30 - 10)\,{}^\circ\mathrm{C} = 20\,{}^\circ\mathrm{C}$

$$\Delta T_{M-\,\mathrm{countercurrent}} = \frac{(40 - 20)\,{}^\circ\mathrm{C}}{\ln\frac{40}{20}} = 28.85\,{}^\circ\mathrm{C}$$

$$L_{\mathbf{countercurrent}} = \frac{31.7\,\mathrm{kW} * \mathrm{m} * {}^\circ\mathrm{C}}{0.2275\,\mathrm{kW} * 0.0975\,\mathrm{m} * \mathrm{m} * 28.85\,{}^\circ\mathrm{C}} = \mathbf{15.8\,m}$$

d) $L = \frac{\dot{Q}}{Kw * d_M * \pi * \Delta T_M}$

$$\Delta T_{M-\mathrm{cocurrent}} = \frac{\Delta T_1 - \Delta T_2}{\ln\frac{\Delta T_1}{\Delta T_2}} \text{ with } \Delta T_1 = (70 - 10)\,{}^\circ\mathrm{C} = 60\,{}^\circ\mathrm{C} \text{ and}$$

$\Delta T_2 = (30 - 30)\,{}^\circ\mathrm{C} = 0\,{}^\circ\mathrm{C}$

$\Delta T_{M-\mathrm{cocurrent}} = \frac{(60-0)\,{}^\circ\mathrm{C}}{\ln\frac{60}{0}} \rightarrow 0\,{}^\circ\mathrm{C}$ ***This makes L_{DC} infinitely long.***

→ ***Result***

a) **The heat flow is 36.7 kW.**
b) **A cooling water flow of 1.36 t/h = 0.377 kg/s is required.**
c) **The necessary length of the heat exchanger in counter current operation is 15.8 m.**
d) **In cocurrent operation, the cooling water and the cooled production mixture would have the same temperature, i.e., no heat would flow due to the 0 °C heat gradient. As a result, an infinitely long length of the exchanger tube would result. This arrangement is therefore not realistic under the given operating conditions.**

Exercise 67

A stirred reactor with a mixture of 2000 L of tetrachlorocarbon (density: 1590 kg/m^3; cp = 0.89 kJ/(kg * °C) and 1000 L of heptane (density: 680 kg/m^3; cp = 2.3 kJ/(kg * °C) is to be brought from 20 °C to 80 °C using saturated steam at 120 °C (heat of vaporization 2300 kJ/kg). The heat transfer area is 4 m^2, the heat transfer coefficient is 500 J/(s * m^2 * °C).

a) How much heat is required for the heating process?
b) How much condensate at 120 °C results?
c) How much time is required for the heating process?

⊗ **Solution**

→ ***Strategy***

a) The required heat Q for heating the mixture from 20 °C to 80 °C is calculated according to formula 20b. The mass of the tetrachlorocarbon and that of the heptane are calculated from volume and density. The corresponding heat capacities are known.
b) The heat Q required to heat the mixture is equal to the heat released by the condensation of the steam. To calculate the amount of condensate, formula 24a is rearranged accordingly.
c) The time required for heating is calculated from the heat Q required to heat the mixture, divided by the heat flow $\dot{Q}$, which results from the heat transfer coefficient Kw, the heat transfer area and the temperature difference steam to mixture according to formula 30. Since the temperature of the mixture increases with time, the aforementioned temperature difference also decreases. No further information is given in the task for this purpose. Therefore, the mean of the initial temperature and the final temperature of the mixture is used for the solution as an approximation.

→ ***Calculation***

a) $Q = \left(m_{\mathrm{Cl_4}} * cp_{\mathrm{CCl_4}} + m_{\mathrm{Hep}} * cp_{\mathrm{Hep}}\right) * (T_{\mathrm{End}} - T_0)$

The mass results from the volume and density in accordance with $m = V * \rho$
It follows: $Q = \left(V_{CCl_4} * \rho_{CCl_4} * cp_{CCl_4} + V_{Hep} * \rho_{Hept} * cp_{Hep}\right) * (T_{End} - T_0)$

$$Q = \left(2\,\text{m}^3 * 1590\frac{\text{kg}}{\text{m}^3} * 0.89\frac{\text{kJ}}{\text{kg} * {}^\circ\text{C}} + 1\,\text{m}^3 * 680\frac{\text{kg}}{\text{m}^3} * 2.3\frac{\text{kJ}}{\text{kg} * {}^\circ\text{C}}\right) * (80 - 20)\,{}^\circ\text{C}$$

$$\mathbf{Q = 263{,}652\,kJ = 263.7\,MJ}$$

b) $m_{\text{Condensat}} = \frac{Q}{\Delta vH} = \frac{263{,}652\,\text{kJ}*\text{kg}}{2300\,\text{kJ}} = \mathbf{114.6\,kg}$

c) $t = \frac{Q}{\dot{Q}}$

$$\dot{Q} = Kw * A * \Delta T$$

$$\Delta T = T_{\text{Steam}} - T_M$$

$$T_M = \frac{T_0 + T_{\text{End}}}{2} = \frac{20 + 80}{2}\,{}^\circ\text{C} = 50\,{}^\circ\text{C}$$

$$\Delta T = (120 - 50)\,{}^\circ\text{C}$$

$$\dot{Q} = 500\frac{\text{J}}{\text{s}*\text{m}^2 * {}^\circ\text{C}} * 4\,\text{m}^2 * 70\,{}^\circ\text{C} = 140{,}000\frac{\text{J}}{\text{s}} = 140\frac{\text{kJ}}{\text{s}}$$

$$t = \frac{263{,}652\,\text{kJ} * \text{s}}{140\,\text{kJ}} = \mathbf{1883.2\,s = 31.4\,Min.}$$

→ ***Result***

a) **The amount of heat required to heat the mixture is 263.7 MJ.**
b) **114.6 kg of steam condensate is produced.**
c) **The heating process takes about 1883 s = 31.4 min.**

Task 68
The enthalpy of reaction for the production of acetic acid by oxidation of ethanol is -496 kJ/mole. Molar masses: ethanol = 46 g/mol, acetic acid = 60 g/mol
The reaction proceeds as follows: $C_2H_5OH + O_2 \rightarrow CH_3COOH + H_2O$

a) What heat generation occurs at a production rate of 0.5 metric t of acetic acid per hour?
b) How large must the area of a heat exchanger (Kw = 500 W/[m² * ° C]) be in order to cope with this heat load at an average temperature difference of 70 ° C?
c) How much cooling water at a temperature of 20 ° C and a specific heat capacity of 4.2 kJ/(kg * ° C) is required per hour if the water outlet temperature may not exceed 50 ° C?

⊗ **Solution**

→ ***Strategy***

The heat generation of the reaction is calculated using formula 27c, whereby the mass flow of acetic acid is converted into the molar flow. The necessary heat exchange surface results from the heat formation using the appropriately rearranged formula 30. The heat the reaction is discharged by the cooling water. Thus, the mass flow of the cooling water is calculated according to formula 19a.

→ ***Calculation***

a) $\dot{Q} = -\dot{n} * \Delta_R H$ with $\dot{n} = \frac{\dot{m}}{M} = \frac{500\,\text{kg} * \text{mol}}{\text{h}*0.060\,\text{kg}} = 8333\frac{\text{mol}}{\text{h}} = \frac{8333\,\text{mol} * \text{h}}{\text{h} * 3600\,\text{s}} = 2.315\frac{\text{mol}}{\text{s}}$

$$\dot{Q} = -2.315\frac{\text{mol}}{\text{s}} * \left(-496\frac{\text{kJ}}{\text{mol}}\right) = \mathbf{1148\frac{kJ}{s}} = \mathbf{1148\,kW}$$

b) $\dot{Q} = Kw * A * \Delta T_M$

$$A = \frac{\dot{Q}}{Kw * \Delta T_M} = \frac{1148\,\text{kW} * \text{m}^2 * {}^\circ\text{C}}{0.5\,\text{kW} * 70\,{}^\circ\text{C}} = \mathbf{32.8\,m^2}$$

c) $\dot{Q} = \dot{m}_W * cp_W * \Delta T_W$

$$\Delta T_W = \frac{\dot{Q}}{cp_W * \Delta T_W} = \frac{1148\,\text{kJ} * \text{kg} * {}^\circ\text{C}}{\text{s} * 4.2\,\text{kJ} * (50 - 20)\,{}^\circ\text{C}} = \mathbf{9.11\frac{kg}{s}} = \mathbf{32.8\frac{t}{h}}$$

→ ***Result***

a) **The heat capacity of the reaction is 1148 kW.**
b) **The surface area required for heat exchange is 32.8 m^2.**
c) **The cooling water flow is 9.11 kg/s = 32.8 t/h.**

Task 69

In a stirred vessel, a 10wt% aqueous solution of ammonium nitrate is to be prepared. For this purpose, 900 kg of water is filled into the vessel, brought from 20°C to 85 °C by directfeed of 130 °C saturated steam, a corresponding amount of ammonium nitrate of a temperature of 15 °C is added and dissolved. The heat capacity of ammonium nitrate is 1.74 kJ/(kg * ° C), that of water 4.2 kJ/(kg * ° C), the condensation heat of the steam 2230 kJ/kg. The solution enthalpy of ammonium nitrate in water is 25.7 kJ/mol, the ammonium nitrate molar mass 80.0 g/mol.

a) How much steam is required for this purpose?
b) Which amount of ammonium nitrate must be added and which total mass of solution results?
c) What is the temperature of the solution after complete dissolution of the salt?

⊗ Solution

→ *Strategy*

a) The amount of heat required to increase the water temperature to 85 °C is determined using Formula 18a. This amount of heat is obtained by the sum of the heat quantities from the condensation of the steam at 130 °C (Formula 24a) and the cooling of the steam condensate from 130 °C to 85 °C (Formula 18a). From the resulting formula of the sum of these heat quantities, the amount of steam required for this purpose is calculated by appropriate rearrangement.
b) The amount of water is now the sum of the introduced water and the steam condensate. With the help of the rule of three, the mass of ammonium nitrate necessary to produce a 10wt% solution is calculated from this.
c) According to Formula 26a, the negative amount of heat that cools the system is calculated from the previously determined amount of ammonium nitrate and the positive solution enthalpy. In addition, the ammonium nitrate is heated from the addition temperature of 15 °C to the final temperature of the solution (Formula 18a). The sum of both heat quantities is provided by the cooling down of the water (Formula 18a). The final temperature can be calculated by the appropriate combination of the equations and their rearrangement.

→ *Calculation*

Indices:

S = steam; W = water; W_o = introduced water; WL = water of the solution; A = ammonium nitrate;

E = End → Finished solution; LW = solution heat

a) *Heat amount heating water:* $Q = m_{W_o} * c_{pW} * (T_{85} - T_{W_o})$
Heat amount by steam introduction: $Q = m_S * \Delta_v H + m_S * c_{pW} * (T_S - T_{85})$

$$m_{W_o} * c_{pW} * (T_{85} - T_{W_o}) = m_S * \Delta_v H + m_S * c_{pW} * (T_S - T_{85})$$

$$m_S = \frac{m_{W_o} * c_{pW} * (T_{85} - T_{W_o})}{\Delta_v H + c_{pW} * (T_D - T_{85})} = \frac{900\,\text{kg} * 4.2\frac{\text{kJ}}{\text{kg} * °\text{C}} * (85 - 20)\,°\text{C}}{2230\frac{\text{kJ}}{\text{kg}} + 4.2\frac{\text{kJ}}{\text{kg} * °\text{C}} * (130 - 85)\,°\text{C}} = \mathbf{101.6\,kg}$$

b) *Total amount of water:* $m_{WL} = (900 + 101.6)\,\text{kg} = 1001.6\,\text{kg}$
90 kg water + 10 kg NH_4NO_3
1001,6 kg water + X kg NH_4NO_3

$$X = \frac{10 * 1001.6}{90}\text{kg} \rightarrow \mathbf{111.3\,kg\ NH_4NO_3}$$

$$m_{Solution} = (1001.6 + 111.3)\,\text{kg} = \mathbf{1112.9\,kg}$$

c) $Q_{\text{Delivered by Water}} + Q_{\text{Consumption by NH4NO3}} + Q_{\text{LW}} = 0$

$$Q_{\text{LW}} = -n_A * \Delta_L H$$

$$n_A = \frac{m_A}{M_A} = \frac{111.3\,\text{kg} * \text{mol}}{0.080\,\text{kg}} = 1391.3\,\text{mol}$$

$$Q_{\text{LW}} = -1391.3\,\text{mol} * 25.7\frac{\text{kJ}}{\text{mol}} = -35{,}755\,\text{kJ}$$

$$Q_{\text{LW}} = Q_{\text{Delivered byWater}} + Q_{\text{Consumed by NH}_4\text{NO}_3}$$

$$Q_{\text{Delivered by Water}} = m_{\text{WL}} * cp_W * (T_{85} - T_E)$$

$$Q_{\text{Consumed by NH}_4\text{NO}_3} = m_A * cp_A * (T_A - T_E)$$

$$m_{\text{WL}} * cp_W * T_{85} - m_{\text{WL}} * cp_W * T_E + m_A * cp_A * T_A - m_A * cp_A * T_E + Q_{\text{LW}} = 0$$

$$T_E = \frac{Q_{\text{LW}} + m_{\text{WL}} * cp_W * T_{85} + m_A * cp_A * T_A}{m_{\text{WL}} * cp_W + m_A * cp_A}$$

$$m_{\text{WL}} * cp_W = 1001.6\,\text{kg} * 4.2\frac{\text{kJ}}{\text{kg} * {}^\circ\text{C}} = 4207\frac{\text{kJ}}{{}^\circ\text{C}}$$

$$m_A * cp_A = 111.3\,\text{kg} * 1.74\frac{\text{kJ}}{\text{kg} * {}^\circ\text{C}} = 194\frac{\text{kJ}}{{}^\circ\text{C}}$$

$$T_E = \frac{-35{,}785\,\text{kJ} + 4207\frac{\text{kJ}}{{}^\circ\text{C}} * 85\,{}^\circ\text{C} + 194\frac{\text{kJ}}{{}^\circ\text{C}} * 15\,{}^\circ\text{C}}{(4207 + 194)\frac{\text{kJ}}{{}^\circ\text{C}}} = \mathbf{73.8\,{}^\circ C}$$

→ ***Result***

a) **101.6 kg of steam are required.**
b) **111.3 kg of ammonium nitrate must be added, resulting in a total mass of the solution of 1112.9 kg ≈1113 kg.**
c) **The final temperature of the 10wt% ammonium nitrate solution is 73.8 °C ≈ 74 °C.**

Task 70

A 15 kg heavy steel forging [cp_{steel} = 460 J/(kg * °C)] of a temperature of 700 °C is given for hardening in a 30 °C warm oil bath [ρ_{oil} = 0.9kg/L; cp_{oil} = 0.95 kJ/(kg * °C)] of a volume of 100 L.

a) Which equal final temperature of steel and oil arises?
b) How much cooling fluid of -5 °C [cp_{cf} = 4.9 kJ/(kg * °C)] is needed to cool the oil bath back to its initial temperature? The cooling fluid warms up to 20 °C in this process.

c) The heat exchanger for cooling the oil bath has an area of 3 m^2 and under the existing operating conditions a heat transfer coefficient of $Kw = 50\ W/(m^2 * °C)$. How long does the cooling process take, if the temperature difference oil to cooling fluid averaged over time is chosen as the difference between the arithmetic mean of the final and the initial temperature of the oil and the mean of the inlet and outlet temperature of the fluid?

d) Which volume flow of fluid ($\rho_{cf} = 1.12$ kg/L) must be set for this purpose?

⊗ **Solution**

→ ***Strategy***

a) At the end of the hardening of the forging, the temperature of the steel piece and the oil bath are ideally identical. The heat emitted by the forging is absorbed by the oil bath. This is quantified in both cases by formula 18a for the steel piece and the oil bath. From this, the end temperature of both results. The heat emitted by the forging into the oil bath is to be discharged by the cooling fluid.

b) With formula 18a, the mass of cooling fluid required is determined from the heat emitted by the oil bath and the temperature difference of the fluid.

c) The heat flow in the heat exchanger results from formula 30. The necessary time for cooling the oil bath results from the heat emitted, divided by the heat flow in the heat exchanger.

d) The volume flow of fluid is the quotient of fluid mass, divided by the density, and the cooling time.

→ ***Calculation***

a) $Q_{\text{steel}} = Q_{\text{oil}}$

$$Q_{\text{steel}} = m_{\text{steel}} * cp_{\text{steel}} * (T_{\text{steel}-o} - T_{\text{end}})$$

$$Q_{\text{oil}} = m_{\text{oil}} * cp_{\text{oil}} * \left(T_{\text{end}} - T_{\text{oil}-o}\right)$$

$$m_{\text{steel}} * cp_{\text{steel}} * T_{\text{steel}-o} + m_{\text{oil}} * cp_{\text{oil}} * T_{\text{oil}-o} = T_{\text{end}} * (m_{\text{steel}} * cp_{\text{steel}} + m_{\text{oil}} * cp_{\text{oil}})$$

$$T_{\text{end}} = \frac{m_{\text{steel}} * cp_{\text{steel}} * T_{\text{steel}-o} + m_{\text{oil}} * cp_{\text{oil}} * T_{\text{oil}-o}}{m_{\text{steel}} * cp_{\text{steel}} + m_{\text{oil}} * cp_{\text{oil}}}$$

$$m_{\text{steel}} * cp_{\text{steel}} = 15\,\text{kg} * 0.460\frac{\text{kJ}}{\text{kg} * °\text{C}} = 6.90\frac{\text{kJ}}{°\text{C}}$$

$$m_{\text{oil}} * cp_{\text{oil}} = 0.9\frac{\text{kg}}{\text{L}} * 100\,\text{L} * 0.95\frac{\text{kJ}}{\text{kg} * °\text{C}} = 85.5\frac{\text{kJ}}{°\text{C}}$$

$$T_{\text{end}} = \frac{6.90\frac{\text{kJ}}{°\text{C}} + 700\,°\text{C} + 85.5\frac{\text{kJ}}{°\text{C}} * 30\,°\text{C}}{(6.90 + 85.5)\frac{\text{kJ}}{°\text{C}}} = \mathbf{80.0\,°C}$$

b) $Q = m_{cf} * cp_{cf} * (T_{cf-ex} - T_{cf-in})$

$$m_{cf} = \frac{Q}{cp_{cf} * (T_{cf-ex} - T_{cf-in})}$$

$$Q = m_{steel} * cp_{steel} * (T_{steel-o} - T_{end}) = 15\,\text{kg} * 0.460\frac{\text{kJ}}{\text{kg} * °\text{C}} * (700 - 80)\,°\text{C} = 4278\,\text{kJ}$$

$$\boldsymbol{m_{cf}} = \frac{4278\,\text{kJ} * \text{kg} * °\text{C}}{4.9\,\text{kJ} * [20 - (-5)]\,°\text{C}} = \mathbf{34.9\,kg}$$

c) $\dot{Q} = Kw * A * \overline{\Delta T}$

$$T_{oil} = (80 + 30)°\text{C}/2 = 55°\text{C} \quad T_{cf} = (-5 + 20)°\text{C}/2 = 7.5°\text{C}$$
$$\overline{\Delta T} = (55 - 7.5)°\text{C} = 60°\text{C}$$

$$\dot{Q} = 50\frac{\text{W}}{°\text{C*m}^2} * 3\,\text{m}^2 * 47.5\,°\text{C} = 7125\,\text{W} = 7.125\frac{\text{kJ}}{\text{s}}$$

$$t = \frac{Q}{\dot{Q}} = \frac{4278\,\text{kJ*s}}{7.125\,\text{kJ}} = \mathbf{600\,s} = \mathbf{10\,min}$$

d) $\dot{m}_{cf} = \frac{m_{cf}}{\text{t}} = \frac{34.9\,\text{kg}}{600\,\text{s}} = 0.0582\frac{\text{kg}}{\text{s}}$

$$\dot{V}_{cf} = \frac{\dot{m}_{cf}}{\rho_{cf}} = \frac{0.0582\text{kg} * \text{L}}{\text{s} * 1.12\,\text{kg}} = \mathbf{0.0520\frac{L}{s}} = \mathbf{0.187\frac{m^3}{h}}$$

→ ***Result***

a) **The final temperature is 80.0 °C**
b) **34.9 kg of cooling fluid is required.**
c) **The cooling process takes 600 s = 10 min.**
d) **The volume flow of cooling fluid is 0.0520 L/s = 0.187 m³/h.**

3.4.6 Reactor Stability Criteria

Task 71

In a batch reactor, a mixture of higher alcohols (C_{14}–C_{18}) is oxidized to the corresponding fatty acid mixture (average molecular mass of the alcohols = 237 g/mol). The reaction enthalpy is $\Delta_R H$ = -145 kJ/mol. The reactor is charged with 400 kg of the alcohol mixture (heat capacity cp_{Alc} = 2.5 kJ/[kg * °C]), 800 kg of the solvent octane (cp_{Oc} = 2.2 kJ/[kg * °C]). 120 L of water phase (density ρ_W = 1026 kg/m^3; cp_W = 4.35 kJ/[kg * °C]), in which the oxidizing agent is dissolved, are added. To get the reaction started well, the feedstock is preheated to 40 °C. The allowed maximum temperature for the operation of the stirred tank is 95 °C.

a) Can the allowed maximum temperature be maintained under the assumption of no heat losses if the reactor cooling fails?
b) How much octane is minimally required for safe operation, even if the reactor cooling fails?

⊗ **Solution**

→ ***Strategy***

a) The final temperature is the sum of the starting temperature (40 °C) and the adiabatic temperature increase and should be below the allowed maximum temperature of 95 °C for safe operation in case of cooling failure. The adiabatic temperature increase is calculated according to formula 32b
b) The minimum amount of octane, needed to maintain reactor stability (i.e. 95 °C maximum in case of cooling failure) can be calculated by the corresponding transformation of formula 32b. The temperature difference between the maximum allowed temperature (95 °C) and the starting temperature (40 °C) stands for the adiabatic temperature increase.

→ ***Calculation***

a) $\Delta T_{\mathrm{ad}} = \frac{-n_i * \Delta_R H}{\sum (m_i * cp_i)} = \frac{-n_{\mathrm{Alc}} * \Delta_R H}{m_{\mathrm{Alc}} * cp_{\mathrm{Alc}} + m_{\mathrm{Oc}} * cp_{\mathrm{Oc}} + m_W * cp_W}$

$$m_W = V_W * \rho_W$$

There are 400 kg of alcohol mixture in the reactor with an average molar mass of 0.237 kg/mol:

$$n_{\mathrm{Alc}} = \frac{m_{\mathrm{Alc}}}{M_{\mathrm{Alc}}} = \frac{400\,\mathrm{kg} * \mathrm{mol}}{0.237\,\mathrm{kg}} = 1688\,\mathrm{mol}$$

$$\Delta T_{\mathrm{ad}} = \frac{-n_i * \Delta_R H}{\sum (m_i * cp_i)} = \frac{-n_{\mathrm{Alc}} * \Delta_R H}{m_{\mathrm{Alc}} * cp_{\mathrm{Alc}} + m_{\mathrm{Ok}} * cp_{\mathrm{Oc}} + V_W * \rho_W * cp_W}$$

$$\boldsymbol{\Delta T_{\mathrm{ad}}} = \frac{-1688\,\mathrm{mol} * \left(-145\,\mathrm{kJ}/\mathrm{mol}\right) * \mathrm{kg} * {}^\circ\mathrm{C}}{400\,\mathrm{kg} * 2.5\,\mathrm{kJ} + 800\,\mathrm{kg} * 2.2\,\mathrm{kJ} + 0.120\,\mathrm{m}^3 * 1026\,\mathrm{kg}/\mathrm{m}^3 * 4.35\,\mathrm{kJ}} = \mathbf{74.2\,{}^\circ C}$$

$$\boldsymbol{T_{\mathrm{End}}} = T_{\mathrm{Start}} + \Delta T_{\mathrm{ad}} = (40 + 74.2)\,{}^\circ\mathrm{C} = \mathbf{114.2\,{}^\circ C}$$

b) $\Delta T_{\mathrm{ad}} = \frac{-n_{\mathrm{Alc}} * \Delta_R H}{m_{\mathrm{Alc}} * cp_{\mathrm{Alc}} + m_{\mathrm{Oc}} * cp_{\mathrm{Oc}} + m_W * cp_W}$

$$\boldsymbol{m_{\mathrm{Oc}}} = \frac{-\frac{n * \Delta_R H}{\Delta T_{\mathrm{ad}}} - m_{\mathrm{Alc}} * cp_{\mathrm{Alc}} - V_W * \rho_W * cp_W}{cp_{\mathrm{Oc}}}$$

$$\Delta T_{\mathrm{ad}} = (95 - 40)\,{}^\circ\mathrm{C} = 55\,{}^\circ\mathrm{C}$$

$$\frac{n_{\text{Alc}} * \Delta_R H}{\Delta T_{\text{ad}}} = -\frac{1688\,\text{mol} * \left(-145\frac{\text{kJ}}{\text{mol}}\right)}{55\,°\text{C}} = 4450\frac{\text{kJ}}{°\text{C}}$$

$$m_{\text{Oc}} = \frac{4450\frac{\text{kJ}}{°\text{C}} - 400\,\text{kg} * 2.5\frac{\text{kJ}}{\text{kg}*°\text{C}} - 0.12\,\text{m}^3 * 1026\,\text{m}^3 * 4.35\frac{\text{kJ}}{\text{kg}*°\text{C}}}{2.2\frac{\text{kJ}}{\text{kg}*°\text{C}}}$$

$$= \frac{4450 - 1000 - 536}{2.2}\text{kg} = \mathbf{1325\,kg}$$

→ ***Result***

a) The adiabatic temperature increase is 74.2 °C, so that the reactor content would increase to 114.2 °C in case of cooling failure and the criterion of a maximum temperature of 95 °C could not be met.
b) By increasing the octane content to 1325 kg, the criterion of a resulting maximum temperature of 95 °C in the case of cooling failure would be met.

Task 72

In a reactor, 1000 mol of substance A are dissolved in 800 kg of hexane. Then 2500 mol of substance B are added. A and B react according to the following formula to C:

$$\text{A} + \text{B} \rightarrow \text{C}$$

Exothermic reaction $\Delta_R H = -200$ kJ/mol

Substance A reacts completely according to the formula. The initial temperature is 15 °C. There are no side reactions. Substances A, B and C have boiling points above 100 °C, while the boiling point of the highly flammable hexane at atmospheric pressure is 68 °C. The absolute pressure in the reactor should not exceed 1 bar.

The heat capacities are known:
$cp_A = 20.0$ J/(mol * °C)
$cp_B = 15.0$ J/(mol * °C)
$cp_{\text{Hexane}} = 2.0$ kJ/(kg * K)

a) Can the reaction still be safely controlled in the event of cooling failure?
b) What happens if only one third of the reactants A and B are used, but the amount of hexane remains unchanged?

⊗ **Solution**

→ ***Strategy***

a) Since a pressure of 1 bar may not be exceeded, the maximum allowed temperature is equal to the boiling point of hexane at 68 °C, i.e. the adiabatic temperature increase may not be greater than the difference between the starting temperature and the boiling point of hexane. The adiabatic temperature increase is calculated according to Formula 32a,b, with the molar amount of the fully reacted component A being used.

b) The solution is analogous to the procedure in the first part of the task.

→ ***Calculation***

a) $\Delta T_{\text{ad}} = \frac{-n_i * \Delta_R H}{\sum (n_i * cp_i)}$

$$\Delta T_{\text{ad}} = \frac{-n_i \ * \ \Delta_R H}{\sum (m_i \ * \ cp_i)}$$

Since cp is given both molar and mass-related, a mixture of both formulas applies:

$$\Delta T_{\text{ad}} = \frac{-n_i \ * \ \Delta_R H}{\sum (n_i \ * \ cp_i) + \sum (m_i \ * \ cp_i)}$$

$$-n_A * \Delta_R H = -1000\,\text{mol} * \left(-200\frac{\text{kJ}}{\text{mol}}\right) = 200{,}000\,\text{kJ}$$

$$\sum (n_i * cp_i) = n_A * cp_A + n_B * cp_B = 1000\,\text{mol} \ * \ 20\frac{\text{J}}{\text{mol} \ * \ °\text{C}} + 2500\,\text{mol} * 15\frac{\text{J}}{\text{mol} \ * \ °\text{C}}$$
$$= 57{,}500\frac{\text{J}}{°\text{C}} = 57.5\frac{\text{kJ}}{°\text{C}}$$

$$\sum (m_i * cp_i) = n_H * cp_H = 800\,\text{kg} * 2\frac{\text{kJ}}{\text{kg} * °\text{C}} = 1600\frac{\text{kJ}}{°\text{C}}$$

$$\boldsymbol{\Delta T_{\text{ad}}} = \frac{200{,}000\,\text{kJ} * °\text{C}}{1657.5\,\text{kJ}} = \mathbf{120.7\,°C}$$

$$\boldsymbol{T_{\text{End}}} = T_{\text{Start}} + \Delta T_{\text{ad}} = (15 + 121)\,°\text{C} = \mathbf{136\,°C}$$

b) $n_A = \frac{1000\,\text{mol}}{3} = 333\,\text{mol}$

$$n_B = \frac{2500\,\text{mol}}{3} = 833\,\text{mol}$$

$$n_A * \Delta_R H = -333\,\text{mol} * \left(-200\frac{\text{kJ}}{\text{mol}}\right) = 66{,}600\,\text{kJ}$$

$$\sum (n_i * cp_i) = n_A * cp_A + n_B * cp_B = 333\,\text{mol} * 20\frac{\text{J}}{\text{mol} *°\text{C}} + 833\,\text{mol} * 15\frac{\text{J}}{\text{mol} *°\text{C}}$$
$$= 19{,}155\frac{\text{J}}{°\text{C}} = 19.16\frac{\text{kJ}}{°\text{C}}$$

$$\sum (m_i * cp_i) = n_H * cp_H = 800\,\text{kg} * 2 = 1600 \frac{\text{kJ}}{°\text{C}}$$

$$\Delta T_{\text{ad}} = \frac{66{,}600\,\text{kJ} * °\text{C}}{1619\,\text{kJ}} = \mathbf{41.1\,°C} \quad .$$

$$T_{\text{End}} = T_{\text{Start}} + \Delta T_{\text{ad}} = (15 + 41.1)\,°\text{C} = \mathbf{56.1\,°C}$$

→ ***Result***

a) **Under the given conditions, an adiabatic temperature increase of 121 °C results, and thus a maximum end temperature of 136 °C in the event of cooling failure. The reaction would not be controllable in this case.**
b) **With only one third of the originally planned reactant amount, but with the same amount of hexane, a lower adiabatic temperature increase of 41.1 °C results, and thus a maximum end temperature of 56 °C in the worst case. In this case, the reactor would be inherently safe.**

Exercise 73
In a stirred tank, a solution of raw material A dissolved in 2000 kg ethanol is mixed with an equivalent amount of raw material B. The addition temperature of all substances used is 20 °C. The reaction heat is -169 kJ/mol based on reactant A. The boiling point of ethanol is with 78 °C (1 bar) significantly lower than that of substances A, B and C. The heat capacity of ethanol is $cp_{\text{Ethanol}} = 2.4$ kJ/(kg * °C), that of raw material A $cp_A = 400$J/(mol * °C), that of raw material B $cp_B = 0.10$ kJ/(mol * °C). The raw material A has a molar mass of 150 g/mol. The molar mass of B is 100 g/mol.

The reaction equation is described as follows: A + 2 B → C

There is no cooling or heating of the reactor. Heat losses are to be neglected. Side reactions do not occur. The conversion of A is 100%.

a) How many kg of raw materials A and B may be used so that the reactor content just do not begin to boil? How large may the weight percent content of A in the initial solution be?
b) How much weight percent of product C does the resulting solution contain?

⊗ **Solution**
→ ***Strategy***

a) According to the reaction equation, in a stoichiometric approach, the molar number of B used is twice that of A ($2n_A = n_B$). Based on this fact, the molar number of B is substituted in the formula 32a,b by the relationship mentioned to A and solved for the molar number of A. The adiabatic temperature increase corresponds to the boiling point of the ethanol minus the starting temperature.

The percentage content of A in the alcoholic solution results from the amounts of substance A used and the ethanol.

b) The mass of product C is the sum of the amounts of substances A and B used. From this, the percentage share of C in the reacted mixture can be calculated with the total mass.

→ *Calculation*

a) $\Delta T_{\mathrm{ad}} = \frac{-n_i * \Delta_R H}{\sum (n_i * cp_i)}$ *and* $\Delta T_{\mathrm{ad}} = \frac{-n_i * \Delta_R H}{\sum (m_i * cp_i)}$

Since cp is specified both per mole and per mass, a mixed form of both formulas applies (Subscript Et = Ethanol):

$\Delta T_{\mathrm{ad}} = \frac{-n_i * \Delta_R H}{\sum (n_i * cp_i) + \sum (m_i * cp_i)} = \frac{-n_A * \Delta_R H}{n_A * cp_A + n_B * cp_B + m_{\mathrm{Et}} * cp_{\mathrm{Et}}}$ *with* $n_B = 2 * n_A$

$$\Delta T_{\mathrm{ad}} = \frac{-n_A * \Delta_R H}{n_A * cp_A + 2 * n_A * cp_B + m_{\mathrm{Et}} * cp_{\mathrm{Et}}} = \frac{-n_A * \Delta_R H}{n_A * cp_A + 2 * n_A * cp_B + m_{\mathrm{Et}} * cp_{\mathrm{Et}}}$$

$$\Delta T_{\mathrm{ad}} = \frac{-n_A * \Delta_R H}{n_A * (cp_A + 2 * cp_B) + m_{\mathrm{Et}} * cp_{\mathrm{Et}}}$$

$$n_A * \Delta_R H = n_A * (cp_A + 2 * cp_B) * \Delta T_{\mathrm{ad}} + m_{\mathrm{Et}} * cp_{\mathrm{Et}} * \Delta T_{\mathrm{ad}}$$

$$n_A = \frac{m_{\mathrm{Et}} * cp_{\mathrm{Et}} * \Delta T_{\mathrm{ad}}}{-\Delta_R H - (cp_A + 2 * cp_B) * \Delta T_{\mathrm{ad}}} \quad \text{with } \Delta T_{\mathrm{ad}} = (78 - 20)\,°\mathrm{C} = 58\,°\mathrm{C}$$

$$n_A = \frac{2000\,\mathrm{kg} * 2.4 \frac{\mathrm{kJ}}{\mathrm{kg} * °\mathrm{C}} * 58\,°\mathrm{C}}{169 \frac{\mathrm{kJ}}{\mathrm{mol}} - \left(0.4 \frac{\mathrm{kJ}}{\mathrm{mol} * °\mathrm{C}} + 2 * 0.1 \frac{\mathrm{kJ}}{\mathrm{mol} * °\mathrm{C}}\right) * 58\,°\mathrm{C}} = 2074.5\,\mathrm{mol}$$

$$\boldsymbol{m_A} = n_A * M_A = 2074.5\,\mathrm{mol} * 0.150 \frac{\mathrm{kg}}{\mathrm{mol}} = \mathbf{311.2\,kg}$$

$$\boldsymbol{m_B} = n_B * M_B = 2 * n_A * M_B = 2 * 2074.5\,\mathrm{mol} * 0.100 \frac{\mathrm{kg}}{\mathrm{mol}} = \mathbf{414.9\,kg}$$

$$m_{\text{Alcoholic } Solution\text{ A}} = m_A + m_{\mathrm{Et}} = (311.2 + 2000)\,\mathrm{kg} = 2311\,\mathrm{kg}$$

$$2311\,\mathrm{kg} \mathrel{\hat{=}} 100\% \quad 311.2\,\mathrm{kg}\,A \mathrel{\hat{=}} x$$

$$x = \frac{311.2\,\mathrm{kg} * 100\%}{2311\,\mathrm{kg}} = \mathbf{13.47\,wt\%}\boldsymbol{A} \cong \mathbf{13.5\,wt\%}\boldsymbol{A}$$

b) $m_C = m_A + m_B \quad m_B = n_B * M_B = 2 * n_A * M_B$

$$\boldsymbol{m_C} = m_A + m_B = (311.2 + 414.9)\,\mathrm{kg} = \mathbf{726.1\,kg}$$

$$m_{\mathrm{total}} = m_{\mathrm{Et}} + m_c = (2000 + 726.1)\,\mathrm{kg} \cong 2726\,\mathrm{kg}$$

$$2726\,\text{kg} \triangleq 100\% \quad 726.1\,\text{kg} \triangleq x$$

$$x = \frac{726.1\,\text{kg} * 100\%}{2726\,\text{kg}} = \mathbf{26.6\,wt\%°C}$$

→ ***Result***

a) **311 kg of raw material A and 415 kg of raw material B may be used. The ethanol solution used may contain a maximum of 13.5% of raw material A.**
b) **The resulting ethanol solution contains 26.6% of product C.**

Task 74
In a CSTR, an ester is to be produced from an acid chloride and an alcohol. The reaction is exothermic. The heat output of the reaction was determined as a function of temperature in the laboratory and extrapolated to the conditions of production reactor 5 in which the reaction is to be carried out on a large scale:

$$\dot{Q}_{\text{Reaction}} = 2.15 * 10^{-6}\text{kW} * e^{0.0535*\left(273+T_{\text{Reactor}}/°\text{C}\right)}$$

The parameters for the cooling are known:
$K_w = 650$ J/(s * °C * m²); cooling surface = 8 m²; average cooling water temperature = 25 °C

a) Determine the lower and upper operating point.
b) What happens, if the heat transfer coefficient decreases by half due to formation of a dirt layer at the cooling surface?
c) What happens if, with a clean cooling surface, the average cooling water temperature rises to 40 °C?

⊗ **Solution**
→ ***Strategy***
The intersections of the heat output curve of the reaction as a function of temperature with the heat removal line represent the operating points (see Sec. 2.4.7). To determine the intersections, a $\dot{Q}$ *vs.T*-diagram is created for both functions in the temperature range of T = 20 °C–100 °C.

→ ***Calculation***

$$\dot{Q}_{\text{Reaction}} = 2.15 * 10^{-3}\text{kW} * e^{0.0535*\left(273+T_{\text{Reactor}}/°\text{C}\right)}$$

Case a) $\dot{Q}_{\text{Cooling}} = Kw * A * (T_{\text{Reactor}} - T_{\text{Water}}) = 0.65\,\text{kW} * 8\,\text{m}^2 * (T_{\text{Reactor}} - 25\,°\text{C})$
Case b) $\dot{Q}_{\text{Cooling}} = Kw * A * (T_{\text{Reactor}} - T_{\text{Water}}) = 0.325\,\text{kW} * 8\,\text{m}^2 * (T_{\text{Reactor}} - 25\,°\text{C})$
Case c) $\dot{Q}_{\text{Cooling}} = Kw * A * (T_{\text{Reactor}} - T_{\text{Wasser}}) = 0.65\,\text{kW} * 8\,\text{m}^2 * (T_{\text{Reactor}} - 40\,°\text{C})$

For the expected temperature range, values of temperature and corresponding heat flows are calculated using the above relationships:

T-reactor °C	Q reactor kW	Case a. Q cooling kW	Case b. Q cooling kW	Case c. Q cooling kW
20	14	-26	-13	-104
40	40	78	39	0
60	117	182	91	104
80	342	286	143	208

The heat-transfer curves and the heat-removal lines are created for the questions a), b), c) from this.

a)

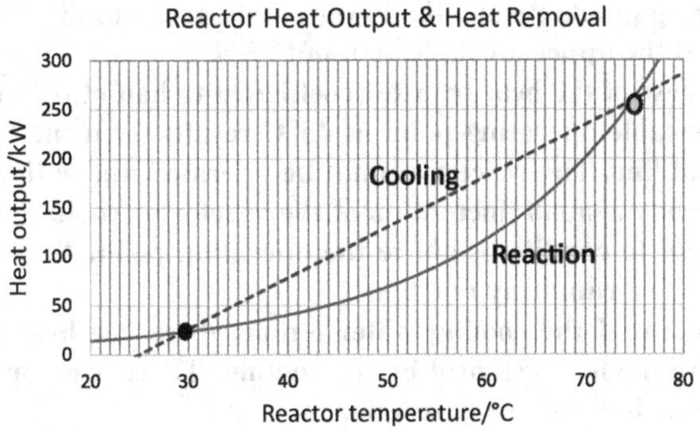

b)

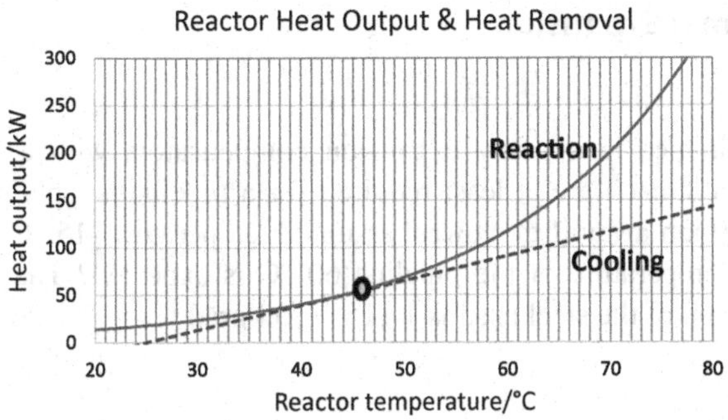

c)

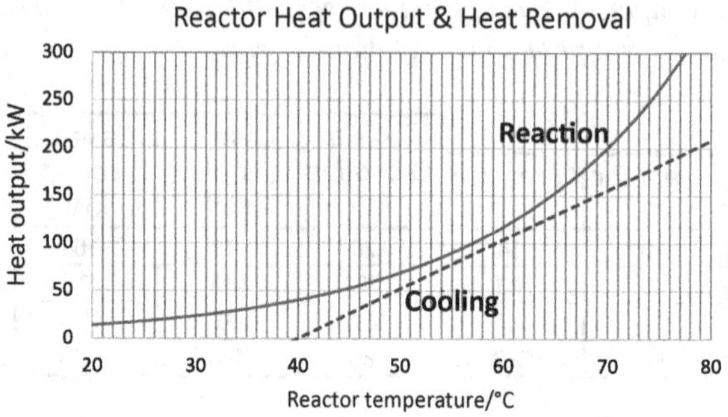

→ *Result*

a) **Two operating points result: the lower stable at about 30 °C reactor temperature and the upper unstable at about 75 °C.**
b) **By the decrease of the heat-transfer coefficient to half of the original value, only one unstable operating point at 45°C results from the tangent of the heat-removal line. The reactor cannot be operated under these conditions because already a small fluctuation of the reactor or cooling water parameters can lead to conditions above the operating point. As a consequence, there would be a run-away reaction..**
c) **By the increase of the cooling water temperature, the heat generation is always above the heat removal by the cooling: There is no operating point → This results in a run-away reaction.**

3.4.7 Thermal Expansion

Task 75

It is to be estimated how many expansion compensators with a compensation capacity of 30 cm each are at least required for a pipeline supplying a chemical plant located 39 km away if the lowest assumed temperature -15 °C would be and as the highest temperature of the pipeline 60 °C is expected? The coefficient of thermal expansion of the steel used is 1.115 * 10^{-5} $grad^{-1}$.

⊗ **Solution**

→ *Strategy*

The maximum length difference of the 39 km long pipeline at a difference of the lower temperature of -15 °C to an upper temperature of 60 °C is calculated using equation 34a. Since each compensator takes up 30 cm of this, the number of compensators results from the ratio of the length difference of the pipeline to the capacity of one compensator.

→ *Calculation*

$$\Delta L = L_o * \alpha * (T_1 - T_o)$$

Given:
L_0= 39,000 m
*α = 1,115 * 10^{-5}/°C*
T_0= −15 °C
T_1= 60°C

Maximum length difference:
$\Delta L = 39{,}000\ m * 1.115 * 10^{-5}\ °C^{-1} * (60 - [-15])\ °C = 32.61\ m$

One compensator can absorb a maximum of 0.3 m. This results in the proportion:
1 compensator = 0.3 m
X compensators = 32.61 m → X = 32.61 m/0.3 m = **108.7 compensators**

→ ***Result***
109 compensators are needed to catch the maximum thermal expansion of the pipeline.

Exercise 76
A cylindrical steel reactor with a wall thickness of 10 mm has an outer length of 2.00 m and an outer diameter of 2.15 m at 20 °C. The specific length expansion of the steel by temperature increase is described as 1.1 * 10^{-5}/°C. The reactor is to be operated at 200 °C.

a) The mountings specify a maximum extension of 30 mm for the reactor length. Are there problems with the operation at 200 °C?
b) What is the reactor volume at 20 °C?
c) What is the volume of the reactor at 200 °C?

⊗ **Solution**
→ ***Strategy***

a) The extension of the reactor length by heating from 20 °C to 200 °C is calculated using formula 34a and compared with the maximum possible value of 30 mm.
b) The volume of the cylinder is calculated from the inner length (outer length minus twice the wall thickness) and the inner diameter (outer diameter minus twice the wall thickness), both at 20 °C.
c) There are two possible pathes of solution:
   1. The inner reactor length and the inner diameter at 200 °C is calculated using formula 34b. Using length and diameter the volume of the cylinder for 200 °C is calculated.
   2. The reactor volume at 200 °C can be approximately calculated using formula 35b from the cubic expansion coefficient (formula 35c) and the volume calculated under point b).

→ ***Calculation***

a) *Length expansion:* $\Delta L = L_o * \alpha * (T_1 - T_o)$
   $\boldsymbol{\Delta L} = 2.00\ \text{m} * 1.1 * 10^{-5}\ ^\circ\text{C}^{-1} * (200 - 20)\ ^\circ\text{C} = 0.00396\ \text{m} \cong \mathbf{4\ mm}$
b) *Volume at* 20 °C: $V = L_i * d_i^2 * \pi/4$
   $V = (2.00\ \text{m} - 2 * 0.01\ \text{m}) * (2.15\ \text{m} - 2 * 0.01\ \text{m})^2 * \pi/4 = \mathbf{7.052\ m^3}$
c) *Volume at* 200 °C
   Alternative 1:
   $V = L_{i\,200\,°C} * d_i^2 * \pi/4$
   $L_{i200\,°C} = L_{i20\,°C} * (1 + \alpha * [200 - 20]\ °C)$
   $= (2.00\ \text{m} - 2*0.01\ \text{m}) * (1 + 1.1 * 10^{-5}\ °C^{-1} * [200 - 20]\ °C) = 1.9839\ \text{m}$
   $d_{i\,200\,°C} = d_{i20\,°C} * (1 + \alpha * [200 - 20]\ °C)$
   $= (2.15\ \text{m} - 2 * 0.01\ \text{m}) * (1 + 1.1 * 10^{-5}\ °C^{-1} * [200 - 20]\ °C) = 2.1342\ \text{m}$
   $\mathbf{V_{200\,°C}} = 1.9839\ \text{m} * 2.1342^2\ \text{m}^2 * \pi/4 = \mathbf{7.0935\ m^3}$
   Alternative 2:
   $V_{200\,°C} = V_{20\,°C} * (1 + 3 * \alpha * [200 - 20]\ °C)$
   $\mathbf{V_{200\,°C}} = 7.052\ \text{m}^3 * (1 + 3 * 1.1 * 10^{-5}/°C * [200 - 20]\ °C) = \mathbf{7.0939\ m^3}$

→ **Result**

a) **Since the length expansion of the reactor is significantly below the maximum permissible of 30 mm at about 4mm, there are no problems with an operation temperature of 200 °C.**
b) **The reactor volume at 20 °C is 7.052 m³.**
c) **The reactor volume at 200 °C is 7.094 m³** according to both alternative solution approaches.

Task 77
The density of water was determined at different temperatures in two measurements:

T = 4 °C
$\rho_{4\,°C} = 1000\ \text{kg/m}^3$
T = 25 °C
$\rho_{25\,°C} = 997\ \text{kg/m}^3$

Calculate the cubic expansion coefficient of water for the range from 4 °C to 25 °C. (Since the expansion coefficient itself is temperature-dependent [it increases with temperature], only the average expansion coefficient for the given temperature range can be calculated using this method with two value pairs of density & temperature).

⊗ **Solution**
→ **Strategy**
The volume difference resulting from a temperature change is described by formula 35b, which contains the cubic expansion coefficient. It can be calculated by rearranging it.

Mass is the product of density and volume, hence for the same mass but different densities and volumes $\rho_{4\,°C} * V_{4\,°C} = m = \rho_{25\,°C} * V_{25\,°C}$.

At 4 °C, $V_{4\,°C} = 1\ m^3$ of water has a mass of 1000 kg. The volume $V_{25\,°C}$ of 1000 kg of water at 25 °C results from rearranging the previous equation. The volume increase ΔV is the difference between $V_{25\,°C}$ and $V_{4\,°C}$.

This volume increase ΔV, the original volume at 4 °C and the temperature are inserted into the rearranged equation 35a to obtain the cubic expansion coefficient.

→ ***Calculation***

$\Delta V = V_0 * \gamma * \Delta T \rightarrow \gamma = \Delta V/(V_0 * \Delta T)$

$\rho_{4\,°C} * V_{4\,°C} = m = \rho_{25\,°C} * V_{25\,°C}$

$V_{25\,°C} = \frac{m}{\rho 25\,°C}$ *with a mass of 1000 kg water, which is at* 4 °C $V_{4\,°C} = 1.000\ m^3$ *corresponds to*

$V_{25\,°C} = \frac{1000\,kg * m^3}{997\,kg} = 1.00301\ m^3$

$\Delta V = V_{25\,°C} - V_{4\,°C} = 1.0026\ m^3 - 1.000\ m^3 = 0.00301\ m^3$

$\gamma = \Delta V/(V_0 * \Delta T) = 0.00301\ m^3/(1.000\ m^3 * [25 - 4]\ °C) = \mathbf{1.43 * 10^{-4}/°C}$

→ ***Result***

The mean cubic expansion coefficient of water between 4 °C and 25 °C is $\gamma = 1.43 * 10^{-4}/°C$.

3.5 Electrochemistry

Task 78

How much energy is needed to produce one metric ton of chlorine (M = 71.1 g/mol) at a decomposition voltage of 4.3 V?

⊗ **Solution**

→ ***Strategy***

First, the molar amount is calculated that corresponds to 1 t of chlorine. With this, the ampere-seconds necessary for the electrolysis are calculated with the rearranged formula 36b. For one molecule of chlorine, two electrons are exchanged: $v_{Cl_2} = 2$. The multiplication with the applied volt-number results in the energy consumption necessary for this according to formula 37b.

→ ***Calculation***

$$n_{Cl_2} = \frac{m_{Cl_2}}{M_{Cl_2}} = \frac{1000\,kg * mol}{0.071\,kg} = 14{,}085\,mol$$

$$I * t = n_{Cl_2} * v_e * F = 14{,}085\,mol * 2 * 96{,}485 \frac{A * s}{mol} = 2.718 * 10^9 A * s$$

$$E = U * I * t = 4.3V * 2.718 * 10^9 \text{A} * \text{s} = 1.169 * 10^{10} Ws = \frac{1.169 * 10^7 \text{kW} * \text{s} * \text{h}}{3600\,\text{s}}$$

$$\mathbf{E = 1.169 * 10^7\,kJ = 3246\,kWh}$$

→ ***Result***

For the electrolytic production of 1 t of chlorine, 3246 kWh or 1.169 * 10^7kJ are consumed under the given conditions.

Exercise 79

In the electrolysis of a zinc chloride solution, a current of 250 A flows. The duration of the electrolysis is 10 h. The atomic weight of zinc is 65.38 g/tom. The voltage the electrolysis cell is operated is 5.3 V.

a) How much zinc is deposited at the cathode in total?
b) How many liters of chlorine gas at 15 °C and 0.7 bar are produced per second?
c) What is the electrical power and how much electrical energy is consumed in the electrolysis process?

$$\text{Zn}^{2+} + 2\text{e} \rightarrow \text{Zn}\ (\text{e} = \text{elektrons})\ 2\text{Cl}^- - 2\text{e} \rightarrow \text{Cl}_2$$

⊗ **Solution**

→ ***Strategy***

The number of moles of zinc and chlorine can be calculated with formula 36b. Where the number of exchanged electrons $\nu_e = 2$. The amount of zinc deposited follows from the multiplication of the number of moles with the molar mass. The volume of chlorine gas is calculated from formula 2. The volume flow is the quotient of chlorine volume and electrolysis time. The electrical power is the product of current and voltage (formula 37a). The consumed electrical energy is the product of power and electrolysis time (formula 37b).

→ ***Calculation***

a) $n = \frac{I*t}{\nu_e*F} = \frac{250\,\text{A}*10\,\text{h}*3600\,\text{s}*\text{mol}}{2*96{,}485\,\text{A}*\text{s}*\text{h}} = 46.64\,\text{mol}$

$$m_{\text{Zn}} = n * M_{\text{Zn}} = 46.64\,\text{mol} * 0.06538\frac{\text{kg}}{\text{mol}} = \mathbf{3.05\,kg}$$

b) $V_{\text{Cl}_2} = \frac{n*R*T}{p} = \frac{46.64\,\text{mol}*8.315*10^{-5}\text{bar}*\text{m}^3*(15+273)\text{K}}{\text{mol}*\text{K}*0.7\,\text{bar}} = 1.60\,\text{m}^3$

$$\dot{V}_{\text{Cl}_2} = \frac{V_{\text{Cl}_2}}{t} = \frac{1600\,\text{L}}{10 * 3600\,\text{s}} = \mathbf{0.0444\frac{L}{s}}$$

c) $\mathbf{P} = U * I = 250\,\text{A} * 5.3\,\text{V} = 1325\,\text{V} * \text{A} = \mathbf{1.325\,kW}$

$$E = U * I * t = 250\,\text{A} * 5.3\,\text{V} * 10\,\text{h} = \mathbf{13.25\,kWh} = \frac{13.25\,\text{kWh} * 3600\,\text{s}}{\text{h}} = \mathbf{47{,}700\,kJ}$$

→ ***Result***

a) **3.05 kg of metallic zinc are deposited.**
b) **At 15°C and a pressure of 0.7 bar, a chlorine volume flow of 44.4 mL/s is produced.**
c) **1.325 kW of power is applied to the electrolysis. The electrical energy used for electrolysis is 13.25 kWh or 47,700 kJ.**

Task 80

Aluminum metal is manufactured by electrolysis of molten aluminum oxide Al_2O_3. A smaller aluminumworks produces 150 metric t of this metal per day. The necessary voltage of the electrolysis cells is 5.1 V. ($M_{aluminum} = 27$ g/tom; $M_{oxygen} = 16$ g/tom)

$$Al^{3+} + 3e \rightarrow Al^{o}$$

a) How much aluminum oxide must be used for such a production performance per day?
b) How many kWh are theoretically used for the production of 1 kg aluminum in the melt electrolysis?
c) What is the necessary electrical power for the melt electrolysis of this site?

⊗ **Solution**

→ ***Strategy***

a) The amount of aluminium oxide needed is calculated from the daily production output of the metal, using the ratio of the molar masses of the metal oxide/metal.

b. & c. In accordance with the rearranged formula 36b and 37b, the corresponding electrical current quantity or corresponding electrical current flow for the production of 1 kg aluminium is calculated using the molar quantity of aluminium metal produced per unit of time and the number of electrons exchanged of $\nu_e = 3$. From this, the electrical power is obtained by multiplication with the applied voltage and the used electrical energy by multiplication with time.

→ ***Calculation***

a) $\dot{m}_{Al_2O_3} = \dot{m}_{Al} * \frac{M_{Al_2O_3}}{2*M_{Al}} = 150\frac{t}{day} * \frac{102}{2*27} = \mathbf{283.3\frac{t}{day}}$

b) *For 1 kg aluminium:*

$$n_{Al} = \frac{m_{Al}}{M_{Al}} = \frac{1\,kg * mol}{0.027\,kg} = 37.04\,mol$$

$$E = U * I * t$$

$$I * t = n_{Al} * \nu_{Al} * F = 37.04\,mol * 3 * 96{,}486\frac{A * s}{mol} = 1.072 * 10^7 A * s$$

$$E = U * I * t = 5.1\,\text{V} * 1.072 * 10^7\text{A*s} = 5.467 * 10^4\text{kW*s} = \frac{5.467 * 10^4\,\text{kW*s*h}}{3600\,\text{s}}$$

$$= \mathbf{15.19\,kWh}$$

c) *150 t Al/day* → $E = 15.19\frac{\text{kWh}}{\text{kg}} * 150 * 1000\,\text{kg} = 2.279 * 10^6\,\text{kWh}$

With 24 h/day → $\boldsymbol{P} = \frac{2.279*10^6\,\text{kWh}}{24\,\text{h}} = \mathbf{9.494 * 10^4\,kW = 94.9\,MW}$

→ ***Result:***

a) **283.3 t aluminium oxide are needed per day.**
b) **15.19 kWh are needed to produce 1 kg aluminium.**
c) **The named aluminium melt electrolysis has a power requirement of 94.9 MW.**

Exercise 81
In a 60 MW wind farm, regularly in the period from 13:00 to 16:00 only 40% of the generated electrical energy can be fed into the electrical supply network. A "power-to-gas" solution is being considered to electrolytically split water into hydrogen and oxygen in this time, store the hydrogen in an empty salt cavern, and then generate electricity from fuel cells at times of high electricity consumption and feed it into the network. The electrolysis voltage is 1.9 V. How large would the necessary volume of such a cavern be to store the hydrogen generated in the period from 13:00 to 16:00 at 25 °C and a pressure of 10 bar? Losses from the electrolysis efficiency are to be neglected.

⊗ **Solution**
→ ***Strategy***
First, the excess electrical energy generated in three hours is calculated as W * s. With the voltage applied to the electrolysis, the corresponding number of amperes * seconds is determined. This results in the number of moles of hydrogen generated with relation 36b. With the gas law (formula 2) follows the volume of the hydrogen to be stored and thus the necessary empty space volume of the salt cavern.

→ ***Calculation***

$$E = P * t = 36\,\text{MW} * 3\,\text{h} = 108\,\text{MWh} = 108 * 10^6\text{W} * \text{h} * 3600\frac{\text{s}}{\text{h}} = 3.888 * 10^{11}\text{W} * \text{s}$$

$$n_{H_2} = \frac{I * t}{\nu_e * F} \quad I * t = \frac{E}{U} = \frac{3.888 * 10^{11}\,\text{W} * \text{s}}{1.9\,\text{V}} = 2.046 * 10^{11}\text{A} * \text{s} \quad \nu_e = 2$$

$$n_{H_2} = \frac{2.046 * 10^{11}\,\text{A} * \text{s} * \text{mol}}{2 * 96{,}485\,\text{A} * \text{s}} = 1.060 * 10^6\,\text{mol}$$

$$V = \frac{n * R * T}{p} = \frac{1.06 * 10^6\,\text{mol} * 8.315 * 10^{-5}\,\text{bar} * \text{m}^3 * (273 + 25)\text{K}}{\text{mol} * \text{K} * 10\,\text{bar}} = \mathbf{2627\,m^3}$$

→ ***Result***

The cavern would have to have a minimum volume of 2627 m³. Such a project would thus be realistic.

Exercise 82

In the electrolytic refining of copper, crude copper plates are used as anodes and a thin sheet of pure copper as a cathode in an acid bath of dilute sulfuric acid. At an applied DC voltage of 0.25 V, the impure copper of the anode dissolves and is deposited as pure copper on the cathode:

$$Cu \rightleftarrows Cu^{2+} + 2e \text{ (atomic mass Cu } = 63.55 \text{g/mol)}$$

a) How much electrical energy does a plant need to produce 20 metric tons of pure copper per day?
b) What is the electrical power flow in this case?

⊗ **Solution**

→ ***Strategy***

Anode reaction: $Cu - 2e \rightarrow Cu^{2+}$
Cathode reaction: $Cu^{2+} + 2e \rightarrow Cu$
$\nu_e=2$

The number of moles is calculated from the daily production of pure copper and the molar mass of copper, inserted into the formula 36b converted to I * t, and multiplied by the cell voltage. The result represents the amount of electrical energy in kWh required. The electrical power is obtained by dividing the energy amount thus calculated by the number of hours in a day.

→ ***Calculation***

a) $n = \frac{m}{M} = \frac{20{,}000 \text{ kg*mol}}{0.06355 kg} = 3.147 * 10^5 \text{ mol}$

$$E = U * I * t$$

$$I * t = n * \nu_e * F = 3.147 * 10^5 \text{ mol} * 2 * 96{,}485 \frac{\text{A} * \text{s}}{\text{mol}} = 6.073 * 10^{10} \text{ A} * \text{s}$$

$$E = 0.25 \text{ V} * 6.073 * 10^{10} \text{ V} * \text{A} * \text{s} = 1.518 * 10^7 \text{kW} * \text{s} = \frac{1.518 * 10^7 \text{ kW} * \text{s} * \text{h}}{3600 \text{ s}} = \mathbf{4217\,kWh}$$

b) $P = \frac{E}{t} = \frac{4217 \text{ kWh}}{24 \text{ h}} = \mathbf{175.7\,kW}$

→ ***Result***

a) **For the production of 20 t of pure copper, an electrical energy amount of 4217 kWh is required.**
b) **The electrolysis unit has a demand for electrical power of 176 kW.**

3.6 Liquid Conveying

Task 83
2 m^3 of water (density 1000 kg/m^3) per minute is pumped from a deep well with a water level of 20 m below ground to an open container 30 m above ground. The friction loss in the piping system is 0.4 bar.

a) What must be the power of the driving electric motor be to accomplish this task, given a pump efficiency of 70 % and a motor efficiency of 90 %?
b) How much electrical energy is required daily for continuous operation?

⊗ **Solution**
→ ***Strategy***

a) The theoretically necessary power consumption of the pump motor is calculated according to Formula 39a. For this purpose, the total height is determined in accordance with Formula 38a. Since both the deep well and the container are atmospheric, there is also no pressure difference and thus no pressure height. The height equivalent of friction is determined with the aid of Formula 38b. The real power consumption of the electric motor results from dividing the theoretical value by the product of the efficiencies of the pump and the motor.
b) The daily pump energy is the product of the real power consumption and 24 h pump time.

→ ***Calculation***

a) $P = \dot{m} * g * H$

$$\dot{m} = \frac{2\,\text{m}^3 * \text{min} * 1000\,\text{kg}}{\text{min} * 60\,\text{s} * \text{m}^3} = 33.3\frac{\text{kg}}{\text{s}}$$

$$H = h_{geo} + h_p + h_f$$

$$h_{geo} = (20 + 30)\text{m} = 50\,\text{m} \quad h_p = 0\,\text{m}$$

$$h_f = \frac{\Delta p}{\rho * g} = \frac{0.4\,\text{bar} * \text{m}^3 * \text{s}^2 * 10^5\,\text{kg}}{1000\,\text{kg} * 9.81\,\text{m} * \text{m} * \text{s}^2} = 4.08\,\text{m} \cong 4.1\,\text{m}$$

$$H = (50 + 4.1)\,\text{m} = 54.1\,\text{m}$$

$$P = \frac{33.3\,\text{kg}\,\text{m}^3 * 9.81\,\text{m} * 54.1\,\text{m}}{\text{s} * \text{s}^2} = 17.67\,\text{kW}$$

$$\boldsymbol{P}_{\textbf{real}} = \frac{P}{\eta_{\text{Pumpe}} * \eta_{\text{Motor}}} = \frac{17.67\,\text{kW}}{0.7 * 0.9} = \mathbf{28\,kW}$$

b) $\boldsymbol{E}_{\textbf{real}} = P_{\text{real}} * t = 28\,\text{kW} * 24\,\text{h} = \mathbf{672\,kWh}$

→ ***Result***

a) **The rated power of the drive motor is 28 kW.**
b) **The daily energy consumption of the pump is 672 kWh.**

Exercise 84

An inland vessel supplies liquid propane with a density of 0.52 kg/L to a tank farm in a harbor. The propane tank of the farm is 15 m above the pier edge. The tank of the ship is 3 m below the pier edge. The pressure in the ship's tank is 4.5 bar, that in the storage tank is 3.5 bar. The unloading of the propane load is carried out at 50 t/h. The friction losses under these conditions are 0.8 bar. In total, 600 metric t of propane are to be pumped.

a) How large must the power of the electric motor of the pump be, if the overall efficiency of the pump and motor is 65%?
b) How much electrical energy in kWh will be required for unloading the ship?

⊗ **Solution**

→ ***Strategy***

a) The pump performance is calculated using formula 39a, the total height required for this purpose from formula 38a with the conversion of the pressure difference from inland vessel to storage tank and the friction pressure loss according to formula 38b to the corresponding heights. This theoretical value is increased by division with the overall efficiency to the real value of the power connection value of the pump motor.
b) The total energy of the pumping process results from the product of the power and the time of the pumping process.

→ ***Calculation***

a) $P = \dot{m} * H * g$
$H = h_{\text{geo}} + h_p + h_f$
$h_{\text{geo}} = (3 + 15)\,\text{m} = 18\,\text{m}$

$$h_p = \frac{\Delta p}{\rho * g} = \frac{p_{\text{tank}} - p_{\text{ship}}}{\rho * g} = \frac{(3.5 - 4.5)\,\text{bar} * \text{m}^3 * \text{s}^2}{520\,\text{kg} * 9.81\,\text{m}}$$
$$= \frac{-1.0\,\text{bar} * 10^5\,\text{kg} * \text{m}^3 * \text{s}^2}{\text{bar} * \text{m} * \text{s}^2 * 520\,\text{kg} * 9.81\,\text{m}} = -19.6\,\text{m}$$

$$h_r = \frac{\Delta p}{\rho * g} = \frac{0.8\,\text{bar} * 10^5\,\text{kg} * \text{m}^3 * \text{s}^2}{\text{bar} * \text{m} * \text{s}^2 * 520\,\text{kg} * 9.81\,\text{m}} = 15.7\,\text{m}$$

$$H = (18 - 19.6 + 15.7)\,\text{m} = 14.1\,\text{m}$$

$$P = \frac{50\,\text{t} * 14.1\,\text{m} * 9.81\text{m}}{\text{h} * \text{s}^2} = \frac{50\,\text{t} * 14.1\,\text{m} * 9.81\text{m} * 1000\,\text{kg} * \text{h}}{\text{h} * \text{s}^2 * \text{t} * 3600\,\text{s}}$$
$$= 1921\frac{\text{kg} * \text{m}^2}{\text{s}^2} = 1.92\,\text{kW}$$

$$\boldsymbol{P}_{\textbf{real}} = \frac{P}{\eta_{\text{pump}} * \eta_{\text{motor}}} = \frac{1.92\,\text{kW}}{0.65} = \mathbf{2.95\,kW \cong 3.0\,kW}$$

b) $E_{\text{real}} = P * t \quad t = \frac{m}{\dot{m}} = \frac{600\,\text{t} * \text{h}}{50\,\text{t}} = 12\,\text{h} \rightarrow \boldsymbol{E}_{\textbf{real}} = 3.0\,\text{kW} * 12\,\text{h} = \mathbf{36\,kWh}$

→ ***Result:***

a) **The power of the pump motor is to be set at 3 kW.**
b) **The total energy for the pumping process is 36 kWh.**

Task 85
A centrifugal pump is to be used to pump 30 m^3 of river water per hour into a high tank. The river is 4 m below the pump, the filling level of the high tank, which does not change, is 10 m above the pump. The air pressure is 1 bar. The pressure inside the high tank is 3 bar. The friction losses correspond to a height of 1 m. The river water has a density of 1000 kg/m^3.

a) How large must the pump power (kW) be if the overall efficiency (pump and motor) is 69%?
b) How much energy (kWh) is required for the pumping process per day?

⊗ **Solution**
→ ***Strategy***

a) Pump performance is calculated using Formula 39c. The total height is obtained from Formula 38a, with Formula 38b used to calculate the pressure height and the friction height.
b) The necessary energy is obtained from the product of power and delivery time.

→ ***Calculation***

a) $P = \frac{\dot{m}*g*H}{\eta_{\text{pump}}*\eta_{\text{motor}}}$

$$\dot{m} = \dot{V} * \rho = 30\frac{\text{m}^3 * \text{h}}{\text{h} * 3600\,\text{s}} * 1000\frac{\text{kg}}{\text{m}^3} = 8.333\frac{\text{kg}}{\text{s}}$$

$$H = h_{\text{geo}} + h_p + h_f$$

$$h_{\text{geo}} = (4 + 10)\,\text{m} = 14\,\text{m} \quad h_f = 1\,\text{m}$$

$$h_p = \frac{\Delta p}{\rho * g} = \frac{(3 - 1)\,\text{bar} * \text{m}^3 * \text{s}^2}{1000\,\text{kg} * 9.81\,\text{m}} = 0.0002039 * \frac{10^5\,\text{kg} * \text{m}^3 * \text{s}^2}{\text{kg} * \text{m} * \text{s}^2 * \text{m}} = 20.4\,\text{m}$$

$$H = (14 + 20.4 + 1)\,\text{m} = 35.4\,\text{m}$$

$$P = \frac{8.333\,\text{kg} * 9.81\,\text{m} * 35.4\,\text{m}}{\text{s} * \text{s}^2 * 0.69} = 4194\frac{\text{kg} * \text{m}^2}{\text{s}^3} = \mathbf{4.194\,kW \cong 4.2\,kW}$$

b) $E = P * t = 4.194\,\text{kW} * 24\,\text{h} = \mathbf{100.7\,kWh}$

→ ***Result***

a) **The connecting power of the pump motor must be 4.2 kW.**
b) **The required daily energy is 100.7 kWh.**

Task 86

A distillation column is supplied with 5 kg/s of a mixture ($\rho = 900$ kg/m ^{3}) by means of a centrifugal pump (characteristic curve is available). The height of the liquid level in the mixture storage tank is 10 m below that of the feed point of the column. The pressure height between the storage tank and the higher pressure at the feed point of the column is 20 m.

The pressure loss by friction was determined empirically according to the following equation:

$$h_f = 0.04 * \dot{V}^2 \text{ with } \dot{V} \text{ as m}^3/\text{h} \ \ h_f \text{ results as m.}$$

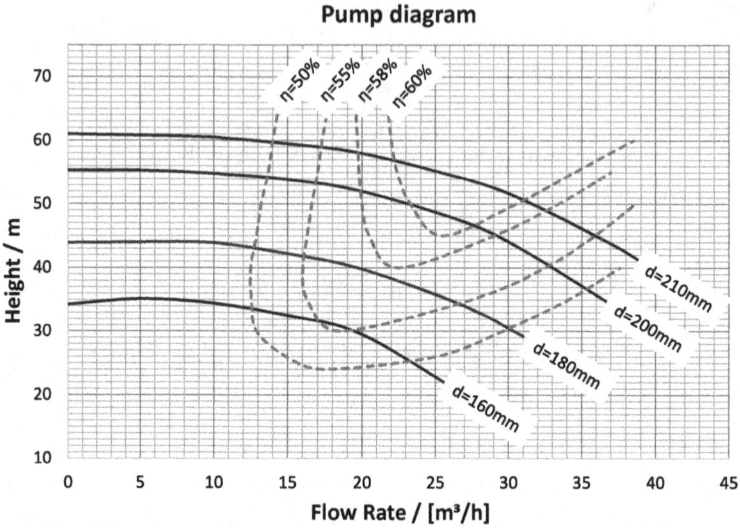

a) What pump impeller diameter should be used and what pump efficiency will result?

b) How large must the power of the electric motor (efficiency: 95%) be, which drives the centrifugal pump.

⊗ Solution

→ *Strategy*

The volumetric flow of the mixture is calculated from the mass flow using the density. The total height is composed of the geodetic height, the pressure height and the height of the friction loss. The latter is calculated from the formula given in the task statement and the determined volumetric flow. From the thus determined point of the total height and the volumetric flow in the pump characteristic curve, the next larger impeller diameter and the corresponding pump efficiency are read off. The connection value of the power of the electric motor is calculated according to formula 39c.

→ *Calculation*

a) $\dot{V} = \frac{\dot{m}}{\rho} = \frac{5\,\text{kg} * \text{m}^3}{\text{s} * 900\,\text{kg}} = 0.00556\frac{\text{m}^3}{\text{s}} = \frac{0.00556\,\text{m}^3 * 3600\,\text{s}}{\text{s} * \text{h}} = 20\frac{\text{m}^3}{\text{h}}$

$$H = h_{\text{geo}} + h_p + h_f$$

$$h_{\text{geo}} = 10\,\text{m} \quad h_p = 20\,\text{m} \quad h_f = 0.04 * 20^2\,\text{m} = 16\,\text{m}$$

$$H = (20 + 10 + 16)\,\text{m} = 46\,\text{m}$$

Diagram → Point H to $\dot{V}$ lies between impeller diameters 180 mm and 200 mm → **The impeller of 200 mm must be used. The corresponding efficiency of the pump is about 57 %.**

b) $P = \frac{\dot{m} * H * g}{\eta_{\text{pump}} * \eta_{\text{motor}}} = \frac{5\,\text{kg} * 46\,\text{m} * 9.81\,\text{m}}{\text{s} * \text{s}^2 * 0.57 * 0.95} = 4167 \frac{\text{kg} * \text{m}^2}{\text{s}^3} \cong \mathbf{4.2\,kW}$

→ ***Result***

a) **The impeller of the diameter of 200 mm must be installed.**
b) **This results in a pump efficiency of 57 %. The electrical power of the pump is 4.2 kW.**

Task 87

A centrifugal pump is to be used to pump 20 m^3 of trichloromethane (density: 1.49 kg/L) from an atmospheric stock tank into a reactor 10 m high. The reactor is under a pressure of 6.5 bar. The friction loss in the pipeline is 0.1 bar.

a) What energy is required for this theoretically?
b) What theoretical power must the pump have, if the pumping process is to be completed within 30 minutes?
c) How much energy and how much power are required for this under real conditions (pump efficiency: 60%; electrical drive efficiency of the pump: 89%)?

⊗ **Solution**

→ ***Strategy***

a) The necessary energy for the pumping process is calculated using Formula 39b. The total height required for this is calculated from Formula 38a with the height for the pressure difference and the pressure loss due to friction determined by Formula 38b.
b) Power is the ratio of energy used to the time required for the pumping process.
c) In order to determine the energy and power consumption required in practice for the pumping process, the theoretical values previously determined under b) are divided by the product of the efficiencies of the pump and the electric drive motor.

→ ***Calculation***

a) $E = m * g * H \quad m = V * \rho = 20\,\text{m}^3 * 1490 \frac{\text{kg}}{\text{m}^3} = 29{,}800\,\text{kg}$

$$H = h_{\text{geo}} + h_p + h_f$$

$$h_{\text{geo}} = 10\,\text{m}$$

$$h_p = \frac{\Delta p}{\rho * g} = \frac{(6.5 - 1.0)\,\text{bar} * \text{m}^3 * \text{s}^2}{1490\,\text{kg} * 9.81\,\text{m}} = \frac{5.5 * 10^5\,\text{kg} * \text{m}^3 * \text{s}^2}{1490\,\text{kg} * 9.81\text{m} * \text{s}^2 * \text{m}} = 37.6\,\text{m}$$

$$h_f = \frac{\Delta p}{\rho * g} = \frac{0.1\,\text{bar} * \text{m}^3 * \text{s}^2}{1490\,\text{kg} * 9.81\,\text{m}} = \frac{0.1 * 10^5\,\text{kg} * \text{m}^3 * \text{s}^2}{1490\,\text{kg} * 9.81\,\text{m} * \text{s}^2 * \text{m}} = 0.68\,\text{m} \cong 0.7\,\text{m}$$

$$H = (10 + 37.6 + 0.7)\,\text{m} = 48.3\,\text{m}$$

$$E = 29{,}800\,\text{kg} * 9.81\frac{\text{m}}{\text{s}^2} * 48.3\,\text{m} = 14{,}119{,}925\,\text{kg} * \text{m}^2/\text{s}^2$$

$$\boldsymbol{E = 14{,}120\,\textbf{kJ} = 14{,}120\,\textbf{kW} * \text{s}} = \frac{14{,}120\,\text{kW} * \text{s} * \text{h}}{3600\,\text{s}} = \mathbf{3.92\,kWh}$$

b) $\boldsymbol{P} = \frac{E}{t} = \frac{14{,}120\,\text{kJ}*\text{min}}{30\,\text{min}*60\,\text{s}} = \mathbf{7.84\,kW}$ alternative : $\boldsymbol{P} = \frac{E}{t} = \frac{3.92\,\text{kWh}}{0.5\,\text{h}} = \mathbf{7.84\,kW}$

c) $\boldsymbol{E}_{\textbf{Real}} = \frac{E}{\eta_{\text{pump}} * \eta_{\text{motor}}} = \frac{14{,}120\,\text{kJ}}{0.6*0.89} = \mathbf{26{,}442\,kJ}$

$$= \frac{3.92\,\text{kWh}}{0.6 * 0.89} = \mathbf{7.34\,kWh}$$

$$\boldsymbol{P}_{\textbf{Real}} = \frac{P}{\eta_{\text{pump}} * \eta_{\text{motor}}} = \frac{7.84\,\text{kW}}{0.6 * 0.89} = \mathbf{14.68\,kW \cong 15\,kW}$$

→ ***Result***

a) **The theoretical energy consumption is 14,120 kJ = 3.92 kWh.**
b) **The theoretical power is 7.84 kW.**
c) **In practical operation, 26,442 kJ = 7.34 kWh are consumed. The electric motor of the pump must have a power of 15 kW.**

Problem 88

In a whisky distillery, 10,000 liters of 60% distillate of a density of 0.9 g/cm^3 is to be pumped from the collection tank into a high tank after each batch distillation. Both containers are atmospheric. The lifting height is 15 m during the entire thirty-minute pumping process. For this purpose, a centrifugal pump with a hydraulic efficiency of 75% is available. Unfortunately, the electric motor has burned out and the type label is no longer legible.

a) What minimum electrical power must a replacement motor have to cope with the pumping task with an electrical efficiency of 90%? Friction losses in the piping system are to be neglected here.
b) In the same distillery, the cooling water for the condenser of the distillation is pumped by means of a submersible pump (centrifugal pump → pump characteristic curve is given below) with a flow of at least 15 m^3/h of water ($\rho = 1000$ kg/m^3) from a well 30 m deep into an open storage tank 10 m above the factory

floor. The small pressure difference and the small pressure loss due to friction are to be neglected. With what wheel diameter must the pump be operated to meet this requirement? What pump efficiency is achieved? What power of the pump drive motor is required at a motor efficiency of 85%?

c) Because of the constantly recurring complex maintenance work on the 30 m deep submersible pump, the proposal is to replace the submersible pump with a centrifugal pump on the factory floor level. What advantages would a piston pump have, which would also be mounted on the factory floor level?

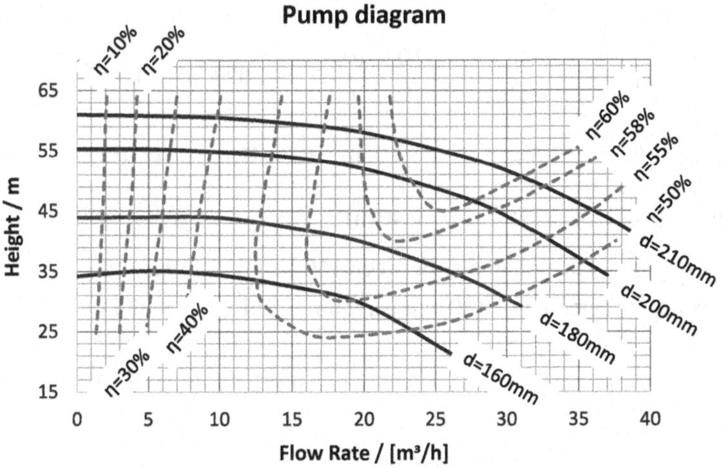

⊗ **Solution**

→ ***Strategy***

a) The necessary power of the liquid flow is calculated with formula 39a. Since, according to the task description, no pressure difference between the distillate collecting tank and the high container exists and the friction losses of the flow can be neglected, the delivery height corresponds to the geodetic. The efficiency of the pump and the drive motor must be taken into account.

b) The flow rate is given and the total delivery height corresponds to the geodetic height, since the small pressure difference and friction differences can be neglected. With these two values, the next larger wheel diameter and the associated pump efficiency are determined in the pump diagram. With the previously calculated theoretical pump performance, the given efficiency of the motor and the determined efficiency of the pump itself, the real value of the electrical power consumption is calculated.

c) A pumping water with a suction height of more than 8m cannot be realized because of the water vapor pressure. This part of the task is therefore nonsense.

→ ***Calculation***

a) $P = \dot{m} * g * H$

$$H = h_{geo} + h_p + h_f \quad h_{geo} = 15\,\text{m} \quad h_p = 0\,\text{m} \quad h_f = 0\,\text{m} \rightarrow H = 15\,\text{m}$$

$$\dot{m} = \frac{\dot{V} * \rho}{t} = \frac{10\,\text{m}^3 * 900\,\text{kg} * \text{min}}{30\,\text{min} * \text{m}^3 * 60\,\text{s}} = 5.0\frac{\text{kg}}{\text{s}}$$

$$P = \frac{5\,\text{kg} * 9.81\,\text{m} * 15\,\text{m}}{\text{s} * \text{s}^2} = 735.8\,\text{W}$$

$$P_{real} = \frac{P}{\eta_{pump} * \eta_{motor}} = \frac{0.736\,\text{kW}}{0.75 * 0.9} = \mathbf{1.09\,kW \cong 1.1\,kW}$$

b) $H = h_{geo} + h_p + h_f \quad h_{geo} = (30 + 10)\,\text{m} \quad h_p = 0\,\text{m} \quad h_f = 0\,\text{m} \rightarrow H = 40\,\text{m}$

Pump-diagramm ($\dot{V} = 15\frac{\text{m}^3}{\text{h}}$; $H = 40\,\text{m}$)

→ pump wheel diameter between 160 mm und 180 mm :
hence an impeller of 180 mm is required and $\eta_{pump} = 0.53$

$$P = \dot{m} * g * H \quad \dot{m} = \dot{V} * \rho = \frac{15\,\text{m}^3 * 1000\,\text{kg} * \text{h}}{\text{h} * \text{m}^3 * 3600\,\text{s}} = 4.17\frac{\text{kg}}{\text{s}}$$

$$P = \frac{4.17\,\text{kg} * 9.81\,\text{m} * 40\,\text{m}}{\text{s} * \text{s}^2} = 1636\frac{\text{kg} * \text{m}^2}{\text{s}^3} = 1.64\,\text{kW}$$

$$P_{real} = \frac{P}{\eta_{pump} * \eta_{motor}} = \frac{1.64\,\text{kW}}{0.53 * 0.85} = \mathbf{3.64\,kW}$$

→ ***Result***

a) **The drive motor of the distillate pump has a power consumption of 1.1 kW.**
b) **The cooling water pump must be equipped with a 180 mm impeller. This results in a hydraulic efficiency of 53 %. The power consumption of the pump motor is 3.64 kW.**
c) **Since it is not possible to achieve a suction lift of more than 8 m due to the vapour pressure of the water, the proposal to install the pump on the factory floor level is senseless, regardless of whether it is a centrifugal or a piston pump.**

Exercise 89

A mixture of carbon tetrachloride (content: 50 wt%, density: 1594 kg/m³), chloroform (30 wt%, 1490 kg/m³), dichloromethane (20 wt%, 1300 kg/m³) is pumped from an atmospheric storage tank to the feed point of a rectification column at a height of 25 m with a centrifugal pump. The absolute pressure in the column at the feed point is 2 bar. The friction losses in the piping system are 0.5 bar. The electric motor driving the centrifugal pump has an efficiency of 85%.

a) What flow rate is achieved according to the attached pump characteristic curve and what is the power of the electric motor of the pump?
b) The flow rate is to be increased to 8 m^3/h. It is approximately assumed that the friction losses only increase slightly by this change and can therefore be neglected. The same pump and the same electric motor of the pump with the same power are to be used. How much must the pressure in the storage tank be increased to achieve this goal?

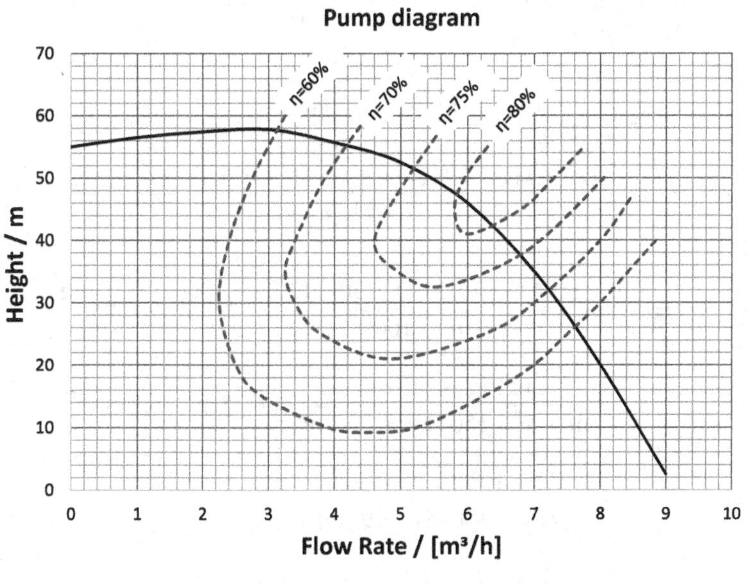

⊗ **Solution**
→ ***Strategy***

a) The total height (formula 38a) can be taken from the pump characteristic curve to determine both the flow rate and the efficiency of the pump. The pressure height and friction height are obtained from formula 38b. For this purpose, the mean density of the flow must be calculated. For the resulting total height, take the corresponding volume flow and efficiency from the characteristic curve. Using formula 39c, the electrical connection value of the power of the pump motor can be determined.
b) From the pump characteristic curve, the total height and the corresponding efficiency are taken for a flow rate of 8 m^3/h. Using formula 38a, the pressure height is calculated and from this, with the appropriately rearranged formula 38b, the pressure difference from column inlet to tank, from which the

necessary tank pressure results. Using formula 39c, it is checked whether the expected similar motor power value is adequate.

→ ***Calculation***

a) $H = h_{geo} + h_p + h_f$

$$h_{geo} = 25\,\text{m}$$

$$h_p = \frac{\Delta p}{\rho * g} \quad \Delta p = (2-1)\,\text{bar} = 1\,\text{bar} = 10^5 \frac{\text{kg}}{\text{m} * \text{s}^2}$$

$$h_f = \frac{\Delta p}{\rho * g} \quad \Delta p = 0.5\,\text{bar} = 0.5 * 10^5 \frac{\text{kg}}{\text{m} * \text{s}^2}$$

$$\overline{\rho} = \frac{\sum (m_i * \rho_i)}{\sum m_i} = \frac{m_{\text{total}} * \sum \left(\frac{\%_i}{100\%} * \rho_i\right)}{m_{\text{total}}}$$
$$= \frac{\%_{CCl_4} * \rho_{CCl_4} + \%_{CHCl_3} * \rho_{CHCl_3} * \%_{CH_2Cl_2} * \rho_{CH_2Cl_2}}{100\%}$$

$$\overline{\rho} = \frac{(50 * 1594 + 30 * 1490 + 20 * 1300)\% * \text{kg}}{\text{m}^3 * 100\%} = 1504 \frac{\text{kg}}{\text{m}^3}$$

$$h_p = \frac{10^5\,\text{kg} * \text{m}^3 * \text{s}^2}{\text{m} * \text{s}^2 * 1504\,\text{kg} * 9.81\,\text{m}} = 6.78\,\text{m} \cong 6.8\,\text{m}$$

$$h_r = \frac{0.5 * 10^5\,\text{kg} * \text{m}^3 * \text{s}^2}{\text{m} * \text{s}^2 * 1504\,\text{kg} * 9.81\,\text{m}} = 3.39\,\text{m} \cong 3.4\,\text{m}$$

$$H = (25 + 6.8 + 3.4)\text{m} = 35.2\,\text{m}$$

The corresponding volume flow of this total height is obtained from the pump characteristic curve in Figure 7 m³/h at a pump efficiency of 0.725. This corresponds to a mass flow of

$$\dot{m} = \dot{V} * \rho = 7{,}0 \frac{\text{m}^3}{\text{h}} * \frac{\text{h}}{3600\,\text{s}} * 1504 \frac{\text{kg}}{\text{m}^3} = \mathbf{2.92 \frac{kg}{s}}$$

$$P = \frac{\dot{m} * g * H}{\eta_{\text{pump}} * \eta_{\text{motor}}} = \frac{2.92\,\text{kg} * 9.81\,\text{m} * 35.2\,\text{m}}{\text{s} * \text{s}^2 * 0.725 * 0.85} = \mathbf{1.64\,kW}$$

b) *From Figure 8 m³/h → H = 20 m;* $\eta_p \cong 0{,}5$

$$H = h_{geo} + h_p + h_f$$

$$h_p = H - h_{geo} - h_f = (20 - 25 - 3.4)\text{m} = -8.4\,\text{m}$$

$$\Delta p = hp * \rho * g \frac{-8.4\,\mathrm{m} * 1504\,\mathrm{kg} * 9.81\,\mathrm{m}}{\mathrm{m}^3 * \mathrm{s}^2} = -123{,}936 \frac{\mathrm{kg}}{\mathrm{m} * \mathrm{s}^2} \cong -123936 \frac{\mathrm{Pa} * \mathrm{bar}}{10^5\,\mathrm{Pa}} \cong -1.24\,\mathrm{bar}$$

$$\Delta p = P_{\mathrm{column}} - P_{\mathrm{tank}}$$

$$\boldsymbol{p_{\mathrm{tank}}} = p_{\mathrm{column}} - \Delta p = 2\,\mathrm{bar} - (-1.24\,\mathrm{bar}) = \mathbf{3.24\,bar}$$

$$\dot{m} = \dot{V} * \rho = 8 \frac{\mathrm{m}^3}{\mathrm{h}} * \frac{\mathrm{h}}{3600\,\mathrm{s}} * 1504 \frac{\mathrm{kg}}{\mathrm{m}^3} = 3.34 \frac{\mathrm{kg}}{\mathrm{s}}$$

$$P = \frac{\dot{m} * g * H}{\eta_{\mathrm{pump}} * \eta_{\mathrm{motor}}} = \frac{3.34\,\mathrm{kg} * 9.81\,\mathrm{m} * 20\,\mathrm{m}}{\mathrm{s} * \mathrm{s}^2 * 0.5 * 0.85} = 1.54\,\mathrm{kW} \rightarrow \cong 1.64\,\mathrm{kW}$$

→ ***Result***

a) **The volume flow is 7 m^{3}per hour. The electrical connection power of the pump motor required for case a) is 1.64 kW.**
b) **The increased delivery capacity at 8 m^3/h with the same pump power can be achieved by increasing the pressure in the storage tank to 3.24 bar. The electrical connection power calculated with the resulting pump efficiency confirms this result.**

3.7 Scale Enlargement

Task 90

A customer requires 2200 kg of 3,4,5-trimethylaniline (TMA). The research laboratory has developed a synthesis for this purpose, in which 3,4,5-trimethylnitrobenzene (TMNB) is reduced to 3,4,5-TMA in the final reaction step by adding iron filings. This reaction step is the slowest in the sequence of synthesis steps and processing procedures.

A laboratory experiment was carried out as follows: In a 1.5-L double-walled reactor, 100 g of TMNB was completely dissolved in 400 mL of toluene within 20 min with stirring and brought to 35 °C. Then, within one hour, 500 mL of 25% by weight hydrochloric acid and 75 g of iron filings were added continuously and the temperature was kept at 35 °C. The mixture was then heated to 60 °C and the reaction was completed within three more hours. The purification of the TMA was carried out by filtration and rectification. Including the preparation of the product, a yield of TMA with respect to TMNB of 85% was obtained.

To satisfy the customer's wishes, the reaction is to be carried out in a pilot plant reactor of 3 m^3. A fill level of 60% is planned. The briefing of the shift before each batch takes 10 min. 20 min are required to fill the reactor with toluene and TMNB. The times of dissolution of the TMNB and the reaction itself, including the addition of the iron filings and the post-reaction at 60 °C, correspond to those of the laboratory batch. Pumping out the reaction mixture requires half an hour, cleaning the reactor 40 min.

a) What is the scale-up factor?
b) What quantities of TMNB, toluene, hydrochloric acid and iron filings are required per pilot plant batch?
c) How much TMA is expected from a pilot plant batch, and how many batches are theoretically required to meet the customer's wishes?
d) The pilot plant is used around the clock in three shifts. In what period of time can the required amount of TMA be produced if, during the entire campaign, one failed batch and about 4 h of maintenance downtime can be expected?

Molar masses in g/mol: C = 12 g; H = 1; N = 14; O = 16; Fe = 58
Density in kg/m^3: TMA = 920; TMNB = 1240; iron = 5580

⊗ **Solution**
→ ***Strategy***

a) The ratio of the active volume of the pilot plant reactor to the final volume of the laboratory batch is formulated as a meaningful scale-up factor.
b) The quantities of the laboratory batch, multiplied by the scale-up factor, correspond to the quantities used for a pilot plant batch.
c) Equation 10a (yield calculation) is rearranged to product molarity (n_{TMA}) and the TMNB mass, used in the pilot plant reactor is converted to moles (equation 7e). The calculated moles of TMA are converted to mass per pilot plant batch.
d) The number of batches required for the scheduled delivery quantity and the TMA mass produced per batch is determined theoretically. This includes a probable failed batch. The number is rounded up to the next whole number. The time required for a batch is calculated from the individual times and the total time required for the batches is calculated from this. In addition, the probably necessary time for maintenance work is added.

→ ***Calculation***

a) $ScF = \frac{V_{PilotReaktor}}{V_{Lab}}$

$$V_{PilotReactor} = 3.0\,\text{m}^3 * \frac{60\%}{100\%} = 1.8\,\text{m}^3 = 1800\,\text{L}$$

$$V_{Lab} = V_{TMNB} + V_{Toluene} + V_{HyrochloricAcid} + V_{Fe} \quad \text{V} = \frac{m}{\rho}$$

$$V_{Lab} = \frac{m_{TMNB}}{\rho_{TMNB}} + V_{Toluene} + V_{HydrochloricAcid} + \frac{m_{Fe}}{\rho_{Fe}}$$

$$V_{Lab} = \frac{0.1\,\text{kg} * \text{m}^3}{1240\,\text{kg}} * \frac{1000\,\text{L}}{\text{m}^3} + 0.4\,\text{L} + 0.5\,\text{L} + \frac{0.075\,\text{kg} * \text{m}^3}{5580\,\text{kg}} * \frac{1000\,\text{L}}{\text{m}^3}$$

$$V_{\text{Lab}} = (0.0806 + 0.4 + 0.5 + 0.0134)\,\text{L} = 0.994\,\text{L}$$

$$\boldsymbol{ScF} = \frac{1800\,\text{L}}{0.994\,\text{L}} = \mathbf{1811}$$

b) $m_{i-\text{Pilot}} = m_{i-\text{Lab}} * ScF \quad V_{i-\text{Pilot}} = V_{i-\text{Lab}} * ScF$

$$\boldsymbol{m}_{\mathbf{TMNB-Pilot}} = 0.1\,\text{kg} * 1811 = 181.1\,\text{kg} \cong \mathbf{181\,kg}$$

$$\boldsymbol{V}_{\mathbf{Toluene-Pilot}} = 0.4\,\text{L} * 1811 = 724.4\,\text{L} \cong \mathbf{0.724\,m^3}$$

$$V_{\text{HydrochloricAcid-Pilot}} = 0.5\,\text{L} * 1811 = 905.5\,\text{L}$$

$$\boldsymbol{m}_{\mathbf{Fe-Pilot}} = 0.075\,\text{kg} * 1811 = 135.8\,\text{kg} = \mathbf{136\,kg}$$

c) *Pilot Reactor*

$$\text{TMNB} \rightarrow \text{C}_9\text{H}_{11}\text{NO}_2$$

$$M_{\text{TMNB}} = (9 * 12 + 11 * 1 + 14 + 2 * 16)\text{g/mol} = 165\,\text{g/mol}$$

$$TMA \rightarrow \text{C}_9\text{H}_{13}\text{N}$$

$$M_{\text{TMA}} = (9 * 12 + 13 * 1 + 14)\text{g/mol} = 135\,\text{g/mol}$$

$$Y_{P/E} = \frac{\upsilon_E * \left(n_{P_o} - n_P\right)}{\upsilon_p * n_{E_o}} \quad Y_{\text{TMA/TMNB}} = \frac{\upsilon_{\text{TMNB}} * \left(n_{\text{TMA}_o} - n_{\text{TMA}}\right)}{\upsilon_{\text{TMA}} * n_{\text{TMNB}_o}}$$

$$\upsilon_{\text{TMNB}} = -1 \quad \upsilon_{\text{TMA}} = +1$$

$$n_{\text{TMA}_o} = 0 \quad n_{\text{TMNB}_o} = \frac{m_{\text{TMNB}_o}}{M_{\text{TMNB}}} = \frac{181\,\text{kg} * \text{mol}}{0.165\,\text{kg}} = 1097\,\text{mol}$$

$$n_{\text{TMA}} = Y_{\text{TMA/TMNB}} * \frac{\nu_{\text{TMA}}}{-\nu_{\text{TMNB}}} * n_{\text{TMNB}} = 0.85 * \frac{+1}{-(-1)} * 1097\,\text{mol} = 919\,\text{mol}$$

$$\boldsymbol{m}_{\mathbf{TMA}} = n_{\text{TMA}} * M_{\text{TMA}} = 919\,\text{mol} * 0.135\frac{\text{kg}}{\text{mol}} = \mathbf{124\,kg}$$

d) Number of Batches $= \frac{2200\,\text{kg}}{124\,\text{kg}} = 17.4 \cong 18$ *additional time reserve for a possible wrong* batch → *19 batches*

Time per batch = Pre-meeting + Refill + Resolve + Addition of hydrochloric acid & iron + After reaction + Pumping + Cleaning

$$\frac{\text{time}}{\text{batch}} = (10 + 20 + 20 + 60 + 3 * 60 + 30 + 40)\text{min} = 360\,\text{min} = 6\,\text{h}$$

$$t_{\text{Total-Production}} = 6\frac{\text{h}}{\text{Batches}} * 19\,\text{Batches} = 114\,\text{h}$$

$$t_{\text{Campaign}} = 114\,\text{h} + 4\,\text{h} = 118\,\text{h} = 4.9\,\text{Days} \cong \mathbf{5\,Days}.$$

→ ***Result***

a) **The scale-up factor is 1811.**
b) **For a batch in the pilot reactor, 181 kg TMNB, 0.724 m^3 toluene, 0.906 m^3 hydrochloric acid and 136 kg iron filings are required.**
c) **124 kg TMA are produced per batch in the pilot reactor.**
d) **5 days must be planned for the production campaign in the pilot reactor.**

Task 91
Using a PFTR with an inner pipe diameter of 150 mm, 100 kg of dry calcium palmitate (Mw = 548.8 g/mol) is to be produced per hour. For this purpose, a 5 wt% aqueous sodium palmitate solution (Mw = 278.4 g/mol; $\rho_{solution}$ = 1030 kg/m^3) is reacted with a 20 wt% calcium chloride solution (Mw = 111.1 g/mol; $\rho_{solution}$ = 1180 kg/m^3) and the resulting precipitate is filtered off. The calcium palmitate with a water content of 5 wt% is dried.

a) Was the stoichiometric ratio of the reactants maintained in the laboratory experiment?
b) What is the scale-up factor?
c) What are the streams of sodium palmitate and calcium chloride solutions that must be fed to the PFR based on the laboratory experiment?
d) How long must the PFR be with a residence time of 5 min?
e) What is the approximate saturated steam flow required to dry the filtered calcium palmitate if only the evaporation heat of the water content of the precipitate is to be considered and the thermal efficiency of the dryer is 85 %?

⊗ **Solution**
→ ***Strategy***

a) For the laboratory approach, the molar feed of sodium palmitate (NaP) and calcium chloride is calculated and compared.
b) The amount of calcium palmitate (CaP) obtained in the laboratory experiment is calculated from the sodium palmitate used by means of the yield formula 10a. The scale-up factor is defined as the ratio of the desired production performance of CaP and that of the laboratory experiment (formula 40).
c) The material flows of NaP solution and $CaCl_2$ solution to the PFR are calculated from the amounts used in the laboratory experiment at a residence time of 5 min and the scale-up factor.

d) The length of the PFR results from the rearranged volume formula of a cylinder. The volume of the PFR is calculated from the sum of the volume flows of the Na-P solution and the $CaCl_2$ solution, multiplied by the residence time of 5 min. The volume flows result from the mass flows, divided by their density.
e) The evaporation heat of the residual water is approximately equal to the condensation heat of the heating steam. Thus, the necessary steam flow for drying corresponds approximately to the amount of water to be removed, divided by the thermal efficiency of the dryer.

→ ***Calculation***

a) 500 g 5 wt%NaP-Solution → 25 g NaP

$$n_{\mathrm{NaP}} = \frac{m_{\mathrm{NaP}}}{M_{\mathrm{NaP}}} = \frac{25\,\mathrm{g} * \mathrm{mol}}{278.4\,\mathrm{g}} = 0.0898\,\mathrm{mol} \cong 0.09\,\mathrm{mol}$$

25 g 20 wt% $CaCl_2$-Solution → 5 g $CaCl_2$

$$n_{\mathrm{CaCl_2}} = \frac{m_{\mathrm{CaCl_2}}}{M_{\mathrm{CaCl_2}}} = \frac{5\,\mathrm{g}{*}\mathrm{mol}}{111.1\,\mathrm{g}} = 0.045\,\mathrm{mol}$$

For 1 mol $CaCl_2$ 2 mol NaP are needed. Therefore, in the laboratory assay, stoichiometric ratios are present.

b) $ScF = \frac{\dot{m}_{\mathrm{CaP-PFR}}}{\dot{m}_{\mathrm{CaP-Lab}}}$

$$Y_{\mathrm{CaP/NaP}} = \frac{\nu_{\mathrm{NaP}} * (n_{\mathrm{CaPo}} - n_{\mathrm{CaP}})}{\nu_{\mathrm{CaP}} * n_{\mathrm{NaPo}}}$$

$$n_{\mathrm{CaP-Lab}} = -\frac{\nu_{\mathrm{CaP}}}{\nu_{\mathrm{NaP}}} * Y_{\mathrm{CaP/NaP}} * n_{\mathrm{NaP}} = 0.5 * 0.975 * 0.09\,\mathrm{mol} = 0.0439\,\mathrm{mol}$$

$$m_{\mathrm{CaP-Labor}} = n_{\mathrm{CaP}} * M_{\mathrm{CaP}} = 0.439\,\mathrm{mol} * 548.8\frac{\mathrm{g}}{\mathrm{mol}} = 24.1\,\mathrm{g}\ \text{within 5 min}$$

$$\rightarrow \dot{m}_{\mathrm{CaP-Lab}} = \frac{24.1\,\mathrm{g}}{5 * 60\,\mathrm{s}} = 0.0803\frac{\mathrm{g}}{\mathrm{s}}$$

$$\dot{m}_{\mathrm{CaP-PFR}} = \frac{100\,\mathrm{kg}}{3600\,\mathrm{s}} = 27.78\frac{\mathrm{g}}{\mathrm{s}}$$

$$\boldsymbol{ScF} = \frac{27.78}{0.0803} = 346$$

c) $\dot{\boldsymbol{m}}_{\mathbf{NaP-Sol-PFR}} = \dot{m}_{\mathrm{NaP-Sol-Lab}} * ScF = \frac{0.5\,\mathrm{kg}}{300\,\mathrm{s}} * 346 = \mathbf{0.576}\frac{\mathbf{kg}}{\mathbf{s}}$

$$\dot{\boldsymbol{m}}_{\mathbf{CaCl_2-Sol-PFR}} = \dot{m}_{\mathrm{CaCl_2-Sol-Lab}} * ScF = \frac{0.025\,\mathrm{kg}}{300\,\mathrm{s}} * 346 = \mathbf{0.0288}\frac{\mathbf{kg}}{\mathbf{s}}$$

d) $L_{PFR} = \frac{V_{PFR}*4}{d^2*\pi}$

$$V_{PFR} = \tau * \dot{V} \quad \tau = 5\,\text{min} = 300\,\text{s} \quad \dot{V} = \frac{\dot{m}}{\rho}$$

$$\dot{V}_{NaP-Sol-PFR} = \frac{0.576\,\text{kg} * \text{m}^3}{\text{s} * 1030\,\text{kg}} = 5.59 * 10^{-4}\frac{\text{m}^3}{\text{s}}$$

$$\dot{V}_{CaCl_2-Sol-PFR} = \frac{0.0288\,\text{kg} * \text{m}^3}{\text{s} * 1180\,\text{kg}} = 2.44 * 10^{-5}\frac{\text{m}^3}{\text{s}}$$

$$V_{PFR} = 300\,\text{s} * (55.9 + 2.44) * 10^{-5}\frac{\text{m}^3}{\text{s}} = 0.175\,\text{m}^3$$

$$\boldsymbol{L_{PFR}} = \frac{0.175\,\text{m}^3 * 4}{(0.15\,\text{m})^2 * \pi} = \mathbf{9.91\,m \cong 10\,m}$$

e) *100 kg wet CaP contains 95 kg CaP and 5 kg water. When 100 kg dry CaP are produced per hour, 5.26 kg/h of water must be removed.*

$$\dot{m}_{Steam} = \frac{5.26\,\text{kg}}{\text{h} * 0.85} = \mathbf{6.2\frac{kg}{h}}$$

→ ***Result***

a) **The laboratory experiment was carried out under stoichiometric conditions of the sodium palmitate and calcium chloride molar number.**
b) **The product stream-related scale-up factor is ScF = 346.**
c) **The feed stream for the PFR is 0.576 kg/s 5 wt% sodium palmitate solution and 0.0288 kg/s 20 wt% calcium chloride solution.**
d) **The necessary length of the tubular reactor is 10 m.**
e) **6.2 kg of steam per hour is required for drying.**

Task 92

In a pilot plant reactor of 200 L total volume, a certain amount of 1,2-epoxybutane (E) is to be produced from 1-chloro-2-hydroxybutane (CHB) by the reaction with sodium hydroxide. The pilot plant reactor may only be filled up to 70 %. For this purpose, 1,2-epoxybutane (EPB; M = 70.0 g/mol) was produced from 1-chloro-2-hydroxybutane (CHB; M = 106.5 g/mol) in a laboratory experiment with a yield of 81 %. 600 g CHB (density 1.05 g/cm^3), 0.7 L of 40 wt% aqueous sodium hydroxide solution and 1 L hexane were used as solvent. There was no volume contraction.

a) What is the scale-up factor?
b) What are the feed values for the pilot plant reactor?
c) What amount of epoxybutane can be expected from the pilot plant batch?

⊗ **Solution**

→ ***Strategy***

a) The production rate is limited by the usable volume of the pilot plant reactor. Therefore, the ratio of the usable volume of the pilot plant reactor to the reaction volume of the laboratory experiment is a meaningful scale-up factor (Formula 40). The reaction volume of the laboratory experiment is calculated from the given amounts used.
b) The amounts used in the laboratory experiment, multiplied by the scale-up factor, give the amounts that must be fed to the pilot plant reactor per batch.
c) By rearranging the yield formula 10a, the amount of EPB produced in the laboratory experiment can be calculated. Multiplied by the scale-up factor, the amount of EPB produced per pilot plant batch results.

→ ***Calculation***

a) $ScF = \frac{V_{\text{Pilot}}}{V_{\text{Lab}}}$

$$V_{\text{Pilot}} = 0.2\,\text{m}^3 * 0.70 = 0.14\,\text{m}^3$$

$$V_{\text{Lab}} = V_{\text{CHB}} + V_{\text{NaOH}} + V_{\text{Hexane}} = \frac{0.6\,\text{kg} * \text{L}}{1.05\,\text{kg}} + 0.7\,\text{L} + 1\,\text{L} = 2.27\,\text{L}$$

$$\boldsymbol{ScF} = \frac{140\,\text{L}^3}{2.27\,\text{L}} = \mathbf{61.64}$$

b) $\boldsymbol{m}_{\mathbf{CHB}} = 61.64 * 0.6\,\text{kg} = \mathbf{37.0\,kg}$

$$\boldsymbol{V}_{\mathbf{NaOH}} = 61.64 * 0.7\,\text{L} = \mathbf{43.15\,L}$$

$$\boldsymbol{V}_{\mathbf{Hexane}} = 61.64 * 1.0\,\text{L} = \mathbf{61.64\,L}$$

c) $Y_{P/E} = \frac{\upsilon_E * (n_{P_0} - n_P)}{\upsilon_p * n_{E_0}}$

Reactant → Subscript E = CHB
Product → Subscript P = EPB

$$Y_{\text{EPB/CHB}} = \frac{\upsilon_{\text{CHB}} * \left(n_{\text{EPB}_0} - n_{\text{EPB}}\right)}{\upsilon_{\text{EPB}} * n_{\text{CHB}_0}}$$

At the beginning of the reaction, no EPB is present, as pure reactants are used only $\rightarrow n_{\text{EPBo}} = 0\,\text{mol}$

CHB + NaOH → EPB + NaCl + H_2O → $\upsilon_{\text{CHB}} = -1 \quad \upsilon_{\text{EPB}} = +1$

$$Y_{\text{EPB/CHB}} = 0.81$$

$$Y_{\text{EPB/CHB}} = \frac{-1 * (-n_{\text{EPB}})}{+1 * n_{\text{CHB}_0}}$$

$$n_{CHB_0} = \frac{m_{CHB_0}}{M_{CHB_0}} = \frac{600\,g * mol}{106.5\,g} = 5.63\,mol$$

Amount of EPB produced in the laboratory experiment:

$$n_{EPB} = Y_{EPB/CHB} * n_{CHB_0} = 0.81 * 5.63\,mol = 4.56\,mol$$

$$m_{EPB} = n_{EPB} * M_{EPB} = 4.56\,mol * \frac{70.0\,g}{mol} = 319.2\,g$$

Amount of EPB expected in the pilot plant reactor:

$$m_{EPB-Pilot} = m_{EPB} * ScF = 0.3192\,kg * 61.64 = \mathbf{19.7\,kg}$$

→ ***Result***

a) **The scale-up factor is 61.64.**
b) **The following amounts are required for the pilot plant reactor: 37.0 kg 1-chloro-2-hydroxybutane (CHB), 43.2 L 40 %NaOH and 61.6 L hexane.**
c) **19.7 kg 1,2-epoxybutane are produced per batch in the pilot plant reactor.**

Task 93

In the design of the production for the manufacture of a plastic additive "Y" (total size of the market about 250 metric t per year) it is assumed that the stirred batch reactor RK8 with a usable volume of 1.2 m^3 can be used 15 days a month. Two batches can be processed per day. Test series were carried out in the research laboratory with the following result:

Standard batch:

Raw material 1: 210 cm^3; Raw material 2: 108 g ($\rho = 0.95$ g/cm^3); Catalyst solution: 8 cm^3; Solvent: 350 cm^3

Yield of "Y": 152 g

a) What is the scale-up factor per batch?
b) Which raw material quantities and which quantity of product "Y" are to be projected per batch in the RK8?
c) How much of additive "Y" can be produced in this way per year, and what is the maximum market share that can be achieved?
d) Make suggestions on how to increase the production capacity.

⊗ **Solution**

→ ***Strategy***

a) To determine the scale-up factor, the comparison of the reaction volume of the laboratory experiment with the usable volume of the RK8 reactor is suitable (Formula 40).

b) The quantities required for RK8 operation are calculated using the scale-up factor from the values of the laboratory scale, as well as the product quantity per batch.
c) The number of batches in RK8 results from the information on operating times and, from this, the annual production with the achieved product quantity per batch. With the market size, the percentage share achievable with RK8 can be determined by the rule of three.

→ ***Calculation***

a) $ScF = \frac{V_{RK8}}{V_{Lab}}$

$$V_{RK8} = 1.2\,\text{m}^3 \quad V_{Lab} = 210\,\text{cm}^3 + \frac{108\,\text{g} * \text{cm}^3}{0.95\,\text{g}} + 8\,\text{cm}^3 + 350\,\text{cm}^3 = 682\,\text{cm}^3 = 0.682 * 10^{-3}\,\text{m}^3$$

$$ScF = \frac{1.2\,\text{m}^3}{0.682 * 10^{-3}\,\text{m}^3} = 1759 \simeq 1760$$

b) $\textbf{Reactant1}: \mathbf{V} = 0.21\,\text{L} * 1760 = \mathbf{370\,L}$
$\textbf{Reactant2}: \boldsymbol{m} = 0.108\,\text{kg} * 1760 \cong \mathbf{190\,kg}$
$\textbf{Catalyst} - \textbf{Solution}: V = 8\,\text{cm}^3 * 1760 = 14{,}080\,\text{cm}^3 \cong \mathbf{14.1\,L}$
$\textbf{Solvent}: V = 0.350\,\text{L} * 1760 = \mathbf{616\,L}$
$\boldsymbol{Mass}\ \textbf{"}\boldsymbol{Y}\textbf{"perBatch} = 0.152\,\text{kg} * 1760 = \mathbf{267.5\,kg}.$

c) $\frac{\text{Batches}}{Year} = \frac{2}{\text{Day}} * \frac{15\,\text{Days}}{\text{Month}} * \frac{12\,\text{Months}}{Year} = 360$

$$\boldsymbol{Mass}\textbf{"}\boldsymbol{Y}\textbf{"}\boldsymbol{perYear} = 360 * 267.5\,\text{kg} = 96{,}300\frac{\text{kg}}{year} = \mathbf{96.3}\frac{\boldsymbol{t}}{\boldsymbol{year}}$$

250t/a →100 %
96,3t/a →X %

$$\textbf{Relative MarketShare X} = \frac{96.3 * 100\%}{250} = \mathbf{38.5\ \%}$$

→ ***Result***

a) **The scale-up factor is 1760.**
b) **The following amounts are used per RK8 batch:**
raw material 1: 370 L, raw material 2: 190 kg, catalyst solution: 14.1 L, solvent: 616 L.
267.5 kg of product Y are produced per batch.
c) **With the annual production of RK8, a market share of 38.5% can be achieved.**
d) **The production volume might be increased by the following actions:**

- **Shorter batch times through faster preparation of the batches (faster filling or emptying of the reactor, faster setting of the reaction temperature).**
- **Shorter batch times through higher reaction rate [higher reaction temperature, higher reactant concentration (lower proportion of solvent), possibly higher stirrer speed (especially in two-phase reactions)].**

Task 94

The addition of an additive solution to a polymer solution is to be implemented on the basis of the results of the production laboratory as a scale-up for a plastic production plant. The polymer solution leaving the reactor at a rate of 100 metric t/h and a temperature of 30 °C is to be mixed with the additive solution at a rate of 50 L/h, which is to be homogeneously mixed in a 2 m long pipe section of an outer diameter of 10″(wall thickness 3 mm).

In the laboratory, it was found on the basis of model tests that a Reynolds number of >3500 is required to achieve a homogeneous mixing of the additive in the 2 m long pipe section. The density of the polymer solution is 1200 kg/m^3 and is to be approximately assumed to be temperature-independent. The viscosity of the polymer solution at 20 °C was determined to be 0.2 Pa * s. Laboratory measurements yielded the following empirical relationship between the viscosity and the temperature of the polymer solution:

$\eta = 21{,}060\, Pa * s * e^{-0.0395 * T}$ with T as absolute temperature

a) What flow conditions would prevail in the pipe section after the reactor under the described conditions?
b) At what temperature of the polymer solution would the mixing criterion according to the Reynolds number be met?

⊗ **Solution**

→ ***Strategy***

The flow of a fluid is characterized by the Reynolds number according to formula 41. First, the inner pipe diameter is calculated. The flow velocity results from the ratio of the volume flow (quotient of mass flow and density) to the pipe cross section.

a) The viscosity of the solution for 30 °C is determined by means of the given empirical formula and from this, together with the previously calculated quantities, the Reynolds number is calculated.
b) The formula for the Reynolds number is turned into the viscosity and this is calculated for a minimum Re = 3500. From the empirical formula of the viscosity as a function of temperature, the corresponding minimum temperature of the polymer solution is calculated.

→ ***Calculation***

$$Re = \frac{w * \rho * d}{\eta}$$

$$d = d_i = 10'' * 0.0254\frac{\text{m}}{''} - 2 * (0.003)\text{m} = 0.248\,\text{m}$$

$$w = \frac{\dot{V}}{A} \quad A = \frac{d_i^2 * \pi}{4} = \frac{(0.248\,\text{m})^2 * \pi}{4} = 0.0483\,\text{m}^2$$

$$\dot{V} = \frac{\dot{m}}{\rho} = \frac{100{,}000\,\text{kg} * \text{m}^3 * \text{h}}{\text{h} * 1200\,\text{kg} * 3600\,\text{s}} = 0.0232\frac{\text{m}^3}{\text{s}}$$

$$w = \frac{0.0232\,\text{m}^3}{\text{s} * 0.0483\,\text{m}^2} = 0.478\frac{\text{m}}{\text{s}}$$

a) $\eta_{30\,°\text{C}} = 21{,}060\,Pa * s * e^{-0.0395 * 303.15} = 0.133\,Pa * s$

$$\boldsymbol{Re} = \frac{0.478\,\text{m} * 1200\,\text{kg} * 0.248\,\text{m} * \text{s} * \text{m}}{\text{s} * \text{m}^3 * 0.133\,\text{kg}} = \mathbf{1070}$$

b) $\eta = \frac{w*\rho*d}{Re} = \frac{0.478\,\text{m}*1200\,\text{kg}*0.248\,\text{m}}{\text{s}*\text{m}^3*3500} = 0.0406\,\text{Pa} * \text{s}$

$$\eta = 21{,}060\,\text{Pa} * \text{s} * e^{-0.0395*T} \rightarrow \ln\frac{\eta}{21{,}060\,\text{Pa} * \text{s}} = -0.0395\,\text{T}$$

$$\boldsymbol{T} = \frac{-\ln\frac{0.0406\,\text{Pa}*\text{s}}{21060\,\text{Pa}*\text{s}}}{0.0395}\text{K} = 333.1\,\text{K} = \mathbf{60\,°C}$$

→ ***Result***

a) **Under the original conditions, a Reynolds number of only 1070 is achieved, i.e. there is laminar but no pronounced turbulent flow and one remains below the criterion necessary for homogenization of Re> 3500.**
b) **To reach the minimum criterion of Re = 3500, the polymer solution must be heated to 60 °C.**

Task 95
In a pilot plant for wastewater treatment, a pipe reactor with an inner diameter of d = 25 mm is flowing with the wastewater to be treated, in which solid particles are present in suspended form. In tests in the pilot plant, it is found that at a flow rate of w ≥ 0.53 m/s there is no sedimentation of solids. How large may the diameter of the pipe reactor of a planned wastewater treatment plant with a projected flow rate of $\dot{V}$= 50 m ^{3} per hour be at most, so that the solids also remains in suspension in this case? The density of the wastewater is ρ = 1.10 kg/L. The viscosity is

η = 1.5 mPa * s = 0.0015 kg/(m * s). Use the dimensionless number Re to this up-scaling, which describes the flow state (laminar or turbulent).

⊗ Solution

→ *Strategy*

A solid deposition process in a liquid is influenced by the flow state: The more turbulent, the lower the tendency to settle. The flow state in the pilot test, in which no solid settles, should therefore also prevail in the large plant, in order to avoid sedimentation. The flow state is described by the Reynolds number (Re → Formula 41). Thus, the Re number of the pilot reactor represents the minimum necessary Re number of the reactor of the wastewater treatment plant, from which there is no sedimentation of the solids. Re of the pilot reactor is calculated and the maximum diameter of the pipe of the wastewater treatment plant is determined from this.

→ *Calculation*

Pilot reactor:

$$Re = \frac{w*d_{Pilot}*\rho}{\eta} = \frac{0.53\,\mathrm{m}*0.025\,\mathrm{m}*1100\,\mathrm{kg}*\mathrm{m}*\mathrm{s}}{\mathrm{s}*\mathrm{m}^3*0.0015\,\mathrm{kg}} = 9717 \rightarrow \text{Turbulence}$$

Wastewater treatment plant:

$$w = \frac{\dot{V}}{A} \text{ with } A = \frac{d^2*\pi}{4} \rightarrow w = \frac{4*\dot{V}}{A*d^2*\pi} \text{ with } \dot{V} = \frac{50\,\mathrm{m}^3}{\mathrm{h}} = \frac{50\,\mathrm{m}^3}{3600\,\mathrm{s}} = 0.01389\frac{\mathrm{m}^3}{\mathrm{s}}$$

The flow velocity w results from the volume flow and the inner cross section of the pipe:

$$\mathrm{Re} = \frac{4*\dot{V}*d*\rho}{d^2*\pi*\eta} = \frac{4*\dot{V}*\rho}{d*\pi*\eta} \text{ and rearranged to the minimum pipe diameter:}$$

$$d = \frac{4*\dot{V}*\rho}{\mathrm{Re}*\pi*\eta} = \frac{4*0.01389\,\mathrm{m}*1100\,\mathrm{kg}*\mathrm{m}*\mathrm{s}^3}{9717*\pi*0.0015\,\mathrm{kg}*\mathrm{s}*\mathrm{m}^3} = \mathbf{1.335\,m}$$

→ *Result*

In order to avoid sedimentation of the solids, a pipe diameter of 1.335 m must not be exceeded in the wastewater treatment plant.

Task 96

In a plant for the production of drugs, an effluent stream of 800 L per hour occurs. On average, the residual content of pharmaceutical active ingredient is 25 ppm. The active ingredient is to be brought to a concentration below the analytical detection limit (i.e. practically completely removed) by means of a cylindrical activated carbon adsorber bed with the effluent being flowing from above,. For these conditions, the adsorber bed to be designed should ensure operation for 30 days, then the loaded activated carbon is to be replaced by fresh activated carbon. The 2 m long water feed pipe is to be dimensioned in a way that laminar flow conditions prevail in it. For this purpose, a Reynolds number of 1000 is specified

The following data are known or have been determined: Bulk density of activated carbon $\rho_K = 440\ kg/m^3$; Density of effluent $\rho_W = 1010\ kg/m^3$; Viscosity of effluent $\eta_W = 0.001\ Pa * s = 0.001\ kg/(s * m)$.

In a pilot test in the laboratory, a pipe of 0.1 m internal diameter was filled with 400 g of the activated carbon selected for the production plant and fed with an effluent stream of 40 L per hour. After 18 h of operation, traces of the pharmaceutical active ingredient were detected in the effluent stream.

a) How much active ingredient must be removed from the effluent stream of the plant in the specified period of 30 days?
b) How much active ingredient was taken up by the carbon bed in the laboratory test until breakthrough, and what was the average loading of the activated carbon at the time of breakthrough?
c) Which formulation of a scale-up factor appears to be reasonable and what value would it have?
d) With how many kg of activated carbon does the adsorber bed have to be stocked in production?
e) What are the dimensions of the adsorber bed in production if similarity in terms of diameter to length is to be maintained with the laboratory arrangement?
f) What diameter must the feed pipe of the adsorber have in order to ensure laminar flow conditions?

⊗ **Solution**

→ ***Strategy***

a) The mass of pharmaceutical active ingredient (drug) that is absorbed by the activated carbon (AC) is the product of the difference in the active ingredient concentration in the water (W) before and after the adsorption bed with the mass flow of the water and the operating time until traces of the ingredient are detected after the carbon bed. From this loading, the breakthrough of active ingredient begins.
b) Proceed as in a). The loading of the activated carbon with active ingredient at the breakthrough point also applies to the adsorption bed in the production plant. It represents the adsorbed mass of active ingredient relative to the mass of activated carbon.
c) A meaningful formulation of the scale-up factor is the ratio of the adsorbed active ingredient mass in the adsorber of the production plant to that in the laboratory experiment.
d) The mass of activated carbon used in the laboratory experiment, multiplied by the scale-up factor, gives the mass of activated carbon in the production plant.
e) From the data of the laboratory reactor, the volume of activated carbon used there is calculated and the length of the bed is derived from this. From this, the ratio of length to diameter of the activated carbon bed results from the volume formula of the cylinder. The volume of the activated carbon bed in the plant results from its mass and the bulk density of the activated carbon bed. From the

boundary conditions of the geometric similarity to the laboratory arrangement (ratio of diameter to length of the bed), the dimensions of the adsorption bed in the plant result from the volume formula of the cylinder.

f) The formula of the Reynolds number is set up (Formula41) and the flow velocity is substituted by the ratio of the volume flow of the water and the pipe cross section. This equation is rearranged to the pipe diameter.

→ ***Calculation***

a) *Maximum amount of drug on AC in the production plant before drug-breakthrough:*

$$m_{\text{Drug-Plant}} = \Delta C_{\text{Drug}} * \dot{m}_{W-\text{Plant}} * t_{\text{Plant}} = \Delta C_{\text{Drug}} * \dot{V}_{W-\text{Plant}} * \rho_W * t_{\text{Plant}}$$

C_{Drug} *after the adsorber bed is practically equal to zero.*

$$\mathbf{m_{Drug-Plant}} = 25\frac{\text{mg}}{\text{kg}} * 0.8\frac{\text{m}}{\text{h}} * 1010\frac{\text{kg}}{\text{m}^3} * 24\frac{\text{h}}{\text{day}} * 30\,\text{days} = \mathbf{1.454 * 10^7\,mg = 14.54\,kg}$$

b) *Maximum amount of Drug on AC in the laboratory test before Drug-breakthrough:*

$$m_{\text{Drug-Lab}} = \Delta C_{\text{Drug}} * \dot{m}_{W-\text{Lab}} * t_{\text{Lab}} = \Delta C_{\text{Drug}} * \dot{V}_{W-\text{Lab}} * \rho_W * t_{\text{Lab}}$$

C_{Drug} *after the adsorber bed is practically equal to zero.*

$$\mathbf{m_{Drug-Lab}} = 25\tfrac{\text{mg}}{\text{kg}} * 0.040\tfrac{\text{m}}{\text{h}} * 1010\tfrac{\text{kg}}{\text{m}^3} * 18\,\text{h} = \mathbf{18{,}180\,mg = 18.2\,g = 0.0182\,kg}$$

$$\textbf{Drug Load on AC at Breakthrough} = \frac{m_{\text{Drug}}}{m_{\text{AC}}} = \frac{18200\,\text{mg Drug}}{400\,\text{g AC}} = \mathbf{45.5\frac{mg\ Drug}{g\ AC} = 0.0455\frac{kg\ Drug}{kg\ AC}}$$

c) $\boldsymbol{ScF} = \frac{14.54\,\text{kg}}{0.0182\,\text{kg}} = \mathbf{798.9 \cong 800}$

d) $\boldsymbol{m_{AC-\text{Plant}}} = m_{AC-\text{Lab}} * ScF = 0.40\,\text{kg} * 800 = \mathbf{320\,kg}$

e) $V_{AC-\text{Lab}} = \frac{m_{AC-\text{Lab}}}{\rho_{AC}} = \frac{0.4\,\text{kg} * \text{m}^3}{440\,\text{kg}} = 9.1 * 10^{-4}\,\text{m}^3$

$$V = \frac{d^2 * \pi}{4} * L \rightarrow L_{AC-Lab} = \frac{4 * V_{AC-Lab}}{d_{AC-Lab}^2 * \pi}$$

$$= \frac{4 * 9.1 * 10^{-4}\,\text{m}^3}{0.1^2 * \text{m}^2 * \pi} = 0.116\,\text{m} \rightarrow L_{AC-Lab} = 1.16 * d_{AC-Lab}$$

$$V_{AC-\text{Plant}} = \frac{m_{AC-\text{Plant}}}{\rho_{AC}} = \frac{320\,\text{kg} * \text{m}^3}{440\,\text{kg}} = 0.727\,\text{m}^3$$

$$V_{AC-\text{Plant}} = \frac{d_{AC-\text{Plant}}^2 * \pi}{4} * L_{AC-\text{Lab}} = \frac{d_{AC-\text{Plant}}^2 * \pi}{4} * 1.16 * d_{AC-\text{Plant}} = \frac{1.16 * \pi}{4} * d_{AC-\text{Plant}}^3$$

$$\boldsymbol{d_{AC-\text{Plant}}} = \sqrt[3]{\frac{4 * V_{AC-\text{Plant}}}{1.16 * \pi}} == \sqrt[3]{1.098 * 0.727\,\text{m}^3} = \mathbf{0.928\,m \cong 0.93\,m}$$

$$L_{AC-\mathbf{Plant}} = 1.16 * d_{AC-\text{Plant}} = 1.16 * 0.928\,\text{m} = \mathbf{1.08\,m} \cong \mathbf{1.1\,m}$$

f) $\text{Re} = \frac{d_{\text{Pipe}} * w_{w-\text{Plant}} * \rho_w}{\eta_w}$

$$w_{w-\text{Plant}} = \frac{\dot{V}_{w-\text{Plant}}}{A_{\text{Pipe}}} \quad A_{\text{Pipe}} = \frac{d_{\text{Pipe}}^2 * \pi}{4} \rightarrow w_{w-\text{Plant}} = \frac{4 * \dot{V}_{w-\text{Plant}}}{d_{\text{Pipe}}^2 * \pi}$$

$$\text{Re} = \frac{d_{\text{Pipe}} * 4 * \dot{V}_{w-\text{Plant}} * \rho_w}{\eta_w * d_{\text{Pipe}}^2 * \pi} = \frac{4 * \dot{V}_{w-\text{Plant}} * \rho_w}{\eta_w * d_{\text{Pipe}} * \pi}$$

$$\boldsymbol{d}_{\mathbf{Pipe}} = \frac{4 * \dot{V}_{w-\text{Plant}} * \rho_w}{\text{Re} * \eta_w * \pi} = \frac{4 * 0.8\,\text{m}^3 * 1010\,\text{kg} * s * \text{m}}{1000 * 0.001\,\text{kg} * \pi * \text{h} * \text{m}^3} * \frac{\text{h}}{3600\,\text{s}} = \mathbf{0.286\,m}$$

→ ***Result***

a) **In the production plant, 14.54 kg of active ingredient must be removed from the wastewater by adsorption in the activated carbon bed within 30 days.**
b) **In the laboratory experiment, 18.2 g of active ingredient were adsorbed, which corresponds to a loading at the breakthrough point of 0.0455 kg of active ingredient per kg of activated carbon.**
c) **The scale-up factor is 800.**
d) **For the adsorber bed in the production plant, 320 kg of activated carbon must be used.**
e) **The diameter of the activated carbon bed in the production plant is 0.93 m, its length 1.1 m.**
f) **The inflow pipe must have an inner diameter of at least 0.286 m.**

3.8 Combined Tasks

Task 97

The trans-isomer of 1,3-dichloropropene (t-DCP) can be converted to a mixture of 55wt% trans and 45wt% cis 1,3-dichloropropene (c-DCP) by adding a catalytic amount of bromine. The heat of reaction is negligible. In a PFR, 1 metric t of t-DCP with a c-DCP content of 2.5% is to be processed per hour. For this purpose, t-DCP containing 2.5wt% c-DCP is mixed with a solution of 5 wt% bromine in 1,2-dichloropropane (PDC) at 25 °C and reacted in a PFR to the mentioned trans-cis equilibrium. The mixture is then fed to a distillation column from which 95% of the t-DCP & c-DCP are distilled off as vapor at 108 °C. The bottom fraction of bromine, PDC and DCP is led to the catalyst preparation unit. The DCP vapor is separated into the trans and cis isomers. The head fraction c-DCP is transferred to the product tank. The middle fraction of t-DCP with 2.5wt% c-DCP is led to the educt tank and recycled as an internal recycle stream together with external feed of the same composition back to the reactor. The sludge of small amounts of PDC and by-products is disposed off.

A plant for the production of 1 metric t c-DCP per hour is to be designed from laboratory data. For this purpose, 500 g of a mixture of t-DCP with 2.5wt% c-DCP were mixed with 25 g of a 5 wt% bromine solution in PDC. After 5 minutes, the equilibrium of trans- and cis-form had been established.

The PFR is to be made from a 6" pipe with a wall thickness of 3 mm.

a) What is the scale-up factor for the PFR with respect to the laboratory data?
b) What length must the PFR have?
c) How much heat is required for heating and for the evaporation process of the reaction mixture?
d) How much c-DCP is produced per hour? How large is the internal recycle stream of t-DCP with 2.5wt% c-DCP to the educt tank? How large is the hourly feed of the entire plant for t-DCP with 2.5wt% c-DCP (external feed)?

The following material data are known:

	Br_2	DCP	PDC
Density in kg/m^3	3120	1230	1180
Heat capacity in kJ/(kg * °C)	3.15	1.4	1.5
Evaporation heat $\Delta_V H$ in kJ/mol		33.3	
Molar mass in g/mol		111	

⊗ Solution

→ *Strategy*

a) The ratio of the DCP feed stream to the PFR to the amount of DCP used in the laboratory test, based on the reaction time, is suitable as a scale-up factor.
b) The length of the PFR results from the correspondingly rearranged formula of the cylinder volume. The inner pipe diameter must be used here. The volume is the product of residence time and volume flow. The volume flow is calculated from the total mass flow, divided by its density. The single feed mass flows result from the laboratory experiment and its reaction time, multiplied by the scale-up factor. The average density of the total feed stream results from the individual mass flows, multiplied by their density, divided by the total mass flow (see Sect. 1.2.5).
c) The heat required to heat the stream leaving the PFR is calculated according to Formula 19a. The method for calculating the average heat capacity of the mixture is similar in principle to the determination of the average density of the mixture described under b). The heat required to vaporize 95% of the DCP contained in the mixture results from Formula 24b.

d) The mass flows requested can be determined by means of a mass balance. For this purpose, a flow sheet of the plant is first set up, the available data of the associated streams are considered and the requested streams are calculated by logical consideration of their relationships.

→ ***Calculation***

a) $ScF = \frac{\dot{m}_{\text{DCP-PFR}}}{\dot{m}_{\text{DCP-Lab}}}$

$$\dot{m}_{\text{DCP-PFR}} = 1\frac{\text{t}}{\text{h}} = \frac{1000\,\text{kg}}{3600\,\text{s}} = 0.278\frac{\text{kg}}{\text{s}}$$
$$\dot{m}_{\text{DCP-Lab}} = \frac{0.50\,\text{kg} * \text{min}}{5\,\text{min} * 60\,\text{s}} = 0.00167\frac{\text{kg}}{\text{s}}$$

$$\boldsymbol{ScF} = \frac{0.278}{0.00167} = \mathbf{166.5}$$

b) $V_{\text{PFR}} = \frac{d_{\text{PFR}}^2 * \pi}{4} * L_{\text{PFR}}$

$$L_{\text{PFR}} = \frac{4 * V_{\text{PFR}}}{d_{\text{PFR}}^2 * \pi} \quad d_{\text{PFR}} = 6'' * 0.0254\frac{\text{m}}{''} - 2 * 0.003\,\text{m} = 0.1464\,\text{m}$$

$$V_{\text{PFR}} = \tau * \dot{V}_{\text{PFR}} \quad \dot{V}_{\text{PFR}} = \frac{\dot{m}_{\text{PFR}}}{\rho_{\text{Feed}}} \quad V_{\text{PFR}} = \tau * \frac{\dot{m}_{\text{PFR}}}{\rho_{\text{Feed}}}$$

$$\rho_{\text{Feed}} = \frac{\sum_i (\dot{m}_i * \rho_i)}{\sum_i \dot{m}_i} = \frac{\dot{m}_{\text{DCP}} * \rho_{\text{DCP}} + \dot{m}_{\text{PDC}} * \rho_{\text{PDC}} + \dot{m}_{\text{Br2}} * \rho_{\text{Br2}}}{\dot{m}_{\text{DCP}} + \dot{m}_{\text{PDC}} + \dot{m}_{\text{Br2}}}$$

$$\dot{m}_{\text{DCP-PFR}} * \rho_{\text{DCP}} = 0.278\frac{\text{kg}}{\text{s}} * 1230\frac{\text{kg}}{\text{m}^3} = 341.94\frac{\text{kg}^2}{\text{s*m}^3}$$

$$\dot{m}_{i\text{-PFR}} = \frac{m_{i\text{-Lab}}}{t} * \text{ScF}$$

$$\dot{m}_{\text{PDC-PFR}} = \frac{0.025\,\text{kg}}{300\,\text{s}} * 166.5 * \frac{95\%}{100\%} = 0.0132\frac{\text{kg}}{\text{s}}$$

$$\dot{m}_{\text{PDC-PFR}} * \rho_{\text{PDC}} = 0.0132\frac{\text{kg}}{\text{s}} * 1180\frac{\text{kg}}{\text{m}^3} = 15.58\frac{\text{kg}^2}{\text{s*m}^3}$$

$$\dot{m}_{\text{Br2-PFR}} = \frac{0.025\,\text{kg}}{300\,\text{s}} * 166.5 * \frac{5\%}{100\%} = 0.0007\frac{\text{kg}}{\text{s}}$$

$$\dot{m}_{\text{Br2-PFR}} * \rho_{\text{Br2}} = 0.0007\frac{\text{kg}}{\text{s}} * 3120\frac{\text{kg}}{\text{m}^3} = 2.84\frac{\text{kg}^2}{\text{s} * \text{m}^3}$$

$$\dot{m}_{PFR} = \dot{m}_{DCP-PFR} + \dot{m}_{PDC-PFR} + \dot{m}_{Br2-PFR} = (0.278 + 0.0132 + 0.0007)\frac{kg}{s} = 0.2919\frac{kg}{s}$$

$$\rho_{Feed} = \frac{(341.9 + 15.58 + 2.84)kg^2 * s}{s * m^3 * 0.2919\,kg} = 1234\frac{kg}{m^3}$$

$$\tau = 5\,min * \frac{60\,s}{min} = 300\,s$$

$$\dot{m}_{PFR} = \frac{m_{Lab}}{t} * ScF = \frac{0.5\,kg + 0.025\,kg}{300\,s} * 166.5 = 0.291\frac{kg}{s}$$

$$\boldsymbol{L_{PFR}} = \frac{4 * \tau * \dot{m}_{DCP-PFR}}{d_{PFR}^2 * \pi * \rho_{Feed}} = \frac{4 * 300\,s * 0.291\,kg * m^3}{s * 0.1464^2\,m^2 * \pi * 1234\,kg} = \mathbf{4.21\,m}$$

c) $\dot{Q}_{Heating} = \dot{m}_{PFR} * cp_{PFR} * \left(T_{BoilingPoint} - T_{PFR}\right)$

$$cp_{PFR} = \frac{\sum_i (\dot{m}_i * cp_i)}{\sum_i \dot{m}_i} = \frac{\dot{m}_{DCP} * cp_{DCP} + \dot{m}_{PDC} * cp_{PDC} + \dot{m}_{Br2} * cp_{Br2}}{\dot{m}_{DCP} + \dot{m}_{PDC} + \dot{m}_{Br2}}$$

$$cp_{PFR} = \frac{(0.278 * 1.4 + 0.0132 * 1.5 + 0.0007 * 3.15)\frac{kg * kJ}{s * kg * °C}}{0.2919\frac{kg}{s}} = 1.41\frac{kJ}{kg * °C}$$

$$\dot{\boldsymbol{Q}}_{\mathbf{Heating}} = 0.2919\frac{kg}{s} * 1.14\frac{kJ}{kg * s} * (108 - 25)\,°C = 34.2\frac{kJ}{s} = \mathbf{34.2\,kW}$$

$$\dot{Q}_{DCP-Evap} = \dot{m}_{DCP-PFR} * \frac{95\%}{100\%} * \Delta_V H$$

$$\dot{\boldsymbol{Q}}_{\mathbf{DCP-Evap}} = 0.278\frac{kg}{s} * 0.95 * \frac{33.3\frac{kJ}{mol}}{0.111\frac{kg}{mol}} = 79.2\frac{kJ}{s} = \mathbf{79.2\,kW}$$

$$\dot{\boldsymbol{Q}}_{\mathbf{Total}} = \dot{Q}_{Heating} + \dot{Q}_{DCP-Evap} = (34.2 + 79.2)\,kW = \mathbf{113.4\,kW}$$

d)

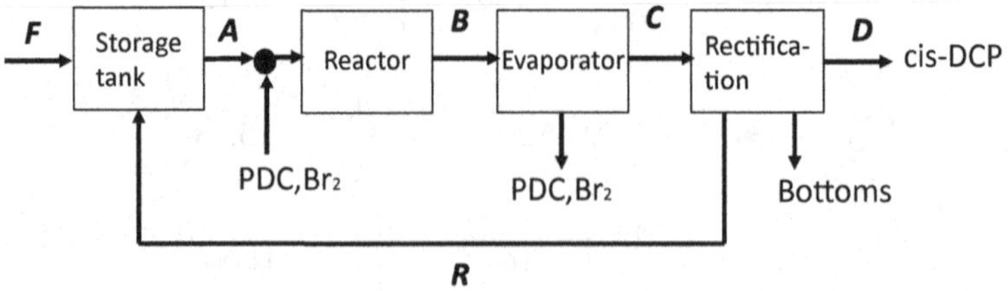

Flow A:

$$\dot{m}_{DCP-A} = 1\frac{t}{h} \quad C_{DCP-trans-A} = 97.5\% \quad C_{DCP-cis-A} = 2.5\%$$

$$\rightarrow \dot{m}_{\text{DCP-trans-}A} = 0.975\frac{\text{t}}{\text{h}}\quad \dot{m}_{\text{DCP-cis-}A} = 0.025\frac{\text{t}}{\text{h}}$$

Flow B:

$$\dot{m}_{\text{DCP-}B} = 1\frac{\text{t}}{\text{h}}\quad C_{\text{DCP-trans-}B} = 55.0\%\quad C_{\text{DCP-cis-}B} = 45.0\%$$

$$\rightarrow \dot{m}_{\text{DCP-trans-}B} = 0.55\frac{\text{t}}{\text{h}}\quad \dot{m}_{\text{DCP-cis-}B} = 0.45\frac{\text{t}}{\text{h}}$$

Flow C:

$$\dot{m}_{\text{DCP-C}} = 0.95\frac{\text{t}}{\text{h}}\quad C_{\text{DCP-trans-}C} = 55\%\quad C_{\text{DCP-cis-}C} = 45\%$$

$$\rightarrow \dot{m}_{\text{DCP-trans-}C} = 0.5225\frac{\text{t}}{\text{h}}\quad \dot{m}_{\text{DCP-cis-}C} = 0.4275\frac{\text{t}}{\text{h}}$$

FlowR:

$$\dot{m}_{\text{DCP-}R} = \dot{m}_{\text{DCP-trans-}R} + \dot{m}_{\text{DCP-cis-}R}$$

$$\dot{m}_{\text{DCP-trans-}R} = \dot{m}_{\text{DCP-trans-}C} = 0.5225\frac{\text{t}}{\text{h}}$$

$$\dot{m}_{\text{DCP-cis-}R} = 2.5\% = 0.5225\frac{\text{t}}{\text{h}} * \frac{2.5\%}{97.5\%} = 0.0134\frac{\text{t}}{\text{h}}$$

$$\dot{m}_{\textbf{DCP-}R} = (0.5225 + 0.0134)\frac{\text{t}}{\text{h}} = \mathbf{0.536\frac{t}{h}}$$

Flow D:

$$\dot{m}_{\textbf{DCP-cis-}D} = \dot{m}_{\text{DCP-cis-}C} - \dot{m}_{\text{DCP-cis-}R} = (0.4275 - 0.0134)\frac{\text{t}}{\text{h}} = \mathbf{0.414\frac{t}{h}}$$

Flow F:

$$\dot{m}_{\textbf{DCP-}F} = \dot{m}_{\text{DCP-}A} - \dot{m}_{\text{DCP-}R} = (1.0 - 0.536)\frac{\text{t}}{\text{h}} = \mathbf{0.464\frac{t}{h}}$$

→ ***Result***

a) **The scale-up factor is 166.5 .**
b) **The PFR is 4.21 m long.**
c) **The heat flow required to heat the reaction mixture from 25 °C to 108 °C is 34.2 kW. The evaporation of 95% of the DCP requires a heat flow of 79.2 kW. The total heat flow required is 113.4 kW.**

d) **0.414 t/h of cis-DCP are produced. The internal recycle stream of 97.5wt% trans-DCP and 2.5wt% cis-DCP is 0.536 t/h. 0.464 t of a trans-cis-DCP mixture of the composition mentioned must be fed to the plant per hour. The difference between the feed stream and the cis-DCP product stream represents the loss by the rectification bottoms flow.**

Task 98

In a production plant, 450 m^3 of exhaust gas (T = 150 °C; p = 2.5 bar) is produced per hour. The gas consists of 82 vol% nitrogen (M = 28 g/mol; cp = 30 J/[mol * °C]), 3.5 vol% hydrogen (M = 2 g/mol; cp = 27.5 J/[mol * °C]), 10 vol% carbon monoxide (M = 28 g/mol; $\Delta_f H = -110.5$ kJ/mol; cp = 29.5 J/[mol * °C]) and 4.5 vol% propane (M = 44.1 g/mol; $\Delta_f H = -103.8$ kJ/mol; cp = 73.9 J/[mol * °C]). The combustible components of the exhaust gas are completely burned to water vapor ($\Delta_f H = -241.8$ kJ/mol) and carbon dioxide ($\Delta_f H = -393.5$ kJ/mol) by adding air at a temperature of 20 °C and a pressure of 2.5 bar with a molar oxygen excess of 15% in a boiler plant consisting of the combustion chamber, followed by an evaporator (air → 21 vol% oxygen [cp = 30 J/[mol * °C]), 79 vol% nitrogen (cp = 30 J/[mol * °C])). The cooled exhaust gases of the combustion leave the evaporator at a temperature of 280 °C. In the evaporator, saturated steam ($\Delta_v H = 2170$ kJ/kg) at a temperature of 130 °C and 2.7 bar is generated. The heat transfer coefficient is Kw = 350 W/(m^2 * °C). The feed temperature of the boiler feed water stored in a tank under atmospheric pressure ($cp_W = 4.2$ kJ/[kg * °C]; $\rho_W = 965$ kg/m^3) is 90°C.

In the calculations, the heat losses through walls are to be neglected to a first approximation.

a) What volume flow of air must be supplied for combustion under these conditions?
b) How much reaction heat is released per unit of time? (The formation enthalpies are assumed to be independent of temperature.)
c) At what temperature does the combustion gas enter the evaporator?
d) What heat transfer area must the evaporator have? (To calculate the mean logarithmic temperature difference, a water-side temperature of 130°C is to be assumed.)
e) What is the mass flow of the saturated steam generated that leaves the evaporator?
f) What is the thermal efficiency of the combined unit of combustion/evaporator, if it is defined as the ratio of the condensation heat output of the 130 °C steam to the reaction heat of the exhaust gas combustion?
g) What power must the boiler feed water pump have if the storage tank is 15 m below the feed of the evaporator, is operated at atmospheric pressure and has a friction loss of 0.15 bar under the operating conditions? The overall efficiency of the pump and motor is 65%.

⊗ **Solution**

→ ***Strategy***

a) The partial volume flows of the combustion gases hydrogen, carbon monoxide, propane and nitrogen result from the total flow of the exhaust gas in accordance with the respective percentage share. Using the general gas law (formula 2), the corresponding molar flows are calculated. From the chemical equations of combustion it follows which molar flow of oxygenis required for the combustion of hydrogen, carbon monoxide and propane with an oxygen excess of 15%. With the gas law (formula 2), the volume flow of oxygen for 20 °C and 2.5 bar is calculated. The volume flow of the combustion air $\dot{V}_{Air}$ is calculated by means of the rule of three for an oxygen content of 21 vol%.

b) The total heat flow from the combustion results from the sum of the heat flows from the combustion of hydrogen, carbon monoxide and propane, which are calculated according to formula 27c from the individual molar flows and the associated reaction enthalpies. The reaction enthalpies result from the formation enthalpies according to formula 28, taking into account that the formation enthalpy of the elements is zero by definition.

c) Both the exhaust gas and the supplied air are heated by the heat output of the combustion reactions $\dot{Q}_{total}$to the desired final temperature = evaporator inlet temperature T_{in}, where the initial temperature of the exhaust gas is at T_{Exh} = 150 °C, while the supplied combustion air has a temperature of T_{Air} = 20 °C. (Since the molar heat capacity of oxygen and nitrogen is practically identical, the total molar number of combustion air can be used here.) These heat flows are calculated according to Formula 21b. The relationship of the sum of the heat quantities absorbed by heating the exhaust gas stream and the combustion air is rearranged to the final temperature T_{in}.

d) The necessary heat transfer area results from the rearranged Formula 30. The heat flow corresponds to the heat, released by cooling the combustion gas according to Formula 21b with the difference in exhaust gas temperature before the evaporator (T_{in}) to that after the evaporator (T_{ex}). The calculation of the mean logarithmic temperature difference is described in Sect. 1.2.5.

e) The heat consumption of the evaporator consists of the amount required to heat the boiler feed water to the temperature of the saturated steam (Formula 19a) and the water vaporization (Formula 24b). The thus composed relationship is transformed to the mass flow of the boiler feed water ($\dot{m}_W$), which corresponds to the mass flow of saturated steam.

f) The thermal efficiency is, according to the definition, the quotient of the condensation heat of the generated steam stream and the total heat generated by combustion per unit of time.

g) The theoretically necessary pump power is calculated according to Formula 39a. The real power required results from the division of this value by the overall efficiency. The total pumping heigth results from Formula 38a, for which the pressureheigth and the friction heigth are calculated by Formula 38b.

→ ***Calculation***

a) *For 2.5 bar and 150 °C:*

$$\dot{n} = \frac{p * \dot{V}}{R * T} = \frac{2.5\,\text{bar} * \text{mol} * \text{K}}{8.315 * 10^{-5}\text{bar} * \text{m}^3 * 423\,\text{K}} * \dot{V} = 71.1\frac{\text{mol}}{\text{m}^3} * \dot{V}$$

$$\dot{V}_{H_2} = 450\frac{\text{m}^3}{\text{h}} * \frac{3.5\%}{100\%} = 15.75\frac{\text{m}^3}{\text{h}}$$

$$\dot{n}_{H_2} = 71.1\frac{\text{mole}}{\text{m}^3} * 15.75\frac{\text{m}^3}{\text{h}} = 1120\frac{\text{mol}}{\text{h}} = 0.311\frac{\text{mol}}{\text{s}}$$

$$\dot{V}_{CO} = 450\frac{\text{m}^3}{\text{h}} * \frac{10\%}{100\%} = 45.0\frac{\text{m}^3}{\text{h}}$$

$$\dot{n}_{CO} = 71.1\frac{\text{mol}}{\text{m}^3} * 45.0\frac{\text{m}^3}{\text{h}} = 3200\frac{\text{mol}}{\text{h}} = 0.889\frac{\text{mol}}{\text{s}}$$

$$\dot{V}_{Prop} = 450\frac{\text{m}^3}{\text{h}} * \frac{4.5\%}{100\%} = 20.25\frac{\text{m}^3}{\text{h}}$$

$$\dot{n}_{Prop} = 71.1\frac{\text{mol}}{\text{m}^3} * 20.25\frac{\text{m}^3}{\text{h}} = 1440\frac{\text{mol}}{\text{h}} = 0.400\frac{\text{mol}}{\text{s}}$$

$$\dot{V}_{N_2exh} = 450\frac{\text{m}^3}{\text{h}} * \frac{82\%}{100\%} = 369.0\frac{\text{m}^3}{\text{h}}$$

$$\dot{n}_{N_2exh} = 71.1\frac{\text{mol}}{\text{m}^3} * 369.0\frac{\text{m}^3}{\text{h}} = 26,236\frac{\text{mol}}{\text{h}} = 7.288\frac{\text{mol}}{\text{s}}$$

$$H_2 + {}^1\!/\!_2O_2 \rightarrow H_2O \rightarrow \frac{1}{2} * \dot{n}_{H_2} = \dot{n}_{O_2}\,\textit{for combustion of hydrogen}$$

$$CO + {}^1\!/\!_2O_2 \rightarrow CO_2 \rightarrow \frac{1}{2} * \dot{n}_{CO} = \dot{n}_{O_2}\,\textit{for combustion of CO}$$

$$C_3H_8 + 5O_2 \rightarrow 3CO_2 + 4H_2O \rightarrow 5 * \dot{n}_{Prop} = \dot{n}_{O_2}\,\textit{for combustion of propane}$$

$$\dot{n}_{O_2-theor} = \frac{1}{2} * \dot{n}_{H_2} + \frac{1}{2} * \dot{n}_{CO} + 5 * \dot{n}_{Prop} = \frac{1}{2} * (1120 + 3200)\frac{\text{mol}}{\text{h}} + 5 * 1440\frac{\text{mol}}{\text{h}} = 9360\frac{\text{mol}}{\text{h}}$$

$$\dot{n}_{O_2-real} = 1.15 * \dot{n}_{O_2-theor} = 10{,}764\frac{\text{mol}}{\text{h}} = 3.0\frac{\text{mol}}{\text{s}}$$

$$\dot{V}_{O_2} = \frac{\dot{n}_{O_2} * R * T_{air}}{p_{air}} = \frac{10{,}764\,\text{mol} * 8.315 * 10^{-5}\,\text{bar} * \text{m}^3 * 293\,\text{K}}{\text{h} * \text{mol} * \text{K} * 2.5\,\text{bar}} = 104.9\frac{\text{m}^3}{\text{h}}$$

$$\dot{V}_{O_2} = 104.9\frac{\text{m}^3}{\text{h}} \rightarrow 21\,\text{vol\%} \quad \dot{V}_L = 100\,\text{vol\%} = 104.9\frac{\text{m}^3}{\text{h}} * \frac{100\%}{21\%} = \mathbf{500\frac{m^3}{h}}$$

b) $\dot{Q}_{total} = \dot{Q}_{H_2} + \dot{Q}_{CO} + \dot{Q}_{Prop} \quad \dot{Q}_i = -\Delta\dot{n}_i * \Delta_R H_i \quad \Delta_R H_i = \sum_i (\nu_i * \Delta_f H_i)$

$$\Delta_R H_{H_2} = \nu_{H_2O} * \Delta_f H_{H_2O} = 1 * \left(-241.8\frac{\text{kJ}}{\text{mol}}\right) = -241.8\frac{\text{kJ}}{\text{mol}}$$

$$\dot{Q}_{H_2} = -\dot{n}_{H_2} * \Delta_R H_{H_2} = -1120\frac{\text{mol}}{\text{h}} * \left(-241.8\frac{\text{kJ}}{\text{mol}}\right) = 270{,}768\frac{\text{kJ}}{\text{h}} = \frac{270{,}768\,\text{kJ} * \text{h}}{\text{h} * 3600\,\text{s}} = 75.2\,\text{kW}$$

$$\Delta_R H_{CO} = \nu_{CO} * \Delta_f H_{CO} + \nu_{CO_2} * \Delta_f H_{CO_2} = -1 * \left(-110.5\frac{\text{kJ}}{\text{mol}}\right) + 1 * \left(-393.5\frac{\text{kJ}}{\text{mol}}\right) = -283.0\frac{\text{kJ}}{\text{mol}}$$

$$\dot{Q}_{CO} = -\dot{n}_{CO} * \Delta_R H_{CO} = -3200\frac{\text{mol}}{\text{h}} * \left(-283.0\frac{\text{kJ}}{\text{mol}}\right) = 905{,}600\frac{\text{kJ}}{\text{h}} = \frac{905{,}600\,\text{kJ} * \text{h}}{\text{h} * 3600\,\text{s}} = 251.6\,\text{kW}$$

$$\Delta_R H_{Prop} = \nu_{Prop} * \Delta_f H_{Prop} + \nu_{CO_2} * \Delta_f H_{CO_2} + \nu_{H_2O} * \Delta_f H_{H_2O}$$

$$\Delta_R H_{Prop} = -1 * \left(-103.8\frac{\text{kJ}}{\text{mol}}\right) + 3 * \left(-393.5\frac{\text{kJ}}{\text{mol}}\right) + 4 * \left(-241.8\frac{\text{kJ}}{\text{mol}}\right) = -2044\frac{\text{kJ}}{\text{mol}}$$

$$\dot{Q}_{Prop} = -\dot{n}_{Prop} * \Delta_R H_{Prop} = -1440\frac{\text{mol}}{\text{h}} * \left(-2044\frac{\text{kJ}}{\text{mol}}\right)$$

$$= 2{,}943{,}000\frac{\text{kJ}}{\text{h}} = \frac{2{,}943{,}000\,\text{kJ*h}}{\text{h} * 3600\,\text{s}} = 817.5\,\text{kW}$$

$$\dot{\mathbf{Q}}_{\mathbf{total}} = (270{,}768 + 905{,}600 + 2{,}943{,}000)\frac{\text{kJ}}{\text{h}} = \mathbf{4119\frac{MJ}{h}} = \mathbf{1144\,kW}$$

c) $\dot{Q}_{total} = \Delta\dot{Q}_{exh} + \Delta\dot{Q}_{air}$

$$\Delta\dot{Q}_{exh} = \left(\dot{n}_{H_2} * cp_{H_2} + \dot{n}_{CO} * cp_{CO} + \dot{n}_{Prop} * cp_{Prop} + \dot{n}_{N_2exh} * cp_{N_2}\right) * (T_{in} - T_{exh})$$

$$\Delta\dot{Q}_{exh} = (0.311 * 27.5 + 0.889 * 29.5 + 0.400 * 73.9 + 7.29 * 30)\frac{\text{J}}{\text{s} * \text{mol} * {}^\circ\text{C}} * (T_{in} - 150\,{}^\circ\text{C})$$

$$\Delta\dot{Q}_{exh} = 283.0\frac{\text{J}}{\text{s} * \text{mol} * {}^\circ\text{C}} * (T_{in} - 150\,{}^\circ\text{C}) = 0.283\frac{\text{kJ}}{\text{s} * \text{mol} * {}^\circ\text{C}} * T_{in} - 42.45\frac{\text{kJ}}{\text{s}}$$

$$\Delta\dot{Q}_{air} = \left(\dot{n}_{O_2} * cp_{O_2} + \dot{n}_{N_2air} * cp_{N_2}\right) * (T_{in} - T_{air})$$

$$\text{with } cp_{O_2} = cp_{N_2} = cp_{air} = 0.03\,\text{kJ}/(\text{mol} * °\text{C}) \rightarrow$$

$$\Delta\dot{Q}_{air} = \left(\dot{n}_{O_2} * cp_{O_2} + \dot{n}_{N_2\text{air}} * cp_{N_2}\right) * (T_{\text{in}} - T_{air}) = \dot{n}_{air} * cp_{air} * (T_{\text{in}} - T_{air})$$

$$\dot{n}_{air} = \dot{n}_{O_2} + \dot{n}_{N_2} \quad \dot{n}_{O_2} = 3.0\frac{\text{mol}}{\text{s}}$$

$$\dot{n}_{N_2} = \dot{n}_{O_2} * \frac{79\%}{21\%} = 11.3\frac{\text{mol}}{\text{s}} \rightarrow \dot{n}_{air} = 14.3\frac{\text{mol}}{\text{s}}$$

$$\Delta\dot{Q}_{air} = 14.3\frac{\text{mol}}{\text{s}} * 0.03\frac{\text{kJ}}{\text{mol} * °\text{C}} * (T_{\text{in}} - 20\,°\text{C}) = 0.429\frac{\text{kJ}}{\text{s} * °\text{C}} * T_{\text{in}} - 8.58\frac{\text{kJ}}{\text{s}}$$

$$1144\frac{\text{kJ}}{\text{s}} = (0.283 + 0.429)\frac{\text{kJ}}{\text{s} * °\text{C}} * T_{\text{in}} - (42.45 + 8.58)\frac{\text{kJ}}{\text{s}} = 0.712\frac{\text{kJ}}{\text{s} * °\text{C}} * T_{\text{in}} - 51\frac{\text{kJ}}{\text{s}}$$

$$\boldsymbol{T_{\text{in}}} = \frac{1195\,\text{kJ} * \text{s} * °\text{C}}{\text{s} * 0.712\,\text{kJ}} = \mathbf{1678\,°C}$$

d) $\Delta\dot{Q}_{\text{evap}} = Kw * A * \overline{\Delta T}_M \quad \rightarrow \quad A = \frac{\Delta\dot{Q}_{\text{evap}}}{Kw * \overline{\Delta T}_M}$

$$\Delta\dot{Q}_{\text{evap}} = \sum_i (\dot{n}_i * cp_i) * (T_{\text{in}} - T_{\text{ex}})$$

$$\Delta\dot{Q}_{\text{evap}} = \left[\left(\dot{n}_{O_2} + \dot{n}_{N_2}\right) * cp_{N2} + \dot{n}_{H_2} * cp_{H_2} + \dot{n}_{\text{CO}} * cp_{\text{CO}} + \dot{n}_{\text{Prop}} * cp_{\text{Prop}}\right] * (T_{\text{in}} - T_{\text{ex}})$$

$$\left(\dot{n}_{O_2} + \dot{n}_{N_2}\right) = \dot{n}_{N_2\text{exh}} + \dot{n}_{air} = (7.3 + 14.3)\frac{\text{mol}}{\text{s}} = 21.6\frac{\text{mol}}{\text{s}}$$

$$\Delta\dot{Q}_{\text{evap}} = (21.6 * 0.03 + 0.311 * 0.0275 + 0.889 * 0.0295 + 0.40 * 0.0739)\frac{\text{mol} * \text{kJ}}{\text{s} * \text{mol} * °\text{C}} * (1678 - 280)\,°\text{C}$$

$$\Delta\dot{Q}_{\text{evap}} = 0.713\frac{\text{kJ}}{\text{s} * °\text{C}} * 1398\,°\text{C} = 997\frac{\text{kJ}}{\text{s}} = 997\,\text{kW}$$

$$\overline{\Delta T}_M = \frac{\Delta T_1 - \Delta T_2}{\ln\frac{\Delta T_1}{\Delta T_2}} \quad \text{with } \Delta T_1 = (1678 - 130)\,°\text{C} = 1548°\text{C und}$$

$$\Delta T_2 = (280 - 130)\,°\text{C} = 150\,°\text{C}$$

$$\overline{\Delta T}_M = \frac{(1548 - 150)\,°\text{C}}{\ln\frac{1548}{150}} = 599\,°\text{C}$$

$$\boldsymbol{A} = \frac{997\,\text{kW} * \text{m}^2 * °\text{C}}{\text{s} * 0.35\,\text{kW} * 599\,°\text{C}} = \mathbf{4.76\,m^2 \cong 5\,m^2}$$

e) $\dot{Q} = \dot{m}_W * [cp_W * (T_{\text{steam}} - T_{W-\text{in}}) + \Delta_V H \rightarrow \dot{Q} = \Delta\dot{Q}_{\text{evap}}$
$\dot{m}_W = \dot{m}_{\text{steam}}$

$$\dot{m}_{\text{steam}} = \frac{\Delta\dot{Q}_{\text{evap}}}{cp_W * (T_{\text{steam}} - T_{W-\text{in}}) + \Delta_V H}$$

$$\boldsymbol{m_{\text{steam}}} = \frac{997\,\text{kW}}{4.2\frac{\text{kJ}}{\text{kg}*{}^\circ\text{C}} * (130 - 90)\,{}^\circ\text{C} + 2170\frac{\text{kJ}}{\text{kg}}} = \mathbf{0.426\frac{kg}{s}} = \mathbf{1534\frac{kg}{h}}$$

f) $\boldsymbol{\eta_{\text{thermal}}} = \frac{\dot{Q}_{\text{steam}}}{\dot{Q}_{\text{total}}}$

$$\dot{Q}_{\text{steam}} = \dot{m}_{\text{steam}} * \Delta_V H = 0.426\frac{\text{kg}}{\text{s}} * 2170\frac{\text{kJ}}{\text{kg}} = 924.5\frac{\text{kJ}}{\text{s}}$$

$$\boldsymbol{\eta_{\text{thermal}}} = \frac{\dot{Q}_{\text{steam}}}{\dot{Q}_{\text{total}}} = \frac{924.5\,\text{kW}}{1144\,\text{kW}} = \mathbf{0.808 \rightarrow 81\,\%}$$

g) $P = \dot{m}_W * g * H \quad \text{with} \quad H = h_{\text{geo}} + h_p + h_f$

$$h_{\text{geo}} = 15.0\,\text{m}$$

$$h_p = \frac{\Delta p_p}{\rho * g} = \frac{(2.7 - 1)\,\text{bar} * \text{m}^3 * \text{s}^2}{965\,\text{kg} * 9.81\,\text{m}} = \frac{1.7 * 10^5\,\text{kg} * \text{m}^3 * \text{s}^2}{\text{s}^2 * \text{m} * 965\,\text{kg} * 9.81\,\text{m}} = 18.0\,\text{m}$$

$$h_f = \frac{\Delta p_f}{\rho * g} = \frac{0.15\,\text{bar} * \text{m}^3 * \text{s}^2}{965\,\text{kg} * 9.81\,\text{m}} = \frac{0.15 * 10^5\,\text{kg} * \text{m}^2 * \text{s}^2}{\text{s}^2 * \text{m} * 965\,\text{kg} * 9.81\,\text{m}} = 1.58\,\text{m} \cong 1.6\,\text{m}$$

$$H = (15.0 + 18.0 + 1.6)\,\text{m} = 34.6\,\text{m}$$

$$P_{\text{theor}} = 0.426\frac{\text{kg}}{\text{s}} * 9.81\frac{\text{m}}{\text{s}^2} * 34.6\,\text{m} = 144.6\frac{\text{kg} * \text{m}^2}{\text{s}^3} \cong 145\,\text{W}$$

$$\boldsymbol{P_{\text{real}}} = \frac{P_{\text{theor}}}{\eta} = \frac{145\,\text{W}}{0.65} = \mathbf{223\,W \cong 0.25\,kW}$$

→ ***Result***

a) **The necessary volume flow of the combustion air (20 °C; 2.5 bar) is 500 m³/h.**
b) **The heat released by the combustion is 4119 MJ/h = 1144 kW.**
c) **The gas inlet temperature into the evaporator is 1678 °C**
d) **The necessary heat transfer area of the steam generator is $4.76\,\text{m}^2 \cong 5\,\text{m}^2$.**
e) **The mass flow of the generated 130 °C saturated steam is 0.426 kg/s = 1534 kg/h.**

f) **The thermal efficiency is 0.808 (≅ 81 %). However, heat losses through the wall/insulation of the plant are not taken into account here, so it would be lower in reality.**
g) **The motor of the boiler feed water pump must have a power input of 223 W ≅ 0.25 kW.**

Exercise 99

A decomposition reaction A → B + C is carried out in a CSTR of a useful volume of 2 m^3 and a total cooling area of 7.0 m^2. The reaction enthalpy is $\Delta_R H$ = -80 kJ/mol. The rate constant at 20 °C is 7.15 * 10^{-5} s^{-1} with an activation energy E_A = 40 kJ/mol. The volumetric flow of the feed stream of 6 m^3/h with a density of 900 kg/m^3 and a temperature of 15 °C has a concentration of A of 4 mol/L. Its heat capacity is 1.6 kJ/(kg * °C). The cooling area is supplied with water (cp = 4.2 kJ/(kg * °C)) at an inlet temperature of 20 °C and a mean temperature of 30 °C. The mean temperature is approximately defined as $\overline{T}_{m-CW} = (T_{\text{in}} + T_{\text{ex}})/2$. The heat transfer coefficient of the cooling area is 600 W/(m^2 * °C).

a) What operating point (temperature and heat flow) is achieved?
b) What is the rate constant in the operating point?
c) What is the concentration of A in the reactor outlet and the corresponding conversion in the operating point?
d) How large must the cooling water flow be in the operating point?

⊗ Solution

→ *Strategy*

The operating point is where the heat power generated by the reaction $\dot{Q}_R$ is equal to the heat power removed by cooling. The removed heat power is the sum of the heat power used to heat the feed flows $\dot{Q}_F$ to the reactor temperature and the heat flow removed by water cooling $\dot{Q}_{CW}$. Since the heat power of the reaction increases exponentially with increasing reactor temperature, while the heat consumption and heat dissipation increase linearly with the reactor temperature, a graphical solution, similar to that described in Sect. 2.4.7, is recommended. For this purpose, it is advisable to create a table that lists the values of the following quantities as a function of the reactor temperature: rate constant k (formula 17b), $\Delta\dot{n}_A$ (formula 13e), $\dot{Q}_R$ (formula 27c), $\dot{Q}_F$ (formula 19a), $\dot{Q}_{CW}$ (formula 30), $\dot{Q}_F + \dot{Q}_{CW}$. From this a diagram is created with $\dot{Q}_R$ as well as $\dot{Q}_F$,$\dot{Q}_{CW}$ and ($\dot{Q}_F + \dot{Q}_{CW}$) as a function of the reactor temperature. The operating point is at the intersection of the two curves.

The conversion of A is calculated using formula 9b with $\Delta\dot{n}_A$ at the operating point.

The necessary amount of cooling water at the operating point results from formula 19a with $\dot{Q}_{CW}$.

→ ***Calculation***

1. *Rate constant and reaction heat power as a function of reactor temperature:*

$$\dot{Q}_R = -\Delta\dot{n}_A * \Delta_R H$$

$$\Delta\dot{n}_A = \dot{n}_{A_o} - \dot{n}_A$$

$$\dot{n} = \dot{V} * c$$

$$\dot{Q}_R = -\dot{V} * \left(c_{A_o} - c_A\right) * \Delta_R H$$

$$c_A = \frac{c_{A_o}}{1 + k * \tau}$$

$$\dot{Q}_R = -\dot{V} * c_{A_o} * \left(1 - \frac{1}{1 + k * \tau}\right) * \Delta_R H$$

$$\dot{V} = \frac{6\mathrm{m}^3}{\mathrm{h}} * \frac{\mathrm{h}}{3600\,\mathrm{s}} = 0.001667\frac{\mathrm{m}^3}{\mathrm{s}}$$

$$c_{A_o} = \frac{4\,\mathrm{mol}}{\mathrm{L}} = 4000\frac{\mathrm{mol}}{\mathrm{m}^3}$$

$$\dot{n}_{A_o} = \dot{V} * c_{A_o} = 0.001667\frac{\mathrm{m}^3}{\mathrm{s}} * 4000\frac{\mathrm{mol}}{\mathrm{m}^3} = 6.67\frac{\mathrm{mol}}{\mathrm{s}}$$

$$\tau = \frac{V_R}{\dot{V}} = \frac{2\,\mathrm{m}^3 * \mathrm{s}}{0.001667\,\mathrm{m}^3} = 1200\,\mathrm{s} \quad \Delta_R H = -80\,\mathrm{kJ/mol}$$

$$\dot{Q}_R = -0.001667\frac{\mathrm{m}^3}{\mathrm{s}} * 4000\frac{\mathrm{mol}}{\mathrm{m}^3} * \left(-80\frac{\mathrm{kJ}}{\mathrm{mol}}\right) * \left(1 - \frac{1}{1 + k * 1200\,\mathrm{s}}\right)$$

$$\dot{Q}_R = 533.4\frac{\mathrm{kJ}}{\mathrm{s}} * \left(1 - \frac{1}{1 + k * 1200\,\mathrm{s}}\right)$$

$$k \text{ as a function of } T \rightarrow k_{T2} = k_{T1} * e^{\frac{E_A}{R} * \left(\frac{1}{T_1} - \frac{1}{T_2}\right)}$$

For $T_1 = 20\,°\mathrm{C} = 293{,}15\,\mathrm{K}$

$$k = 7.15 * 10^{-5}\,\mathrm{s}^{-1} * e^{\frac{40\,\mathrm{kJ} * \mathrm{mol} * \mathrm{K}}{\mathrm{mol} * 0.008315 * \mathrm{kJ}} * \left(\frac{1}{293.15\,\mathrm{K}} - \frac{1}{T_R}\right)}$$

$$\dot{Q}_R = 533.4\frac{\mathrm{kJ}}{\mathrm{s}} * \left(1 - \frac{1}{1 + 0.0858 * e^{4810.6\,\mathrm{K} * \left(\frac{1}{293.15\,\mathrm{K}} - \frac{1}{T_R}\right)}}\right)$$

With this, k and $\dot{Q}_R$ can be calculated as a function of the reactor temperature T_R.

2. *Heat consumption by the feed as a function of reactor temperature:*

$$\dot{Q}_{\text{Feed}} = \dot{m}_{\text{Feed}} * cp_{\text{Feed}} * (T_R - T_{\text{Feed}})$$

$$\dot{m}_{\text{Feed}} = \dot{V} * \rho_{\text{Feed}} = 0.001667\frac{\text{m}^3}{\text{s}} * 900\frac{\text{kg}}{\text{m}^3} = 1.50\frac{\text{kg}}{\text{s}}$$

$$\dot{Q}_{\text{Feed}} = 1.50\frac{\text{kg}}{\text{s}} * 1.6\frac{\text{kJ}}{\text{kg} * {}^\circ\text{C}} * (T_R - 15\,^\circ\text{C}) = 2.4\frac{\text{kJ}}{\text{s} * {}^\circ\text{C}} * (T_R - 15\,^\circ\text{C})$$

With this, the amount of heat absorbed by the feed can be calculated as a function of the reactor temperature T_R.

3. *Heat removal by cooling water:*

$$\dot{Q}_{CW} = Kw * A * (T_R - T_{CW}) = 0.6\frac{\text{kJ}}{\text{s} * \text{m}^2 * {}^\circ\text{C}} * 7\,\text{m}^2 * (T_R - 30\,^\circ\text{C}) = 4.2\frac{\text{kJ}}{\text{s} * {}^\circ\text{C}} * (T_R - 30\,^\circ\text{C})$$

With this, the amount of heat removed by the cooling water can be calculated as a function of the reactor temperature T_R.

A table of the reactor temperature (optionally 20°C – 80°C with 20°C steps) with the associated rate constant k, concentration c_A and the heat flows $\dot{Q}_R$, $\dot{Q}_{\text{Feed}}$, $\dot{Q}_{KW}$ is prepared and a diagram $\dot{Q}_R$ as well as the sum $(\dot{Q}_{\text{Feed}} + \dot{Q}_{CW})$ *as a function of temperature is created. The operating point is at the intersection of the two curves.*

Example of a corresponding table calculation sheet:

k-20°C/(1/s)	7.15E-05	Ea/(kJ/mol)	40
nAo/(mol/s)	6.67	ΔRH/(kJ/mol)	80
Mass flowFeed/(kg/s)	1.5	Time used τ/s	1200
T-Feed/°C	15	cp feed/(kJ/[kg*°C])	1.6
Kw/(kW/[m2 *°C])	0.6	A-cooling/m²	7.0
Tm cooling water/°C	30		

T	T	k	ΔnA	Q-Reaction	Q-Feed	Q-Cooling	Q-Discharge
°C	K	1/s	mol/s	kJ/s	kJ/s	kJ/s	kJ/s
10	283.15	4.01E-05	0.31	24	-12	-84	-96
30	303.15	1.23E-04	0.86	69	36	0	36
40	313.15	2.04E-04	1.31	105	60	42	102
50	323.15	3.28E-04	1.88	151	84	84	168
60	333.15	5.13E-04	2.54	203	108	126	234
70	343.15	7.81E-04	3.23	258	132	168	300
80	353.15	1.16E-03	3.88	311	156	210	366

Associated heat capacity vs. reactor temperature diagram:

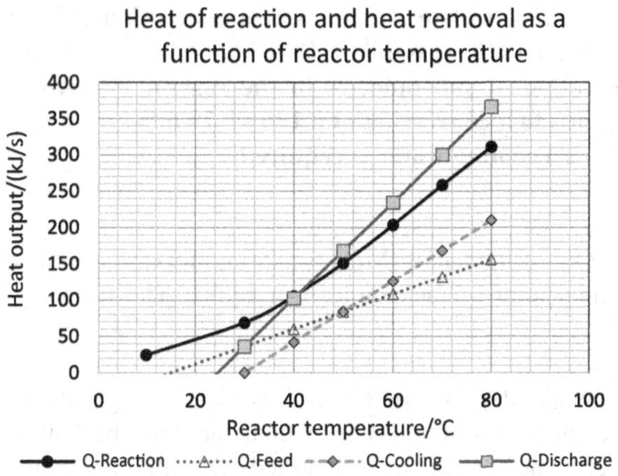

a) *The intersection of the heat of reaction curve and the total heat removal curve is at a heat generation of* ***100 kW*** *and a reactor temperature of* ***40 °C***. *This is the operating point.*

b) *The rate constant at a reactor temperature of 40 °C can be calculated using the previously given formula, but was already calculated in the table with* $\boldsymbol{k = 2.04 * 10^{-4} s^{-1}}$.

c) From the table, $\Delta\dot{n}_A = 1.31$ mol/s can be taken for the operating point at 40 °C.

$$c_A = \frac{n_A}{\dot{V}} = \frac{\dot{n}_{A_0} - \Delta\dot{n}_A}{\dot{V}} = \frac{(6.67 - 1.31)\,\text{mol} * \text{s}}{\text{s} * 0.001667\,\text{m}^3} = 3215\frac{\text{mol}}{\text{m}^3} = \mathbf{3.215\frac{mol}{L}}$$

$$X_A = \frac{\dot{n}_{A_0} - \dot{n}_A}{\dot{n}_{A_0}} = \frac{\Delta\dot{n}_A}{\dot{n}_{A_0}}$$

$$X_A = \frac{1.31\,\text{mol} * \text{s}}{\text{s} * 6.67\,\text{mol}} = \mathbf{0.196 \rightarrow 19.6\,\%}$$

d) $\dot{Q}_{CW} = \dot{m}_{CW} * cp_{CW} * (T_{CW-\text{ex}} - T_{CW-\text{in}})$

$$\overline{T}_{CW-m} = \frac{T_{CW-\text{in}} + T_{CW-\text{ex}}}{2}$$

$$T_{CW-\text{ex}} = 2 * \overline{T}_{CW-m} - T_{CW-\text{in}}$$

$$\dot{m}_{CW} = \frac{\dot{Q}_{CW}}{2 * cp_{CW} * \left(\overline{T}_{CW-m} - T_{CW-\text{in}}\right)}$$

With the heat flow
$\dot{Q}_{CW} = 42\frac{\text{kJ}}{\text{s}}$ for the CoolingWaterStream in the Operating Point of 40 °C *(see table):*

$$\dot{m}_{CW} = \frac{42\,\text{kJ} * \text{kg} * °\text{C}}{\text{s} * 2 * 4.2\,\text{kJ} * (30 - 20)\,°\text{C}} = \mathbf{0.5\frac{kg}{s}} = \mathbf{1.8\frac{t}{h}}$$

→ ***Result***

a) **The operating point is at 40 °C with a heat generation of 100 kW.**
b) **The rate constant at 40 °C is 2.04 * $10^{-4}s^{-1}$.**
c) **The concentration of substance A in the reactor effluent is 3.215 mol/L, this corresponds to a conversion of 0.196 → 19.6 %.**
d) **0.5 kg/s = 1.8 t/h cooling water is required.**

Task 100

In a reactor, 400 kg of methanol (M_{Met} = 32.0 g/mol) are esterified with benzoic acid at the boiling point of methanol under reflux. The methanol conversion is 65%.

a) How many kg of ester (M_E = 136.2 g/mol) do you get if no side reactions occur, and how much kg of non-reacted methanol must be distilled off?
b) The non-reacted methanol is to be distilled off within one hour (Kp = 68 °C; ΔvH = 1100 kJ/kg). The evaporator is heated by means of a natural gas burner (methane) (formation enthalpies in kJ/mol: CH_4 = -74.9; CO_2 = -393.5; H_2O = -241.8). How much natural gas is consumed at a thermal efficiency of 80%?
c) The non-reacted methanol is to be recovered in a condenser of a cooling area of 1 m^2 and Kw = 1300 W(m^2 * K). The average temperature difference is 35 °C. Is the heat exchanger area sufficient for this task?

⊗ **Solution**

→ ***Strategy***

a) The amount of methanol not reacted is calculated using Formula 9a. The mass of ester formed results from the molar difference between the amount of reacted methanol and the amount of methanol not reacted, multiplied by the molar mass of the ester.
b) The heat flow required for the vaporization of the unreacted methanol within one hour is calculated according to Formula 24b. The reaction equation of combustion of methane is set up for the calculation of heat generation required for this purpose and the reaction enthalpy is determined using Formula 28. With Formula 27c, the corresponding molar flow of methane results and, from this by means of the molar mass of methan, its mass flow is calculated. This theoretical mass flow, which would correspond to a process without heat loss, is converted into the real value by division by the thermal efficiency.
c) The heat flow required to vaporize the excess methanol has already been calculated. It is now compared with the data of the heat exchanger using Formula 30. Alternatively, the necessary heat exchanger area for this energy transport can also be calculated.

→ ***Calculation***

Subscripts: Met = Methanol; E = Ester

$CH_3OH + R\text{-}COOH \rightarrow R\text{-}COO\text{-}CH_3 + H_2O$

R = Benzyl group

a) $m_E = n_E * M_E$ with $n_E = n_{Meto} - n_{Met}$

$$X_{Met} = \frac{n_{Meto} - n_{Met}}{n_{Meto}}$$

$$n_{Met} = n_{Meto} * (1 - X_{Met})$$

$$n_{Meto} = \frac{m_{Meto}}{M_{Met}} = \frac{400\,\text{kg} * \text{mol}}{0.032\,\text{kg}} = 12{,}500\,\text{mol}$$

$$n_{Met} = 12{,}500\,\text{mol} * (1 - 0.65) = 4375\,\text{mol}$$

$$\boldsymbol{m_{Met}} = n_{Met} * M_{Met} = 4375\,\text{mol} * 0.032\frac{\text{kg}}{\text{mol}} = \mathbf{140\,kg}$$

$$n_E = (12{,}500 - 4{,}375)\,\text{mol} = 8{,}125\,\text{mol}$$

$$\boldsymbol{m_E} = 8125\,\text{mole} * 0.1362\frac{\text{kg}}{\text{mole}} = \mathbf{1107\,kg}$$

b) *Necessary heat flow for methanol evaporation:*

$$\dot{Q} = \dot{m}_{Met} * \Delta_V H_{Met} = 140\frac{\text{kg}}{\text{h}} * 1100\frac{\text{kJ}}{\text{kg}} = 154{,}000\frac{\text{kJ}}{\text{h}} = 15.400\frac{\text{kJ}}{\text{h}} * \frac{\text{h}}{3600\,\text{s}}$$
$$= 42.78\frac{\text{kJ}}{\text{s}} = 42.78\,\text{kW}$$

$CH_4 + 2O_2 \rightarrow CO_2 + 2H_2O$

$$\dot{Q} = -\dot{n}_{CH_4} * \Delta_R H \rightarrow \dot{n}_{CH_4} = \frac{\dot{Q}}{-\Delta_R H}$$

with $\Delta_R H = \sum_i (\upsilon_i * \Delta_f H_i) = \nu_{CH_4} * \Delta_f H_{CH_4} * \nu_{CO_2} * \Delta_f H_{CO_2} + \nu_{H_2O} * \Delta_f H_{H_2O} \quad \{\Delta_f H_{O_2} = 0\}$

$$\Delta_R H = [-1 * (-74.9) + 1 * (-393.5) + 2 * (-241.8)]\frac{\text{kJ}}{\text{mol}} = -802.2\frac{\text{kJ}}{\text{mol}}$$

$$\dot{n}_{CH_4} = \frac{15{,}400\,\text{kJ*mol}}{\text{h} * 802.2\,\text{kJ}} = 192.0\frac{\text{mol}}{\text{h}}$$

$$\dot{m}_{CH_4} = \dot{n}_{CH_4} * M_{CH_4} = 192.0\frac{\text{mol}}{\text{h}} * 0.016\,\text{kg/mol} = 3.07\frac{\text{kg}}{\text{h}}$$

$$\boldsymbol{\dot{m}_{CH_4-Real}} = \frac{\dot{m}_{CH_4}}{\eta} = \frac{3.07\,\text{kg}}{\text{h} * 0.8} = 3.84\frac{\mathbf{kg}}{\mathbf{h}}$$

c) *Maximum heat flow of the condenser :*
$\dot{Q}_{max} = Kw * A * \Delta T = 1300 \frac{W}{m^2 * °C} * 1\,m^2 * 35\,°C = \mathbf{45.5\,kW}$
Alternative calculation of the minimum heat exchanger area: $\boldsymbol{A} = \frac{\dot{Q}}{Kw * \Delta T} = \frac{42.78\,kJ*m^2*°C}{s*1.3\,kW*35\,°C} = 0.941\,m^2$

→ ***Result***

a) **1107 kg of ester are produced. The excess methanol amount is 140 kg.**
b) **3.84 kg/h of methane must be burned to evaporate the excess methanol within one hour.**
c) **The heat exchanger area is sufficient: The maximum heat flow capacity of the condenser is 45.5 kW and thus above the necessary heat flow of the methanol evaporation of 42.8 kW. Alternatively: The minimum heat exchanger area is 0.941 m²and thus below the existing one.**

Task 101

A gas wash column operated at 25 °C with 10wt% aqueous NaOH is to be used to completely free a gas stream of 1200 m³/h of highly toxic phosgene. The phosgene content in the gas (gas density: 1.26 kg/m³) is 2.5 wt% phosgene ($M_{phosgene}$ = 99 g/mol).

a) How many kilograms of sodium chloride (M = 58.5 g/mol) and sodium carbonate (106 g/mol) are formed per hour in this process?
b) How many kilograms of 40wt% NaOH (M = 40 g/mol) must be added per hour to compensate for the losses due to the reaction?
c) What heat generation must be removed to avoid a temperature increase in the gas washer? (Formation enthalpies in kJ/mol: $COCl_2$ = -219.1; water = -286.3; NaCl = -409.8; NaOH = -427.2; Na_2CO_3 = -1132.6)
d) 15% of this heat generation is compensated by water evaporation during the gas wash (evaporation enthalpy of water $\Delta_v H$ = 2250 kJ/kg)? How much water evaporates in this way?
e) What cooling water flow is required to remove the remaining 85% of the resulting heat output ($cp_{cooling\ water}$ = 4.23 kJ/[kg*°C])? The cooling water temperature may only rise by 10 °C in this process.
f) The cooling water (density = 1000 kg/kg) is taken from a well 20 m below the plant floor. The water feed point of the gas washer heat exchanger is 8 m above the plant floor. The pressure loss in the pipe system is 0.2 bar. The feed pressure into the heat exchanger is 1.5 bar absolute. The efficiency of the centrifugal pump is 60%. What power should an electric motor be chosen to drive the pump if its electrical efficiency is assumed to be 80%?

⊗ **Solution**

→ ***Strategy***

a. & b. First, the reaction equation is set up and the stoichiometric factors are taken from it. Then the molar flow of the removed phosgene is calculated. Using the phosgene molar flow and the molar masses of sodium chloride and sodium carbonate, the corresponding mass flows of these salts are calculated. The same applies to the feed stream of the sodium hydroxide solution. Here, the concentration of 40wt% must also be taken into account.

c) The heat generation of the reaction results from Formula 27c, wherein the reaction enthalpy is calculated according to Formula 28.

d) With the rearranged Formula 24b, the amount of water evaporated can be calculated from 15% of the heat flow determined under c).

e) The mass flow of cooling water is determined by the corresponding rearranged Formula 19a for 85% of the heat generation of the reaction.

f) The theoretically necessary pump power results from the total pumping height according to Formula 39a, wherein the geodetic height results from the position of the well and the heat exchanger and the pressure and friction heights are calculated according to Formula 38b. From this, the calculation of the real power required by the motor is carried out by means of both specified efficiencies.

→ ***Calculation***

	$COCl_2$	NaOH	Na_2CO_3	NaCl	H_2O
ν_i	−1	−4	+1	+2	+2

$$\dot{n}_{COCl_2} = \frac{\dot{m}_{COCl_2}}{M_{COCl_2}}$$

$$\dot{m}_{COCl_2} = \dot{m}_{exhaust} * 0.01 * \%\, COCl_2$$

$$\dot{m}_{exhaust} = \dot{V}_{exhaust} * \rho_{exhaust}$$

$$\dot{n}_{COCl_2} = \frac{\dot{V}_{exhaust} * \rho_{exhaust} * 0.025}{M_{COCl_2}} = \frac{1200\,\text{m}^3 * 1.26\,\text{kg} * 0.025\,\text{mol}}{\text{h} * \text{m}^3 * 0.099\,\text{kg}} = 381.8\frac{\text{mol}}{\text{h}} = 0.106\frac{\text{mol}}{\text{s}}$$

a) $\dot{m}_{NaCl} = \dot{n}_{Nacl} * M_{NaCl} = 2 * \dot{n}_{COCl_2} * M_{NaCl} = 2 * 381.8\frac{\text{mol}}{\text{h}} * 0.0585\frac{\text{kg}}{\text{mol}} = \mathbf{44.67\frac{kg}{h}}$

$$\dot{m}_{Na_2CO_3} = \dot{n}_{Na_2CO_3} * M_{Na_2CO_3} = \dot{n}_{COCl_2} * M_{Na_2CO_3} = 381.8\frac{\text{mol}}{\text{h}} * 0.106\frac{\text{kg}}{\text{mol}} = \mathbf{40.47\frac{kg}{h}}$$

b) $\dot{m}_{NaOH} = \dot{n}_{NaOH} * M_{NaOH} = 4 * \dot{n}_{COCl_2} * M_{NaOH}$

$$= 4 * 381.8\frac{\text{mol}}{\text{h}} * 0.040\frac{\text{kg}}{\text{mol}} = \mathbf{61.09\frac{kg}{h}}$$

1 kg 40% NaOH → 0.40 kg NaOH
X kg 40 % NaOH = 61.09 kg NaOH

$$\rightarrow \dot{m}_{40\%\,\mathrm{NaOH}} = \frac{1\,\mathrm{kg} * 61.09\,\mathrm{kg}}{\mathrm{h} * 0.40\,\mathrm{kg}} = 152.7\frac{\mathbf{kg\ 40\ \%\ NaOH}}{\mathbf{h}}$$

c) $\dot{Q} = -\dot{n}_{\mathrm{COCl_2}} * \Delta_R H$

$$\Delta_R H = \sum_i \left(\nu_i * \Delta_f H_{io}\right)$$

$$\Delta_R H = \nu_{\mathrm{COCl_2}} * \Delta_f H_{\mathrm{COCl_2}} + \nu_{\mathrm{NaOH}} * \Delta_f H_{\mathrm{NaOH}} + \nu_{\mathrm{NaCl}} * \Delta_f H_{\mathrm{NaCl}} + \nu_{\mathrm{Na_2CO_3}} * \Delta_f H_{\mathrm{Na_2CO_3}} + \nu_{\mathrm{H_2O}} * \Delta_f H_{\mathrm{H_2O}}$$

$$\Delta_R H = \{-1 * (-219.1) - 4 * (-427.2) + 1 * (-1132.6) + 2 * (-409.8) + 2 * (286.3)\}\frac{\mathrm{kJ}}{\mathrm{mol}} = -596.9\frac{\mathrm{kJ}}{\mathrm{mol}}$$

$$\dot{Q} = -0.106\frac{\mathrm{mol}}{\mathrm{s}} * \left(-596.9\frac{\mathrm{kJ}}{\mathrm{mol}}\right) = \mathbf{63.27\frac{kJ}{s} = 63.27\,kW \cong 63.3\,kW}$$

d) $\dot{Q}_{15\%} = 63.27\frac{\mathrm{kJ}}{\mathrm{s}} * 0.15 = 9.49\frac{\mathrm{kJ}}{\mathrm{s}}$

$$\dot{Q} = \dot{m}_{W\mathrm{evap}} * \Delta_V H$$

$$\dot{m}_{\mathbf{Wevap}} = \frac{\dot{Q}}{\Delta_V H} = \frac{9.49\,\mathrm{kJ} * \mathrm{kg}}{\mathrm{s} * 2250\,\mathrm{kJ}} = \mathbf{0.00422\frac{kg}{s}} = 0.00422\frac{\mathrm{kg}}{\mathrm{s}} * 3600\frac{\mathrm{s}}{\mathrm{h}} = \mathbf{15.2\frac{kg}{h}}$$

e) $\dot{Q}_{85\%} = 63.27\frac{\mathrm{kJ}}{\mathrm{s}} * 0.85 = 53.78\frac{\mathrm{kJ}}{\mathrm{s}}$

$$\dot{Q}_{85\%} = \dot{m}_W * cp_W * \Delta T_W$$

$$\dot{m}_W = \frac{\dot{Q}_{85\%}}{cp_W * \Delta T_W} = \frac{53.78\,\mathrm{kJ} * \mathrm{kg} * {}^\circ\mathrm{C}}{\mathrm{s} * 4.23\,\mathrm{kJ} * 10\,{}^\circ\mathrm{C}} = \mathbf{1.27\frac{kg}{s}} = \mathbf{4572\frac{kg}{h}}$$

f) $P = \dot{m} * h * g$

$$H = h_{\mathrm{geo}} + h_p + h_f$$

$$h_{\mathrm{geo}} = (20 + 8)\,\mathrm{m} = 28\mathrm{m}$$

$$h_p = \frac{\Delta p}{\rho * g} = \frac{(1.5 - 1.0)\mathrm{bar} * \mathrm{m}^3 * \mathrm{s}^2}{1000\,\mathrm{kg} * 9.81\,\mathrm{m}} = \frac{(1.5 - 1.0) * 10^5\mathrm{kg} * \mathrm{m}^3 * \mathrm{s}^2}{\mathrm{m} * \mathrm{s}^2 * 1000\,\mathrm{kg} * 9.81\,\mathrm{m}} = 5.1\,m$$

$$h_f = \frac{\Delta p_f}{\rho * g} = \frac{0.2\,\mathrm{bar} * \mathrm{m}^3 * \mathrm{s}^2}{1000\,\mathrm{kg} * 9.81\,\mathrm{m}} = \frac{0.2 * 10^5\,\mathrm{kg} * \mathrm{m}^3 * \mathrm{s}^2}{\mathrm{m} * \mathrm{s}^2 * 1000\,\mathrm{kg} * 9.81\,\mathrm{m}} = 2.04\,\mathrm{m}$$

$$H = (28 + 5.1 + 2.04)\,\mathrm{m} = 35.14\,\mathrm{m}$$

$$P_{\text{theor}} = 1.27\frac{\text{kg}}{\text{s}} * 35.14\,\text{m} * 9.81\frac{\text{m}}{\text{s}^2} = 437.8\frac{\text{kg} * \text{m}^2}{\text{s}^3} \cong 438\,\text{W}$$

$$\boldsymbol{P_{\text{real}}} = \frac{P_{\text{theor}}}{\eta_{\text{pump}} * \eta_{\text{motor}}} = \frac{438\,\text{W}}{0.6 * 0.8} = \mathbf{912.5\,W \rightarrow 1\,kW\ Motor}$$

→ ***Result***

a) **44.7 kg of sodium chloride and 40.5 kg of sodium carbonate are formed per hour.**
b) **61.1 kg of sodium hydroxide per hour is required, which corresponds to 152.7 kg/h of a 40% aqueous sodium hydroxide solution.**
c) **The heat generation of the reaction is 63.3 kW.**
d) **15.2 kg of water evaporate per hour.**
e) **1.27 kg/s = 4.57 t/h of cooling water is required.**
f) **The required power consumption of the pump motor is 0.91 kW, so about 1 kW.**

Task 102

The water losses due to evaporation in a cooling tower of a cooling capacity of 4000 kW are compensated by continuous pumping of well water to the system. For this purpose, water is pumped from a reservoir 3.5 m below the factory floor to the feed point of water sprinkling 23 m above the factory floor under atmospheric pressure with a submersible centrifugal pump. The friction loss is 0.5 bar (= 50,000 Pa = 50,000 kg/[m * s ^{2}]). The water has a density of 1000 kg/m ^{3}. The pump efficiency is 80%, the motor efficiency is 92%. The well water has a pH of 6 and contains 4 mmol/L calcium hydrogen carbonate and 0.15 mmol/L iron (II) hydroxide. For corrosion protection reasons, it is brought to a pH of 10.5. This is done by adding 5wt% sodium hydroxide solution. Here, the calcium precipitates as $CaCO_3$ * H_2O (M = 118 g/mol). The iron (II) is oxidized to iron (III) and precipitates as FeO(OH) (M = 88.9 g/mol). It is approximately assumed that there is a complete precipitation. These solids are withdrawn as a 10% suspension.

a) How much water is evaporated to provide the required cooling capacity? (Water vaporization heat $\Delta_v H = 2250$ kJ/kg)
b) What is the power consumption of the pump motor?
c) How much mixture of $CaCO_3$ * H_2O & FeO(OH) precipitates per day, and what is the total sludge flow per day?

⊗ **Solution**

→ ***Strategy***

a) The heat flow discharged by water evaporation is described by Formula 24b, which is rearranged to the mass flow of water for the purpose of solution.

b) The theoretically required pump power results from Formula 39a. The total pumping height is calculated using Formula 38a, with the friction loss height resulting from Formula 38b. The pressure height is zero. The real power of the pump motor results from the theoretical value, divided by the product of both specified efficiencies.
c) The molar flow of calcium and iron compounds precipitated in the cooling tower is calculated from the concentration of calcium and iron in the well water, multiplied by the volume flow of water. With the molar masses, the corresponding mass flow of solids results. The mass flow of the sludge to be discharged is determined by percentage calculation.

→ ***Calculation***

a) $\dot{Q} = \dot{m} * \Delta_V H$

$$\dot{m}_w = \frac{\dot{Q}}{\Delta_V H_w} = \frac{4000\,\text{kW} * \text{kg}}{2250\,\text{kJ}} = \frac{4000\,\text{kJ} * \text{kg}}{\text{s} * 2250\,\text{kJ}} = \mathbf{1.78\frac{kg}{s}} = 1.78\frac{\text{kg}}{\text{s}} * \frac{3600\,\text{s}}{\text{h}} = \mathbf{6.41\frac{t}{h}}$$

b) $P = \dot{m} * g * H$
$H = h_{geo} + h_p + h_f$

$$h_{\text{geo}} = (3.5 + 23)\,\text{m} = 26.5\,\text{m}$$

$$h_f = \frac{\Delta p}{\rho * g} = \frac{0.5\,\text{bar} * \text{m}^3 * \text{s}^2}{1000\,\text{kg} * 9.81\,\text{m}} * \frac{10^5\,\text{kg}}{\text{bar} * \text{m} * \text{s}^2} = 5.1\,\text{m}$$

$$H = (26.5 + 5.1)\,\text{m} = 31.6\,\text{m}$$

$$P_{\text{theor}} = 1.78\frac{\text{m}^3}{\text{s}} * 9.81\frac{\text{m}}{\text{s}^2} * 31.6\,\text{m} = 552\frac{\text{kg} * \text{m}^2}{\text{s}^3} = 0.552\,\text{kW}$$

$$\boldsymbol{P_{\textbf{real}}} = \frac{P_{\text{theor}}}{\eta_{\text{pump}} * \eta_{\text{motor}}} = \frac{0.552kW}{0.80 * 0.92} = \mathbf{0.75\,kW}$$

c) $\dot{m}_i = \dot{n}_i * M_i \quad \dot{n}_i = \dot{V}_w * c_i$

$$\dot{n}_{\text{Ca}} = 1.78\frac{\text{L}}{\text{s}} * 4\frac{\text{mmol}}{\text{L}} = 7.12\frac{\text{mmol}}{\text{s}}$$

$$\dot{m}_{\text{CaCO}_3 * \text{H}_2\text{O}} = 7.12\frac{\text{mmol}}{\text{s}} * 0.118\frac{\text{g}}{\text{mmol}} = 0.84\frac{\text{g}}{\text{s}} = 0.84\frac{\text{g}}{\text{s}} * 3600\frac{\text{s}}{\text{h}} * 24\frac{\text{h}}{\text{day}} = \mathbf{72.6\frac{kg}{day}}$$

$$\dot{n}_{\text{Fe}} = 1.78\frac{\text{L}}{\text{s}} * 0.15\frac{\text{mmol}}{\text{L}} = 0.27\frac{\text{mmol}}{\text{s}}$$

$$\dot{m}_{\mathrm{FeO(OH)}} = 0.27\frac{\mathrm{mmol}}{\mathrm{s}} * 0.0889\frac{\mathrm{g}}{\mathrm{mmol}} = 0.024\frac{\mathrm{g}}{\mathrm{s}}$$
$$= 0.024\frac{\mathrm{g}}{\mathrm{s}} * 3600\frac{\mathrm{s}}{\mathrm{h}} * 24\frac{\mathrm{h}}{\mathrm{day}} = \mathbf{2.1}\frac{\mathbf{kg}}{\mathbf{day}}$$

***Total solids per day* = (72.6 + 2.1) kg = 74.7 kg ≅ 75 kg**

$$\dot{m}_{\mathrm{Suspension}} = 75\frac{\mathrm{kg}}{\mathrm{day}} * \frac{100\%}{10\%} = \mathbf{750}\frac{\mathbf{kg}}{\mathbf{day}}$$

→ ***Result***

a) **The cooling tower must be supplied with 1.78 kg of water per second, or 6.4 t per hour.**
b) **The pump motor must have a power of 0.75 kW.**
c) **72.6 kg of $CaCO_3$* H_2O and 2.1 kg of FeO(OH) are formed per day, or about 75 kg in total. This gives a daily sludge flow of about 750 kg.**

Task 103

The capacity of a plant for the production of propanol is determined by the final rectification column. The limiting part is the condenser of the column head with an exchange surface of 5.6 m^2 and a heat transfer coefficient of 1500 W/(kg * °C). The cooling water inlet is 15 °C, the outlet temperature is 50 °C.

Data propanol: Condensation heat: 754 kJ/kg; Boiling point: 97.3 °C
cp-water = 4.2 kJ/(kg * °C)

a) How large is the average temperature difference in the condenser?
b) How large is the cooling water flow for these conditions?
c) As a rule, a reflux ratio of 0.2 is required for the purity specified for all customers. What is the production rate of propanol under these conditions? RefluxRatio $v_r = \frac{\dot{m}_{\mathrm{Reflux}}}{\dot{m}_{\mathrm{Product}}}$ $\dot{m}_{\mathrm{Condenser}} = \dot{m}_{\mathrm{Reflux}} + \dot{m}_{\mathrm{Product}}$
d) A customer requires propanol of a higher purity, which can only be achieved with a return ratio of 0.4. By what percentage does this reduce the production performance in comparison to the normal production with the return ratio 0.2?

⊗Solution

→ ***Strategy***

a) The average logarithmic temperature difference is calculated according to the corresponding formula given in Sec. 1.2.5.
b) The heat performance of the condenser results from Formula 30. This amount of heat must be dissipated by the heating of the cooling water according to Formula 19a. To calculate the cooling water flow, it is rearranged accordingly.

c) The total amount of propanol liquefied in the condenser is calculated using Formula 24b. The heat performance of the condenser calculated in the previous part of the task is used here. The mass flow of produced propanol is obtained by means of the reflux ratio.

d) To determine the mass flow of product, proceed in the same way as in part c) of the task. The percentage value of the decrease in production rate is calculated by means of the proportion.

→ Calculation

Subscript: Cond = condensation, W = cooling water, P = propanol , R = return, Prod = product

a) $\overline{\Delta T_M} = \frac{\Delta T_1 - \Delta T_2}{\ln \frac{\Delta T_1}{\Delta T_2}}$

$$\Delta T_1 = (97.3 - 15)\,°\text{C} = 82.3\,°\text{C} \quad \Delta T_2 = (97.3 - 50)\,°\text{C} = 47.3\,°\text{C}$$

$$\overline{\Delta T_M} = \frac{(82.3 - 47.3)\,°\text{C}}{\ln \frac{82.3}{47.3}} = \mathbf{63.2\,°C}$$

b)

$$\dot{Q}_{Cond} = K_W * A * \overline{\Delta T_M} = 1500 \frac{\text{W}}{\text{m}^2 * °\text{C}} * 5.6\,\text{m}^2 * 63.2\,°\text{C}$$
$$= 530.9\,\text{kW} = 530.9\,\frac{\text{kJ}}{\text{s}}$$

$$\dot{Q}_{Cond} = \dot{Q}_W = \dot{m}_W * cp_W * \Delta T_W$$

$$\dot{m}_W = \frac{\dot{Q}_{Cond}}{cp_W * \Delta T_W} = \frac{530.9\text{kJ} * \text{kg} * °\text{C}}{\text{s} * 4.2\,\text{kJ} * (50 - 15)\,°\text{C}} = \mathbf{3.61\frac{kg}{s}} = 3.61 \frac{\text{kg} * \text{t} * 3600\,\text{s}}{1000\,\text{kg} * \text{h}} = \mathbf{13.0\frac{t}{h}}$$

c) $\dot{Q}_{Cond} = \dot{m}_{\text{total}-P} * \Delta_V H_P$

$$\dot{m}_{\text{total}-P} = \frac{\dot{Q}_{Cond}}{\Delta_V H_P} = \frac{530.9\,\text{kJ} * \text{kg}}{\text{s} * 754\,\text{kJ}} = 0.704 \frac{\text{kg}}{\text{s}}$$

$$\dot{m}_{\text{total}-P} = \dot{m}_{\text{Prod}} + \dot{m}_R$$

$$\dot{m}_R = \nu_R * \dot{m}_{\text{Prod}}$$

$$\dot{m}_{\text{total}-P} = \dot{m}_{\text{Prod}} * (1 + \nu_R)$$

$$\dot{m}_{\mathbf{Prod}} = \frac{\dot{m}_{\text{total}-P}}{1 + \nu_R} = \frac{0.704\,\text{kg}}{\text{s} * (1 + 0.2)} = \mathbf{0.587\frac{kg}{s}} = 0.587 \frac{\text{kg} * 3600\,\text{s} * \text{t}}{\text{s} * \text{h} * 1000\,\text{kg}} = \mathbf{2.11\frac{t}{h}}$$

d)
$$\dot{m}_{\mathbf{Prod}} = \frac{\dot{m}_{\text{total}-P}}{(1+\nu_R)} = \frac{0.704\,\text{kg}}{s*(1+0.4)} = \mathbf{0.502\frac{kg}{s}}$$
$$= 0.502\frac{\text{kg} * 3600\,\text{s} * \text{t}}{\text{s} * \text{h} * 1000\,\text{kg}} = \mathbf{1.81\frac{t}{h}}$$

2.11 t/h →100 %
1.81 t/h →X

$$X = \frac{100\% * 1.81\,\text{t/h}}{2.11\,\text{t/h}} = 85.8\%$$

*Production decrease: 100 % - 85.8 % = **14.2 %***

→ ***Result***

a) **The average logarithmic temperature difference in the condenser is 63.2 °C.**
b) **13.0 t of cooling water is needed per hour.**
c) **The production rate of propanol at a reflux ratio of 0.2 is 2.11 t per hour.**
d) **The production rate of propanol at a reflux ratio of 0.4 is 1.81 t per hour. This is 14.2% lower than the production performance at a reflux ratio of 0.2.**

Task 104

A plant for the production of 2.5 metric t of chlorine per hour with an integrated natural gas-fired combined heat and power plant is to be roughly pre-projected.

- The chlorine is produced by a chlorine-alkali electrolysis with a cell voltage of 4.4 V.
- The gas power plant is to have an electrical efficiency of 55% and an overall efficiency of 70%, resulting in 30% higher electrical output than the power consumption of the chlorine-alkali electrolysis.
- To generate saturated steam at 140 °C, feedwater at 115 °C is supplied.
- 40% of the unused waste heat of the power plant is to be discharged by water evaporation in a cooling tower operated with river water. For this purpose, a centrifugal pump with an overall efficiency of 45% from pump & motor is used. The level of the river is 30 m below the point of addition into the cooling tower. The friction resistances of the supply pipe system are 0.5 bar.

a) How much 30wt% sodium hydroxide solution is produced as a by-product per hour?
b) How much standard cubic meters of hydrogen (Std-m^3 → 1.013 bar; 0 °C) are produced as a by-product per hour?

c) What diameter must a spherical pressure gas tank have for the hydrogen at a pressure of 10 bar and a temperature of 15 °C in order to catch a three-hour production?
d) How much power does the chlorine-alkali electrolysis require?
e) How much electrical power does the power plant require, and how many Std-m^3 or kg of natural gas (methane) are required for the operation of the power plant per hour?
f) How much saturated steam at 140 °C is produced per hour?
g) How much water must be supplied to the cooling tower per hour, and what pump power is required for this?

Molar masses in g/mol: H_2: 2.0; CH_4: 16.0; Cl_2: 35.5; NaOH: 40.0
Low caloric value of natural gas: 8580 kcal/Std-m^3
Temperature-dependent evaporation heat of water in kJ/kg (steam generator: 2175; cooling tower: 2450)
Heat capacity of water: 4.2 kJ/(kg * °C)
Density of water: 1000 kg/m^3

⊗ **Solution**
→ ***Strategy***

a. & b. First, the molar flow of chlorine generated is calculated from the required mass flow and the molar mass. The molar flow of sodium hydroxide is twice as high as that of chlorine. The mass flow of aqueous sodium hydroxide results from its molar flow, multiplied by the molar mass, taking into account the 30wt% content. The molar flow of formed hydrogen is equal to that of formed chlorine. The gas law according to formula 2 is converted to the volume flow and the previously calculated molar flow as well as standard temperature and pressure are used.

c) The hydrogen volume for a three-hour production time is calculated from the previously calculated hydrogen volume flow under standard conditions (formula 1). Alternatively, the tank volume can be determined from the hydrogen molar number and the pressure and temperature prevailing in the tank using formula 2. The diameter of the corresponding sphere is calculated from the formula of the volume of a sphere.

d) The electrical power consumption is calculated using formula 37a. The current required for this purpose results from the appropriately converted formula 36a, into which the molar flow of chlorine is inserted.

e) The electrical power of the power plant should be 1.3 times higher than the amount required for chlorine production. The electrical efficiency of the power plant is 55%. So the total energy generated by natural gas combustion must correspond to the following amount:
$Q_{total} = 1.3$ * Energy Chloralkali-Elektrolysis/0.55
Accordingly, the volume flow of natural gas (methane) under normal conditions is the quotient of total energy and the low calorific value LCV. The mass flow

of methane can be calculated from its molar flow and the general gas law using the molar mass.

f) The overall efficiency, i.e. electrical power and usable thermal power, is 70%. With an electrical efficiency of 55% therefore 15% of the combustion power of methane are used for steam generation, i.e. 15% of Q_{total}. This is the sum of the heat for heating the boiler feed water to steam temperature (formula 19a) and the amount required for evaporating the water (formula 24b). The amount of steam is obtained by reversing the relationship formed to the mass flow of water.

g) The loss of non-usable power is 30% of the total power. 40% of this should be dissipated by evaporation of water in the cooling tower. The water flow to be evaporated and thus supplied to the cooling tower results from the reversed formula 24b. The required pump power results from formula 39a. The total delivery height is calculated using formula 38a, the friction height is determined using formula 38b. The overall efficiency of the pump & electric motor must be taken into account.

→ ***Calculation***

$2NaCl + H_2O \rightarrow Cl_2 + 2NaOH + H_2$ *exchange of two electrons* $\rightarrow \nu_e = 2$

a) $\dot{m}_{\text{NaOH}} = \dot{n}_{\text{NaOH}} * M_{\text{NaOH}}$

$$\dot{n}_{\text{NaOH}} = 2 * \dot{n}_{\text{Cl}_2}$$

$$\dot{n}_{\text{Cl}_2} = \frac{\dot{m}_{\text{Cl}_2}}{M_{\text{Cl}_2}} = \frac{2500\,\text{kg} * \text{mol}}{\text{h} * 0.071\,\text{kg}} = 35{,}211\frac{\text{mol}}{\text{h}} = \frac{35{,}211\,\text{mol} * \text{h}}{\text{h} * 3600\,\text{s}} = 9.78\frac{\text{mol}}{\text{s}}$$

$$\dot{m}_{\text{NaOH}} = 2 * 35{,}211\frac{\text{mol}}{\text{h}} * 0.040\frac{\text{kg}}{\text{mol}} = 2817\frac{\text{kg}}{\text{h}}$$

30 kg NaOH → 100 kg solution
2817 kg NaOH/h → X

$$\dot{m}_{30wt\%\ \text{NaOH}} = \frac{100\,\text{kg} * 2817\,\text{kg}}{\text{h} * 30\,\text{kg}} = \mathbf{9390\frac{kg}{h}} \cong \mathbf{9.4\frac{t}{h}}$$

b) $p * \dot{V} = \dot{n} * R * T \quad \dot{n}_{\text{H}_2} = \dot{n}_{\text{Cl}_2} = 35{,}211\frac{\text{mol}}{\text{h}}$

Standard Conditions : $p_{Std} = 1.013\,\text{bar} \quad T_{Std} = 273.15\,\text{K}$

$$\dot{V}_{\mathbf{H_2-Std}} = \frac{\dot{n}_{\text{Cl}_2} * R * T_{Std}}{p_{Std}} = \frac{35{,}211\,\text{mol} * 8.315 * 10^{-5}\,\text{bar} * \text{m}^3 * 273.15\,\text{K}}{\text{h} * \text{mol} * \text{K} * 1.013\,\text{bar}} = \mathbf{789.5\frac{m^3}{h}}$$

c) $V_{\text{Tank}} = \frac{d^3 * \pi}{6} \rightarrow d = \sqrt[3]{\frac{V_{\text{Tank}} * 6}{\pi}}$

$$V_{\text{H}_2-\text{Std}} = \dot{V}_{\text{H}_2-\text{Std}} * t = 769.5\frac{\text{m}^3}{\text{h}} * 3\,\text{h} = 2368.5\,\text{m}^3$$

$$\frac{V_{H_2-Std} * p_{Std}}{T_{Std}} = \frac{V_{Tank} * p_{Tank}}{T_{Tank}}$$

$$V_{Tank} = \frac{V_{H_2-Std} * p_{Std} * T_{Tank}}{T_{Std} * p_{Tank}} = \frac{2368.5\,\mathrm{m}^3 * 1.013\,\mathrm{bar} * (273.15 + 15)\,\mathrm{K}}{273.15\,\mathrm{K} * 10\,\mathrm{bar}} = 253.1\,\mathrm{m}^3$$

Alternatively:

$$V_{Tank} = \frac{n_{H_2} * R * T_{Tank}}{p_{Tank}}$$

$$n_{H_2} = \dot{n}_{H_2} * t = 35{,}211\frac{\mathrm{mol}}{\mathrm{h}} * 3\,\mathrm{h} = 105{,}633\,\mathrm{mol}$$

$$V_{Tank} = \frac{105{,}633\,\mathrm{mol} * 8.315 * 10^{-5}\,\mathrm{bar} * \mathrm{m}^3 * (273{,}15 + 15)\,\mathrm{K}}{\mathrm{mole} * \mathrm{K} * 10\,\mathrm{bar}} = 253.1\,\mathrm{m}^3$$

$$\boldsymbol{d} = \sqrt[3]{\frac{253.1\,\mathrm{m}^3 * 6}{\pi}} = \mathbf{7.85\,m}$$

d) $P_{Electrolysis} = U * I$

$$U = 4.4\,\mathrm{V}$$

$$I = \dot{n}_{Cl_2} * \nu_e = 35{,}211\frac{\mathrm{mol}}{\mathrm{h}} * 2 * 96{,}485\,\mathrm{A} * \mathrm{s} = \frac{35{,}211\,\mathrm{mol} * \mathrm{h}}{\mathrm{h} * 3600\,\mathrm{s}} * 2 * 96{,}485\,\mathrm{A} * \mathrm{s} = 1.887 * 10^6\,\mathrm{A}$$

$$\boldsymbol{P}_{\mathbf{Electrolysis}} = 4.4\,\mathrm{V} * 1.887 * 10^6\,\mathrm{A} = 8.305 * 10^6\,\mathrm{W} \cong \mathbf{8.3\,MW}$$

e) $\boldsymbol{P}_{\mathbf{PowerPlant-electrical}} = 1.3 * 8.305\,\mathrm{MW} = 10.797\,\mathrm{MW} \cong \mathbf{10.8\,MW}$
With an electrical efficiency of 55 % → η_{electr} *= 0.55, the total heat required by natural gas combustion is:*

$$\dot{Q}_{Total} = \frac{P_{Power\ Plant-electrical}}{\eta_{electr}} = \frac{10.8\,\mathrm{MW}}{0.55} = 19.64\,\mathrm{MW} = 19.64\frac{\mathrm{MJ}}{\mathrm{s}}$$

$$\dot{Q}_{Total} = \dot{V}_{CH_4-Std} * LCV$$

$$LCV = 8580\frac{\mathrm{kcal}}{\mathrm{Std\text{-}m}^3}$$

$$1\,\mathrm{kcal} = 4.19\,\mathrm{kJ} \rightarrow LCV = 8580\frac{\mathrm{kcal}}{\mathrm{Std\text{-}m}^3} * 4.19\frac{\mathrm{kJ}}{\mathrm{kcal}} = 35{,}950\frac{\mathrm{kJ}}{\mathrm{Std\text{-}m}^3}$$

$$\dot{V}_{\mathbf{CH_4-Std}} = \frac{\dot{Q}_{\text{Total}}}{LCV} = \frac{19.64\,\text{MJ} * \text{Std-m}^3}{\text{s} * 3590\,\text{kJ}} = 0.546\frac{\text{Std-m}^3}{\text{s}} = \frac{0.546\,\text{Std-m}^3 * 3600\,\text{s}}{\text{s} * \text{h}}$$
$$= \mathbf{1967\frac{Std-m^3}{h}}$$

$$\dot{n} = \frac{p * \dot{V}}{R * T}$$

$$\dot{m} = \dot{n} * M$$

$$\dot{m}_{\mathbf{CH_4}} = \frac{1.013\,\text{bar} * 1967\,\text{m}^3 * 0.016\,\text{kg} * \text{mol} * \text{K}}{\text{h} * \text{mol} * 8.315 * 10^{-5}\,\text{bar} * \text{m}^3 * 273.15\,\text{K}} = \mathbf{1403.7\frac{kg}{h}} \cong \mathbf{1.4\frac{t}{h}}$$

f) *Subscript: SG = steam generation; S = steam; WF = water feed*

$$\dot{Q}_{SG} = \dot{Q}_{\text{Total}} * \eta_S = 19{,}640\frac{\text{kJ}}{\text{s}} * 0.15 = 2946\,\text{kW}$$

$$\dot{Q}_{SG} = \dot{Q}_{WF} + \dot{Q}_{\text{Evap}}$$

$$\dot{Q}_{SG} = \dot{m}_S * cp_{WF} * (T_S - T_{WF}) + \dot{m}_S * \Delta_V H$$

$$\dot{m}_S = \frac{\dot{Q}_{SG}}{cp_{WF} * (T_S - T_{WF}) + \Delta_V H}$$

$$\dot{m}_S = \frac{2946\,\text{kJ}}{\text{s} * 4.2\frac{\text{kJ}}{\text{kg}*^\circ\text{C}} * (140 - 115)\,^\circ\text{C} + 2175\frac{\text{kJ}}{\text{kg}}} = \mathbf{1.290\frac{kg}{s}}$$
$$= 1.290\frac{\text{kg} * 3600\,\text{s}}{\text{s} * \text{h}} = 4652\frac{\text{kg}}{\text{h}} \cong \mathbf{4.65\frac{t}{h}}$$

g) *Subscript: CT = cooling tower; Loss= loss; W = water*

$$\dot{Q}_{CT} = \dot{m}_W * \Delta_V H_W \rightarrow \dot{m}_W = \frac{\dot{Q}_{CT}}{\Delta_V H_W}$$

$$\dot{Q}_{CT} = \dot{Q}_{Loss} * 0.4.$$

$$\dot{Q}_{Loss} = \dot{Q}_{\text{Total}} * 0.3 = 1964\,\text{MW} * 0.3 = 5892\frac{\text{kJ}}{\text{s}}$$

$$\dot{Q}_{CT} = 5892\frac{\text{kJ}}{\text{s}} * 0.4 = 2357\frac{\text{kJ}}{\text{s}}$$

$$\dot{m}_W = \frac{5892\,\text{kJ} * \text{kg}}{\text{s} * 2450\,\text{kJ}} = 0.962\frac{\mathbf{kg}}{\mathbf{s}} = 0.962\frac{\text{kg} * 3600\,\text{s}}{\text{s} * \text{h}} = 3.46\frac{\mathbf{t}}{\mathbf{h}}$$

Pump power:

$$P_{\text{theor}} = \dot{m} * g * H$$

$$P_{\text{real}} = \frac{P_{\text{theor}}}{\eta_{\text{pump-total}}}$$

$$H = h_{\text{geo}} + h_p + h_f \quad h_{\text{geo}} = 30\,\text{m} \quad h_p = 0$$

$$h_f = \frac{\Delta p}{\rho * g} = \frac{0.5\,\text{bar} * 10^5\,\text{kg} * \text{m}^3 * \text{s}^2}{\text{bar} * \text{m} * \text{s}^2 * 1000\,\text{kg} * 9.81\,\text{m}} = 5.097\,\text{m} \cong 5.1\,\text{m}$$

$$H = 35.1\,\text{m}$$

$$P_{\text{theor}} = 0.962\frac{\text{kg}}{\text{s}} * 9.81\frac{\text{m}}{\text{s}^2} * 35.1\,\text{m} = 331.2\frac{\text{kg} * \text{m}^2}{\text{s}^3} \cong 331\,\text{W}$$

$$\boldsymbol{P_{\text{real}} = \frac{331\,\text{W}}{0.45} = 735.6\,\text{W} = 0.75\,\text{kW}}$$

→ ***Result***

a) **As a by-product, 9.4 t 30wt% sodium hydroxide solution is manufactured per hour.**
b) **789.5 standard-m^3 hydrogen is produced per hour.**
c) **The diameter of the hydrogen sphere tank is 7.85 m.**
d) **The power consumption of the chlor-alkali electrolysis unit is 8.3 MW.**
e) **The power plant must be designed for an electrical power of 10.8 MW. The gas power plant must be supplied with 1,967 standard-m^3 i.e. 1.4 metric t of natural gas (methane) per hour.**
f) **4.65 metric t of steam is produced per hour.**
g) **The mass flow of water evaporating in the cooling tower is 3.46 t/h. The pump motor power required to supply this mass flow is at least 735.6 W, a motor with a power input of 0.75 kW would be reasonable.**

Task 105

To produce 1.6 metric t of caproic acid chloride ($C_5H_{11}COCl$; Mw = 134.6 g/mol; boiling point = 151 °C; Δ_f H = −575 kJ/mol) per day from caproic acid ($C_5H_{11}COOH$; Mw = 116.2 g/mol; density = 910 kg/m^3; boiling point = 206 °C; Δ_f H = −585 kJ/mol) and thionyl chloride ($SOCl_2$; Mw = 119.0 g/mol; density = 1640 kg/m^3; boiling point = 76 °C; Δ_f H = −246 J/mol) in the presence of the inert solvent n-heptane (C_7H_{16}; density = 700 kg/m; boiling point = 98 °C) a rough preliminary design is to be created. The reaction is to be carried out in a batch reactor equipped with a reflux condenser, which is operated with water at 60 °C. The reaction proceeds according to the following equation:

$C_5H_{11}COOH + SOCl_2 \rightarrow C_5H_{11}COCl + HCl + SO_2$ ($\Delta_f H_{HCl}$ = −92 kJ/mol; $\Delta_f H_{SO_2}$ = −297 kJ/mol)

For this purpose, the following procedure was found to be optimal by the research department by synthesis trials in a three-necked flask:

- 800 mL of heptane in which 200 g of caproic acid were dissolved and heated to 80 °C.
- Addition of thionyl chloride with a molar excess of 5%, followed by a 5-hour reaction phase, during which the formed hydrogen chloride and sulfur dioxide escape. Subsequently, the hexane and the remaining thionyl chloride were distilled off.
- The conversion of caproic acid was 100%. The yield of caproic acid chloride based on caproic acid used was 95%.
- HCl, SO_2 and unreacted $SOCl_2$ leave the reactor as gases.

a) What is the scale-up factor?
b) How much of the reactants and heptane are required per batch in production?
c) What is the size of the production reactor vessel required for the desired production rate if the maximum fill level is 75%, the time to fill with heptane/caproic acid is 30 min, the reaction time is 5 h, the time to transfer the mixture after the reaction to the storage tank for further processing is 10 min, and the time to rinse the reactor for cleaning is 20 min? The reactor vessel is available for 24 hours per day.
d) How many standard cubic meters (T = 0 °C; p = 1.013 bar) of HCl and SO_2 are released per production batch?
e) How much heat of reaction must be removed per batch? How much cooling water (cp = 4.19 kJ/kg * °C), which may be heated from 15 °C to 40 °C, is required for this if the heat removal by the reflux condenser is neglected?

⊗ **Solution**

→ ***Strategy***

a) The scale-up factor is defined as the ratio of the quantities of caproic acid chloride per batch in the production reactor to the laboratory reactor (Formula 40). The total time required for a production batch and, therefore, the number of batches per day can be calculated from the times for preparation and post-processing as well as the reaction time.
b) The quantities used in production per batch are calculated from the data of the laboratory batch, multiplied by the scale-up factor.
c) The total volume supplied to the production reactor consists of the volumes of the reactants (mass divided by density), plus the volume of heptane. This corresponds to 75% of the total volume of the production reactor.
d) The amount of HCl and SO_2 released is equal to the amount of caproic acid used. With the ideal gas law (Formula 2), the volume of HCl and SO_2 under standard conditions can be calculated.
e) The heat to be removed is calculated according to Formula 27a from the amount of caproic acid used and completely reacted, and the reaction enthalpy,

which in turn is determined by Formula 29 from the formation enthalpies. This amount of heat corresponds to the amount of cooling water to be calculated according to Formula 18a.

→ ***Calculation***

Subscript: CC = Caproic acid chloride; CA = Caproic acid; H = Heptan; TC = Thionylchloride; Lab = Laboratory, Prod = Production

a) $ScF = \frac{m_{CC-Prod}}{m_{CC-Lab}}$

Production:

1.6 metric *t per day*

Time required per batch: 30 min filling, 5 h reaction, 10 min pumping, 20 min cleaning →6 h

→ 4 batches per day → m_{CC} = 400 kg Caproic acid chloride per batch

Laboratory:

$$m_{CA-Lab} = 200\,g \quad n_{CA-Lab} = \frac{m_{CA-Lab}}{M_{CA}} = \frac{200\,g{*}mol}{116.2\,g} = 1.72\,mol$$

$$n_{CC-Lab} = Y_{\frac{CC}{CS}} * n_{CS} = 0.95 * 1.75\,mol = 1.634\,mol$$

$$m_{CC-Lab} = n_{CC-Lab} * M_{cc} = 1.634\,mol * 134.6\frac{g}{mol} = 220\,g$$

$$\boldsymbol{ScF} = \frac{400\,kg}{0.22\,kg} = \mathbf{1818}$$

b) $m_{Prod} = m_{Lab} * ScF$

$$\boldsymbol{m_{CA-Prod}} = 0.2\,kg * 1818 = \mathbf{363.6\,kg}$$

$$m_{TC-Lab} = n_{TC-Lab} * M_{TC} \quad n_{TC-Lab} = n_{CA-Lab} * 1.05$$

$$m_{TC-Lab} = 1.72\,mol * 1.05 * 0.119\frac{kg}{mol} = 0.215\,kg$$

$$\boldsymbol{m_{TC-Prod}} = 0.215\,kg * 1818 = \mathbf{391.0\,kg}$$

$$\boldsymbol{V_{H-Prod}} = V_H * ScF = 0.8\,L * 1818 = \mathbf{1.454\,m^3}$$

c) $V_{Educts+H} = V_{CA} + V_{TC} + V_H = \frac{m_{CA}}{\rho_{CA}} + \frac{m_{TC}}{\rho_{TC}} + V_H = \frac{363\,kg * m^3}{910\,kg} + \frac{391\,kg * m^3}{1640\,kg} + 1.454\,m^3 = 2.092\,m^3$

$$2.092\,m^3 = 75\% \quad V_{Reactor} = 100\% \rightarrow \boldsymbol{V_{Reactor}} = \mathbf{2.8\,m^3}$$

d) $V = \frac{n * R * T}{p}$

$$n_{HCl-Prod} = n_{SO_2-Prod} = n_{CA-Prod} = n_{CA-Lab} * ScF = 1.72\,mol * 1818 = 3127\,mol$$

$$V = \frac{3127\ \text{mol} * 8.315 * 10^{-5}\ \text{bar} * \text{m}^3 * 273.15\ \text{K}}{1.013\ \text{bar} * \text{mol} * \text{K}} = 70.1\ \text{m}^3$$

$$\mathbf{V_{HCl} = V_{SO_2} \cong 70\ m^3}$$

e) $Q = -n * \Delta_R H = -n_{\text{CA-Prod}} * \Delta_R H$

$$\Delta_R H = \sum_i (\nu_i * \Delta_f H_i)$$

$$\Delta_R H = \nu_{CA} * \Delta_f H_{CA} + \nu_{TC} * \Delta_f H_{TC} + \nu_{CC} * \Delta_f H_{CC} + \nu_{HCl} * \Delta_f H_{HCl} + \nu_{SO_2} * \Delta_f H_{SO_2}$$

$$\nu_{CA} = -1 \quad \nu_{TC} = -1 \quad \nu_{CC} = +1 \quad \nu_{HCl} = +1 \quad \nu_{SO_2} = +1$$

$$\Delta_R H = \{-1 * (-585) - 1 * (-246) + 1 * (-575) + 1 * (-92) + 1 * (-297)\} \frac{\text{kJ}}{\text{mol}}$$

$$= -133 \frac{\text{kJ}}{\text{mol}}$$

$$Q = -3127\ \text{mol} * (-133) \frac{\text{kJ}}{\text{mol}} = 415{,}891\ \text{kJ}$$

$$Q = m_{\text{Water}} * cp_{\text{Water}} * \Delta T_{\text{Water}}$$

$$m_{\text{Water}} = \frac{Q}{cp_{\text{Water}} * \Delta T_{\text{Water}}} = \frac{415{,}891 \text{kJ} * \text{kg} * {}^\circ\text{C}}{4.19\ \text{kJ} * (40 - 15)\ {}^\circ\text{C}} = \mathbf{3970\ kg \cong 4.0\ t}$$

→ ***Result***

a) **The scale-up factor is based on the mass of caproic acid chloride produced per batch and is ScF = 1818.**
b) **363.6 kg of caproic acid, 391.0 kg of thionyl chloride and 1.454 m^3 heptane are required per production batch.**
c) **The total volume of the production reactor is 2.8 m^3.**
d) **In the production reactor, 70 m^3 of hydrogen chloride and the same volume of sulfur dioxide are released per batch under standard conditions (0°C, 1.013 bar).**
e) **4.0 metric t of cooling water is required per batch.**

Task 106

A wastewater stream [heat capacity: 4.3 kJ/(kg * °C)] of 60 metric t/h with a temperature of 55 °C and a potassium hydroxide content of 0.8 wt% is to be neutralized with another wastewater stream of 20 °C [heat capacity: 3.9 kJ/(kg * °C)], which 5 wt% sulfuric acid contains. Subsequently, the neutralized wastewater stream is cooled in a tube bundle exchanger (outside diameter single tube 200 mm, wall thickness 5 mm) of 20 m in length in countercurrent flow to 25 °C. The heat transfer coefficient was estimated to be 400 W/(m ^{2} * °C). For this purpose, cooling water [heat capacity: 4.2 kJ/(kg * °C); density = 1000 kg/m [3]] of 15 °C is available. The effluent cooling water may not exceed a temperature of 40 °C. The cooling water is pumped from a well (atmospheric pressure) of 20 m depth to the height of the tube bundle exchanger of 5 m. The absolute inlet pressure of the cooling water into the heat exchanger is about 2 bar, the friction pressure losses 0.2 bar.

Formation enthalpies in kJ/mol: $\Delta_f H_{KOH} = -425$; $\Delta_f H_{H_2SO_4} = -811$; $\Delta_f H_{K_2SO_4} = -1438$; $\Delta_f H_{H_2O} = -286$, molar masses in g/mol: KOH = 56; $H_2SO_4 = 98$

a) How large must the mass flow of 5 wt% sulfuric acid be?
b) What is the temperature of the mixture after neutralization? The reaction enthalpy is to be estimated by means of the formation enthalpies.
c) What cooling waterflow is required?
d) How many pipes does the heat exchanger have to be constructed?
e) What power of the cooling water pump motor is required at an overall efficiency of pump and motor of 65%?

⊗ **Solution**

→ ***Strategy***

$2\,KOH + H_2SO_4 \rightarrow K_2SO_4 + 2H_2O$

a) The molar flow of KOH is calculated in the wastewater. The molar flow of H_2SO_4 is half as large. This results in the mass flow of 5wt% sulfuric acid. Formulas 7d and 7e are used.
b) First, the mixing temperature of the wastewater is calculated under the hypothetical assumption that no reaction heat would be generated. The amount of heat released by the wastewater is equal to the amount of heat absorbed by the sulfuric acid solution. For this, formula 19a is used and the balance formula for the temperature of the mixture is solved.
The reaction heat is calculated from formula 27c using the reaction enthalpy determined by formula 28. Formula 19b results in the final temperature of the neutralized wastewater.
c) The cooling water flow results from the combination of formula 19a for the wastewater and cooling water flow converted to the mass flows.
d) The amount of heat transferred from the wastewater stream to the cooling water stream has already been calculated in point c). From the inverted formula 30, the total area of the heat exchanger is calculated with the mean temperature difference. With the heat exchange area of the individual tube, calculated with the mean tube diameter, the number of tubes of the tube bundle exchanger results.
e) The theoretically necessary power of the pump is calculated from formula 39a, whereby the pressure and friction heights result from formula 38b. With the overall efficiency, the electrical power consumption of the motor is calculated.

→***Calculation***

Subscripts:

ww = wastewater, H_2SO_4aq = sulfuric acid solution, R = reaction
M = mixture of wastewater&sulfuric acid without chemical reaction
ME = mixture of wastewater&sulfuric acid with chemical reaction
cw = cooling water, in = inlet, ex = outlet

a) $\dot{m}_{\text{KOH}} = \dot{m}_{\text{ww}} * \frac{\%\,\text{KOH}}{100\%} = \frac{60\,\text{t}*0.8\%}{\text{h}*100\%} = 0.48\frac{\text{t}}{\text{h}}$

$$\dot{n}_{\text{KOH}} = \frac{\dot{m}_{\text{KOH}}}{M_{\text{KOH}}} = \frac{480\,\text{kg} * \text{mol}}{\text{h} * 0.056\,\text{kg}} = 8571\frac{\text{mol}}{\text{h}} = 2.38\frac{\text{mol}}{\text{s}}$$

$$\dot{n}_{\text{H}_2\text{SO}_4} = \frac{\dot{n}_{\text{KOH}}}{2} = \frac{2.38\,\text{mol}}{2\,\text{s}} = 1.19\frac{\text{mol}}{\text{s}}$$

$$\dot{m}_{\text{H}_2\text{SO}_4} = \dot{n}_{\text{H}_2\text{SO}_4} * M_{\text{H}_2\text{SO}_4} = 1.19\frac{\text{mol}}{\text{s}} * 0.098\frac{\text{kg}}{\text{mol}} = 0.117\frac{\text{kg}}{\text{s}}$$

$$\dot{\boldsymbol{m}}_{\mathbf{H_2SO_4-solution}} = \frac{\dot{m}_{\text{H}_2\text{SO}_4} * 100\%}{\%\ \text{H}_2\text{SO}_4} = \frac{0.117\,\text{kg} * 100\%}{\text{s} * 5\%} = \mathbf{2.34\frac{kg}{s}} = \mathbf{8.4\frac{t}{h}}$$

b) $\Delta\dot{Q}_{\text{ww}} = \dot{m}_{\text{ww}} * cp_{\text{ww}} * (T_{\text{ww}} - T_M) = \Delta\dot{Q}_{\text{H}_2\text{SO}_4\text{aq}} = \dot{m}_{\text{H}_2\text{SO}_4\text{aq}} * cp_{\text{H}_2\text{SO}_4\text{aq}} * \left(T_M - T_{\text{H}_2\text{SO}_4\text{aq}}\right)$

$$T_M = \frac{\dot{m}_{\text{ww}} * cp_{\text{ww}} * T_{\text{ww}} + \dot{m}_{\text{H}_2\text{SO}_4\text{aq}} * cp_{\text{H}_2\text{SO}_4\text{aq}} * T_{\text{H}_2\text{SO}_4\text{aq}}}{\dot{m}_{\text{ww}} * cp_{\text{ww}} + \dot{m}_{\text{H}_2\text{SO}_4\text{aq}} * cp_{\text{H}_2\text{SO}_4\text{aq}}}$$

$$\dot{m}_{\text{ww}} * cp_{\text{ww}} = 60{,}000\frac{\text{kg}}{\text{h}} * 4.3\frac{\text{kJ}}{\text{kg} * °\text{C}} = 258{,}000\frac{\text{kJ}}{\text{h} * °\text{C}} = 71.7\frac{\text{kJ}}{\text{s} * °\text{C}}$$

$$\dot{m}_{\text{H}_2\text{SO}_4\text{aq}} * cp_{\text{H}_2\text{SO}_4\text{aq}} = 2.34\frac{\text{kg}}{\text{s}} * 3.9\frac{\text{kJ}}{\text{kg} * °\text{C}} = 9.13\frac{\text{kJ}}{\text{s} * °\text{C}}$$

$$T_M = \frac{(71.7 * 55 + 9.13 * 20)\frac{\text{kJ}}{\text{s}}}{(71.7 + 9.13)\frac{\text{kJ}}{\text{s}*°\text{C}}} = 51.1\,°\text{C}$$

$$\dot{Q} = \left(\dot{m}_{\text{ww}} * cp_{\text{ww}} + \dot{m}_{\text{H}_2\text{SO}_4\text{aq}} * cp_{\text{H}_2\text{SO}_4\text{aq}}\right) * (T_{ME} - T_M) \text{ with } \dot{Q} = -\Delta\dot{n} * \Delta_R H$$

$$\Delta_R H = \sum(\nu_i * \Delta_f H_i) = \nu_{\text{KOH}} * \Delta_f H_{\text{KOH}} + \nu_{\text{H}_2\text{SO}_4} * \Delta_f H_{\text{H}_2\text{SO}_4} + \nu_{\text{K}_2\text{SO}_4} * \Delta_f H_{\text{K}_2\text{SO}_4} + \nu_{\text{H}_2\text{O}} * \Delta_f H_{\text{H}_2\text{O}}$$

$$\Delta_R H = (-2 * [-425] - 1 * [-811] + 1 * [-1438] + 2 * [-286])\frac{\text{kJ}}{\text{mol}} = -349\frac{\text{kJ}}{\text{mol}}$$

$$\dot{Q} = -1.19\frac{\text{mol}}{\text{s}} * \left(-349\frac{\text{kJ}}{\text{mol}}\right) = 415.3\frac{\text{kJ}}{\text{s}}$$

$$T_{ME} = T_M + \frac{\dot{Q}}{\dot{m}_{\mathrm{ww}} * cp_{\mathrm{ww}} + \dot{m}_{\mathrm{H_2SO_4aq}} * cp_{\mathrm{H_2SO_4aq}}} = \frac{415.3\,\mathrm{kJ} + \mathrm{s} * {}^\circ\mathrm{C}}{\mathrm{s} * (71.7{+}9.13)\,\mathrm{kJ}} = 51.1\,{}^\circ\mathrm{C} + 5.14\,{}^\circ\mathrm{C}$$

$$= \mathbf{56.2\,{}^\circ C}$$

c) $\dot{Q}_M = \left(\dot{m}_{\mathrm{ww}} * cp_{\mathrm{ww}} + \dot{m}_{\mathrm{H_2SO_4aq}} * cp_{\mathrm{H_2SO_4aq}}\right) * (T_{\mathrm{ME}} - T_{\mathrm{ME-ex}})$

$$\dot{Q}_M = (71.7 + 9.13)\frac{\mathrm{kJ}}{\mathrm{s} * {}^\circ\mathrm{C}} * (56.2 - 25.0)\,{}^\circ\mathrm{C} = 2522\frac{\mathrm{kJ}}{\mathrm{s}}$$

$$\dot{Q}_{cw} = \dot{m}_{cw} * cp_{cw} * (T_{cw-\mathrm{ex}} - T_{cw-\mathrm{in}}) = \dot{Q}_M$$

$$\dot{m}_{cw} = \frac{\dot{Q}_M}{cp_{cw} * (T_{cw-\mathrm{ex}} - T_{cw-\mathrm{in}})} = \frac{2522\,\mathrm{kJ} * \mathrm{kg} * {}^\circ\mathrm{C}}{\mathrm{s} * 4.2\,\mathrm{kJ} * (40 - 15)\,{}^\circ\mathrm{C}} = \mathbf{24.0\frac{kg}{s}} = \mathbf{86.4\frac{t}{h}}$$

d) $\dot{Q}_M = Kw * A * \Delta T_M$

$$\Delta T_M = \frac{\Delta T_1 - \Delta T_2}{ln\frac{\Delta T_1}{\Delta T_2}} \quad \Delta T_1 = (56.3 - 40.0)\,{}^\circ\mathrm{C} \quad \Delta T_2 = (25 - 15)\,{}^\circ\mathrm{C}$$

$$\Delta T_M = \frac{(16.3 - 10.0)\,{}^\circ\mathrm{C}}{ln\frac{16.3\,{}^\circ\mathrm{C}}{10.0\,{}^\circ\mathrm{C}}} = 12.9\,{}^\circ\mathrm{C}$$

$$A = \frac{\dot{Q}_M}{Kw * \Delta T_M} = \frac{25{,}22\,\mathrm{kJ} * \mathrm{m}^2 * {}^\circ\mathrm{C}}{\mathrm{s} * 0.4\,\mathrm{kW} * 12.9\,{}^\circ\mathrm{C}} = 487.4\,\mathrm{m}^2$$

$$A = n_{\mathrm{pipes}} * d_{M-\mathrm{pipe}} * \pi * L\ , d_{M-\mathrm{pipe}} = \frac{d_a + d_i}{2} = \frac{(200 + 190)\,\mathrm{mm}}{2} = 0.195\,\mathrm{m}$$

$$n_{pipes} = \frac{A}{d_{M\text{-pipe}} * \pi * L} = \frac{487.4\,\mathrm{m}^2}{0.195\,\mathrm{m} * \pi * 20\,\mathrm{m}} = 39.8 \cong \mathbf{40}$$

e) $P = \dot{m}_{cw} * g * H = \dot{m}_{cw} * g * \left(h_{\mathrm{geo}} + h_p + h_f\right)$

$$h_{\mathrm{geo}} = (20 + 5)\mathrm{m} = 25\,\mathrm{m}$$

$$h_p = \frac{\Delta p}{\rho * g} = \frac{1\,\mathrm{bar} * \mathrm{m}^3 * \mathrm{s}^2}{1000\,\mathrm{kg} * 9.81\,\mathrm{m}} = \frac{10^5}{1000 * 9.81}\mathrm{m} = 10.2\,\mathrm{m}$$

$$h_f = \frac{\Delta p}{\rho * g} = \frac{0.2\,\mathrm{bar} * m^3 * \mathrm{s}^2}{1000\,\mathrm{kg} * 9.81\,\mathrm{m}} = \frac{2 * 10^4}{1000 * 9.81}\mathrm{m} = 2.04\,\mathrm{m}$$

$$P_{\mathrm{theor}} = 24.1\frac{\mathrm{kg}}{\mathrm{s}} * 9.81\frac{\mathrm{m}}{\mathrm{s}^2} * (25 + 10.2 + 2.04)\,\mathrm{m} = 8795\frac{\mathrm{kg} * \mathrm{m}^2}{\mathrm{s}^3} = 8.8\,\mathrm{kW}$$

$$P_{\text{real}} = \frac{8.8\,\text{kW}}{0.65} = 13.5\,\text{kW}$$

→ ***Result***

a) **The mass flow of 5% sulfuric acid required for neutralization is 2.34 kg/s = 8.4 t/h.**
b) **The temperature of the wastewater mixture after neutralization is 56.2 °C.**
c) **24.0 kg/s = 86.4 t/h cooling water is required.**
d) **The heat exchanger for the total wastewater flow must be equipped with 40 tubes.**
e) **The power consumption of the pump motor is 13.5 kW.**

Task 107

A reactant solution containing 32% by weight of reactant A (M_A = 85 g/mol) is reacted in a CSTR at 50 °C with reactant B (M_B = 56 g/mol) to form product C (M_C = 152 g/mol). The reaction proceeds according to A + B → 2C + D. Due to side reactions, the yield of desired product C with respect to reactant A is 86%. 3.5% of this is lost during subsequent purification of the product.

In order to recover energy, the reactant solution of A {cp = 1.9 kJ/(kg * °C)} is to be heated with hot waste water from 15 °C to 50 °C. The temperature of the waste water [cp = 4.4 kJ/(kg * °C)]; density = 1160 kg/m^3) is to be lowered by this from 80 °C to 30 °C. For this purpose, a tube bundle heat exchanger 5 m long and 30 tubes of an outer diameter of 20 mm and 2 mm wall thickness is available, which is to be operated in countercurrent mode. Its heat transfer coefficient has been estimated for these conditions at about Kw = 550 W/(m^2 * °C).

The hot water with a pressure of 2.1 bar is pumped into the foot of the 2 m higher located vertical heat exchanger and then led into the wastewater treatment, the feed point of which is 12 m above the heat exchanger outlet. The wastewater treatment operates at atmospheric pressure. The pressure drop in the piping system and in the exchanger is 0.5 bar. The efficiency of the centrifugal pump is 70%, that of the associated electric motor is 92%.

a) What is the effective (mean logarithmic) heat transfer area?
b) What is the mean logarithmic temperature difference?
c) What is the heat flow from the wastewater to the stream of reactant solution A?
d) What is the maximum possible flow of reactant solution A and the reactant A itself under the given conditions?
e) What is the maximum possible wastewater mass and volume flow under the given conditions?
f) What must be the power value of the electric motor of the pump for the wastewater flow at least?
g) What is the maximum daily production of C under the given conditions?

⊗ **Solution**

→ ***Strategy***

a) The heat transfer area for a pipe corresponds to a cylinder jacket and is calculated from its length and the mean logarithmic pipe diameter (see Sect. 1.2.5).
b) The calculation formula for the mean logarithmic temperature difference is also described in Sect. 1.2.5.
c) The heat flow in the exchanger results according to Formula 30, wherein the area and temperature difference calculated under a) and b) are used.
d. & e. The heat flow calculated under c) is taken up by the feed of solution A, as given off by the wastewater. Thus, the size of both mass flows can be calculated by appropriately rearranging Formula 19a.
e) The necessary theoretical power of the wastewater pump results from Formula 39a. The total height is calculated with Formula 38a, wherein the pressure heigt and the friction height are determined according to Formula 38b. With the efficiencies of the pump and the electric drive motor, the real electrical power connection of the motor follows.
g Formula 10b for yield is rearranged for the molar flow of product C. The mass flow of product is calculated using the molar mass and, taking into account the loss in the treatment step, the daily production quantity is determined.

→ ***Calculation***

Subscript: A = reactant A, SolA =solution of reactant A, C = product C, W = waste water

a) $A = n_{\text{pipes}} * \overline{d}_M * \pi * L$

$$\overline{d}_M = \frac{d_a - d_i}{\ln \frac{d_a}{d_i}}$$

$$d_a = 20\,\text{mm} \quad d_i = (20 - 2 * 2)\,\text{mm} = 16\,\text{mm}$$

$$\overline{d}_M = \frac{(20 - 16)\text{mm}}{\ln \frac{20\,\text{mm}}{16\,\text{mm}}} = 17.93\,\text{mm} = 0.0179\,\text{m}$$

$$A = 30 * 0.0179\,\text{m} * \pi * 5\,\text{m} = \mathbf{8.43\,m^2}$$

b) $\overline{\Delta T_M} = \frac{\Delta T_1 - \Delta T_2}{\ln \frac{\Delta T_1}{\Delta T_2}}$

$$\Delta T_1 = (30 - 15)\,°\text{C} = 15\,°\text{C} \quad \Delta T_2 = (80 - 50)\,°\text{C} = 30\,°\text{C}$$

$$\overline{\Delta T_M} = \frac{(15 - 30)\,°\text{C}}{\ln \frac{15\,°\text{C}}{30\,°\text{C}}} = \mathbf{21.6\,°C}$$

c)
$$\dot{Q} = K_W * A * \overline{\Delta T}_M = 550\frac{\text{J}}{\text{s} * \text{m}^2 * {}^\circ\text{C}} * 8.43\,\text{m}^2 * 21.6\,{}^\circ\text{C}$$
$$= 100.150\frac{\text{J}}{\text{s}} = \mathbf{100.15\,kW}$$

d) $\dot{Q} = \dot{Q}_{\text{SolA}} = \dot{m}_{\text{SolA}} * cp_{\text{SolA}} * \Delta T_{\text{SolA}}$

$$\dot{m}_{\text{SolA}} = \frac{\dot{Q}_{\text{SolA}}}{cp_{\text{SolA}} * \Delta T_{\text{SolA}}} = \frac{100.15\,\text{kJ} * \text{kg} * {}^\circ\text{C}}{\text{s} * 1.9\,\text{kJ} * (50 - 15)\,{}^\circ\text{C}} = \mathbf{1.51\frac{kg}{s}} = 1.51\frac{\text{kg} * 3600\,\text{s}}{\text{s}} * \text{h} = \mathbf{5.44\frac{t}{h}}$$

$$\dot{m}_{Ao} = \frac{32\%}{100\%} * \dot{m}_{\text{SolA}} = 0.32 * 1.51\frac{\text{kg}}{\text{s}} = \mathbf{0.483\frac{kg}{s}} = \mathbf{1.74\frac{t}{h}}$$

e) $\dot{Q} = \dot{Q}_W = \dot{m}_W * cp_W * \Delta T_W$

$$\dot{m}_W = \frac{\dot{Q}_W}{cp_W * \Delta T_W} = \frac{100.15\,\text{kJ} * \text{kg} * {}^\circ\text{C}}{\text{s} * 4.4\,\text{kJ} * (80 - 30)\,{}^\circ\text{C}} = \mathbf{0.455\frac{kg}{s}} = 0.455\frac{\text{kg} * 3600\,\text{s}}{\text{s}} * \text{h} = \mathbf{1.64\frac{t}{h}}$$

$$\dot{V}_W = \frac{\dot{m}_W}{\rho_W} = \frac{1.64\,\text{t} * \text{m}^3}{\text{h} * 1.16\,\text{t}} = \mathbf{1.41\frac{m^3}{h}}$$

f) $P = \dot{m} * g * H$

$$H = h_{\text{geo}} + h_p + h_f$$

$$h_{\text{geo}} = h_{\text{feed point heat exchanger}} + h_{\text{heat exchanger}} + h_{\text{feed point waste water plant}} = (2 + 5 + 12)\,\text{m} = 19\,\text{m}$$

$$h_p = \frac{\Delta p}{\rho * g} = \frac{(1 - 2.1)\,\text{bar} * \text{m}^3 * \text{s}^2}{1160\,\text{kg} * 9.81\,\text{m}} = \frac{-1.1 * 10^5\,\text{kg} * \text{s}^2 * \text{m}^3}{\text{m} * \text{s}^2 * 1160\,\text{kg} * 9.81\text{m}} = -9.67\text{m}$$

$$h_f = \frac{\Delta p_f}{\rho * g} = \frac{0.5\,\text{bar} * \text{m}^3 * \text{s}^2}{1160\,\text{kg} * 9.81\,\text{m}} = \frac{0.5 * 10^5\,\text{kg} * \text{s}^2 * \text{m}^3}{\text{m} * \text{s}^2 * 1160\,\text{kg} * 9.81\,\text{m}} = 4.39\,\text{m}$$

$$H = (19 - 9.67 + 4.39)\,\text{m} = 13.72\,\text{m}$$

$$P = 0.455\frac{\text{kg}}{\text{s}} * 9.81\frac{\text{m}}{\text{s}^2} * 13.72\,\text{m} = 61.24\frac{\text{kg} * \text{m}^2}{\text{s}^3} = 61.24\,\text{W}$$

$$P_{\text{Real}} = \frac{P}{\eta_{\text{pump}} * \eta_{\text{Motor}}} = \frac{61.24\,\text{W}}{0.7 * 0.92} = \mathbf{95.1\,W \cong 0.1\,kW}$$

g)
$$Y_{C/A} = \frac{\nu_A * (\dot{n}_{C_o} - \dot{n}_C)}{\nu_C * \dot{n}_{A_o}} \quad \nu_A = -1 \quad \nu_C = +2 \quad \dot{n}_{C_o} = 0\frac{\text{mol}}{\text{s}}$$
$$\dot{n}_{A_o} = \frac{\dot{m}_{A_o}}{M_A} = \frac{0.483\,\text{kg} * \text{mol}}{\text{s} * 0.085\,\text{kg}} = 5.682\frac{\text{mol}}{\text{s}}$$

$$\dot{n}_C = Y_{C/A} * 2 * \dot{n}_{A_o} = 0.86 * 2 * 5.582\frac{\text{mol}}{\text{s}} = 9.77\frac{\text{mol}}{\text{s}}$$

$$\dot{n}_{C-\text{Real}} = \frac{(100 - 3.5)\%}{100\%} * \dot{n}_C = 0.965 * 9.77\frac{\text{mol}}{\text{s}} = 9.43\frac{\text{mol}}{\text{s}}$$

$$\dot{m}_{C-\textbf{Real}} = \dot{n}_{C-\text{Real}} * M_C = 9.43\frac{\text{mol}}{\text{s}} * 0.152\frac{\text{kg}}{\text{mol}} = 1.433\frac{\text{kg}}{\text{s}} = 5.16\frac{\text{t}}{\text{h}}$$

$$= \mathbf{123.85}\frac{\mathbf{t}}{\mathbf{day}} \cong \mathbf{124}\frac{\mathbf{t}}{\mathbf{day}}$$

→ ***Result***

a) **The heat exchanger transfer area is 8.43 m^2.**
b) **The average logarithmic temperature difference in the heat exchanger is 21.6 °C.**
c) **The heat performance of the exchanger is 100.15 kW ≅ 100 kW.**
d) **The mass flow of the reactant solution A is 5.44 t/h, this corresponds to a mass flow of reactant A of 1.74 t/h.**
e) **The wastewater flow is 1.64 t/h, this is 1.41 m^3/h.**
f) **The motor of the pump for conveying the wastewater has a power consumption of 100 W.**
g) **The daily production rate of product C is 123.85 t ≅ 124 t.**

Task 108

In the graduation tower of a salt work, brine of a concentration of 4.5wt% sodium chloride is to be brought to a concentration of 15 wt% sodium chloride by evaporation of water. For this purpose, salt water of a concentration of 4.5 wt% is pumped by means of a centrifugal pump designed as a submersible pump from a 20 m deep brine well below the factory floor into a tank of a level at 10 m above the factory floor, from where it is supplied to the trickle bed in which the corresponding water fraction evaporates. The brine well is atmospheric, while the tank is under an absolute pressure of 2.52 bar. The desired flow rate is 20 m^3 per hour. The pressure drop in the pipeline system is 0.62 bar. The efficiency of the pump under the given conditions is 60 %.

The salt solution leaving the graduation tower has a temperature of 15 °C. To obtain solid salt, the water of this enriched salt solution is completely evaporated at 105 °C. The resulting saturated steam of 105 °C is used to heat the salt solution, to be fed to the evaporator, from 15 °C to 90 °C. For this purpose, a bundle tube exchanger with 50 tubes (d_a = 50 mm; wall thickness = 3 mm; heat conductivity stainless steel λ = 20.5 W/[m * °C]) is operated in countercurrent mode. The heat transfer coefficient of the salt solution side is 400W/(m^2 * °C), that of the steam side 1500 W/(m^2 * °C). The evaporator with a thermal efficiency of 85% is operated with a mixture of 75% natural gas and 25% hydrogen (low heating values LHV_{H2}= 10,840 kJ/m^3;LHV_{CH_4}= 36,040 kJ/m^3 for 0 °C & 1 bar).

The following data are known:
Relationship NaCl content and density of the brine:

$$c_{NaCl} = 0.132 * \rho_{Brine} - 131.6 \quad \text{with } c_{NaCl} \text{ as Gew\% and } \rho_{Brine} \text{ as kg/m}^3$$

Densities, heat capacities, evaporation heat:
NaCl: $\rho_{NaCl} = 2161$ kg/m³; $cp_{NaCl} = 0.87$ kJ/(kg * °C)
Water: = 1000 kg/m³; = 4200 J/(kg * °C); $\Delta_v H = 2250$ kJ/kg

a) How much solid salt is produced by this process in continuous operation per day, and how much water is evaporated per day in the graduation tower?
b) Design of the brine pump
ba. What is the power input of the driving electric motor if its efficiency is 90%?
bb. How much energy is consumed (kWh) per day for pumping?
c) Design of the heat exchanger in front of the evaporator (the mean arithmetic pipe diameter is to be used)
ca. How large must the steam flow be to preheat the 15wt% salt water to 90 °C, and what percentage is this in comparison to the total flow of 105 °C hot steam, generated in the evaporator?
cb. How long must the heat exchanger be?
d) Design of the evaporator
da. What heat flow must be supplied to the evaporator?
db. What is the volumetric flow of the necessary natural gas - hydrogen mixture at a pressure of 5 bar and a temperature of 10 °C?
dc. What is the mass flow of the natural gas - hydrogen mixture?

⊗ **Solution**
→ ***Strategy***

a) The mass flow of NaCl is the product of the mass flow of the brine and its concentration. The mass flow of the brine is calculated from its volume flow and the given relationship between NaCl concentration and density of the brine. The mass flow of water in the 15wt% brine is the difference between the total flow of the brine and that of NaCl. The mass flow of water, evaporating in the gradation plant, is the difference between the water mass flow in the 4.5wt% brine and that of the 15% brine. The latter is determined by means of appropriate percentage calculation.
b) The theoretical pump performance results from Formula 39a with the total height determined from Formula 38a. For this purpose, the pressure difference and the friction pressure loss are transformed into the corresponding height by means of Formula 38b. To calculate the real pump performance, the theoretical value is divided by the product of the pump and motor efficiencies. The daily required pump energy is determined by multiplying the pump performance by the number of daily hours.

c) The heat flow for heating the brine is calculated by Formula 19a from the mass flow of NaCl and water, each multiplied by the corresponding heat capacities and the temperature difference. This results in the steam flow of 105 °C with the formula converted to mass flow 25b. The total steam quantity of 105 °C is equal to the water content of the 15wt% brine stream.
To calculate the length of the heat exchanger, Formula 30 is converted to area and from this, together with the number of pipes and the formula for the cylinder jacket area converted accordingly, the length is calculated. For this purpose, the heat flow previously calculated for preheating the brine, the heat transfer coefficient determined by Formula 31a and the logarithmic temperature difference are used.

da) The necessary amount of heat to evaporate the water of the 15wt% brine is the sum of the heat quantities of heating from the feed temperature to the boiling point of the brine (105 °C) and the heat of vaporization of the brine water. For this, Formula 19a is combined with 25b.

db) This heat flow must be generated by the combustion of a certain amount of the hydrogen - natural gas mixture. The thermal efficiency must be taken into account here. The heat output, generated by combustion for this purpose, is the product of the low heating value and the volume flow (see Sect. 2.4.4). It thus results from the hydrogen stream, multiplied by the corresponding heating value, and the methane stream, multiplied by the corresponding heating value. The individual volume flows are calculated percentage-wise from the total volume flow. The combination of these facts, taking into account the gas laws (Formula 1a), leads to the total volume flow at 5 bar and 10 °C.

dc) To calculate the necessary mass flow of the fuel gas mixture, the molar flows of hydrogen and methane are determined by means of the corresponding formula converted and multiplied by the molar mass.

→ ***Calculation***

a) $\dot{m}_{\mathrm{NaCl}} = \dot{m}_{\mathrm{Brine\,4.5\%}} * C_{\mathrm{NaCl}}$

$$\dot{m}_{\mathrm{Brine}} = \dot{V}_{\mathrm{Brine}} * \rho_{\mathrm{Brine}} \quad \rho_{\mathrm{Brine\,4.5\%}} = \frac{(C_{\mathrm{NaCl}} + 131.6)\mathrm{kg}}{0.132\,\mathrm{m}^3} = \frac{(4.5 + 131.6)\mathrm{kg}}{0.132\,\mathrm{m}^3} = 1031\frac{\mathrm{kg}}{\mathrm{m}^3}$$

$$\dot{m}_{\mathrm{Brine\,4.5\%}} = \frac{20\,\mathrm{m}^3 * 1031\,\mathrm{kg}}{\mathrm{h} * \mathrm{m}^3} = 20,620\frac{\mathrm{kg}}{\mathrm{h}}$$

$$\dot{m}_{\mathrm{NaCl}} = 20\frac{\mathrm{m}^3}{\mathrm{h}} * 1031\frac{\mathrm{kg}}{\mathrm{m}^3} * \frac{4.5\%}{100\%} = 927.9\frac{\mathrm{kg}}{\mathrm{h}} = 0.258\frac{\mathrm{kg}}{\mathrm{s}}$$

$$= \mathbf{22,270\frac{kg}{Day}} \cong \mathbf{22.3\frac{t}{Day}}$$

$$\Delta\dot{m}_{\mathrm{H_2O\;Brine\;4.5-15\%}} = \dot{m}_{\mathrm{H_2O\;Brine4.5\%}} - \dot{m}_{\mathrm{H_2O\;Brine15\%}}$$

$$\dot{m}_{\text{H}_2\text{O Brine4.5\%}} = \dot{m}_{\text{Brine4.5\%}} - \dot{m}_{\text{NaCl}} = (20{,}620 - 928)\frac{\text{kg}}{\text{h}} = 19{,}692\frac{\text{kg}}{\text{h}} = 5.47\frac{\text{kg}}{\text{s}} = 473\frac{\text{t}}{\text{Day}}$$

15% solution:

$$15\frac{\text{kg NaCl}}{\text{h}} \rightarrow 85\frac{\text{kg H}_2\text{O}}{\text{h}} \quad 927.9\frac{\text{kg NaCl}}{\text{h}} \rightarrow x\frac{\text{kg H}_2\text{O}}{\text{h}}$$

$$x = \dot{m}_{\text{H}_2\text{O Brine15\%}} = \frac{85 * 927.9\,\text{kg}}{15\,\text{h}} = 5258\frac{\text{kg}}{\text{h}} = 1.46\frac{\text{kg}}{\text{s}}$$

$$\Delta\dot{m}_{\mathbf{H_2O\ Brine}\ 4.5-15\%} = (19{,}692 - 5258)\frac{\text{kg}}{\text{h}} = \mathbf{14,434\frac{kg}{h}} = \mathbf{4.01\frac{kg}{s}} = \mathbf{346\frac{t}{Day}}$$

b) $\dot{P}_{\text{theor}} = \dot{m}_{\text{Brine4.5\%}} * g * H = \dot{V} * \rho_{\text{Brine}} * \text{g*H}$

$$H = h_{\text{geo}} + h_p + h_f$$

$$h_{\text{geo}} = (20 + 10)\text{m} = 30\,\text{m}$$

$$h_p = \frac{\Delta p}{\rho_{\text{Brine4.5\%}} * g} = \frac{(2.52 - 1.0)\text{bar} * \text{m}^3 * \text{s}^2 * 10^5\,\text{kg}}{1031\,\text{kg} * 9.81\,\text{m} * \text{m} * \text{s}^2 * \text{bar}} = 15.0\,\text{m}$$

$$h_f = \frac{\Delta p_r}{\rho_{\text{Brine4.5\%}} * g} = \frac{0.62\,\text{bar} * \text{m}^3 * \text{s}^2 * 10^5\text{kg}}{1031\,\text{kg} * 9.81\,\text{m} * \text{m} * \text{s}^2 * \text{bar}} = 6.1\,\text{m}$$

$$H = (30 + 15 + 6.1)\text{m} = 51.1\,\text{m}$$

$$P_{\text{theor}} = 20\frac{\text{m}^3 * \text{h}}{\text{h} * 3600\,\text{s}} * 1031\frac{\text{kg}}{\text{m}^3} * 9.81\frac{\text{m}}{\text{s}^2} * 51.1\,\text{m} = 2871\frac{\text{kg} * \text{m}^2}{\text{s}^3} = 2.87\,\text{kW}$$

$$\boldsymbol{P}_{\mathbf{real}} = \frac{P_{\text{theor}}}{\eta_{\text{Pumpe}} * \eta_{\text{Motor}}} = \frac{2.87\,\text{kW}}{0.6 * 0.9} = \mathbf{5.32\,kW}$$

$$\boldsymbol{E} = P_{\text{real}} * t = 5.32\,\text{kW} * 24\,\text{h} = \mathbf{127.7\,kWh} \cong \mathbf{130\,kWh}$$

ca.

$$\Delta\dot{Q}_{\text{Brine15\%}} = \left(\dot{m}_{\text{NaCl}} * cp_{\text{NaCl}} + \dot{m}_{\text{H}_2\text{O Brine15\%}} * cp_{\text{H}_2\text{O}}\right) * (T_{\text{Brine-ex}} - T_{\text{Brine-in}})$$

$$\Delta\dot{Q}_{\text{Brine15\%}} = \left(0.258\frac{\text{kg}}{\text{s}} * 0.87\frac{\text{kJ}}{\text{kg} * {}^\circ\text{C}} + 1.46\frac{\text{kg}}{\text{s}} * 4.2\frac{\text{kJ}}{\text{kg} * {}^\circ\text{C}}\right) * (90 - 15)\,{}^\circ\text{C}$$

$$= 476.7\frac{\text{kJ}}{\text{s}} = 476.7\,\text{kW}$$

$$\dot{Q}_{\text{Steam}} = \dot{m} * \Delta_v H_{\text{H}_2\text{O}} = \Delta\dot{Q}_{\text{Brine15\%}}$$

$$\dot{m}_{\text{Steam}} = \frac{\Delta\dot{Q}_{\text{Brine15\%}}}{\Delta_v H_{\text{H}_2\text{O}}} = \frac{476.7\,\text{kJ} * \text{kg}}{\text{s} * 2250\,\text{kJ}} = \mathbf{0.212\frac{kg}{s}} = \mathbf{763\frac{kg}{h}}$$

$$\dot{m}_{\text{Steam-total}} = \dot{m}_{\text{H}_2\text{O Brine15\%}} = 1.46\frac{\text{kg}}{\text{s}} \quad \dot{m}_{\text{Steam}} \rightarrow \mathbf{14.4\ \%}$$

$$\dot{Q} = Kw * A * \overline{\Delta T_M} \quad A = \frac{\dot{Q}}{Kw * \overline{\Delta T_M}} \quad A = n_{\text{Pipes}} * \overline{d_M} * \pi * L$$

$$L = \frac{A}{n_{\text{Pipes}} * \overline{d_M} * \pi}$$

$$\dot{Q} = 476{,}7\,\text{kW}$$

$$\overline{\Delta T_M} = \frac{\Delta T_1 - \Delta T_2}{\ln\frac{\Delta T_1}{\Delta T_2}} \quad \Delta T_1 = (105 - 15)\,°\text{C} = 90\,°\text{C} \quad \Delta T_2 = (105 - 90)°\text{C} = 15\,°\text{C}$$

$$\overline{\Delta T_M} = \frac{(90 - 15)\,°\text{C}}{\ln\frac{90}{15}} = 41.85\ °\text{C}$$

$$\frac{1}{Kw} = \frac{1}{\alpha_1} + \frac{s}{\lambda} + \frac{1}{\alpha_2} = \frac{\text{m}^2 * °\text{C}}{400\,\text{W}} + \frac{0.003\,\text{m} * \text{m} * °\text{C}}{20.5\,\text{W}} + \frac{\text{m}^2 * °\text{C}}{1500\,\text{W}} = 0.00332\frac{\text{m}^2 * °\text{C}}{\text{W}}$$

$$Kw = 301.3\frac{\text{W}}{\text{m}^2 * °\text{C}}$$

$$\overline{d_M} = \frac{d_a + d_i}{2} = \frac{0.050 + 0.044}{2}\text{m} = 0.047\,\text{m}$$

$$A = \frac{476.7\,\text{kW} * \text{m}^2 * °\text{C}}{0.3013\,\text{kW} * 41.85\,°\text{C}} = 37.7\,\text{m}^2$$

$$\mathbf{L} = \frac{37.7\,\text{m}^2}{50 * 0.047\,\text{m} * \pi} = \mathbf{5.11\,m}$$

da. $\dot{Q}_{\text{Evaporator}} = \left(\dot{m}_{\text{NaCl}} * cp_{\text{NaCl}} + \dot{m}_{\text{H}_2\text{OBrine 15\%}} * cp_{\text{H}_2\text{O}}\right) * \left(T_{\text{Evaporator}} - T_{\text{Brine-in}}\right) + \dot{m}_{\text{H}_2\text{O Brine15\%}} * \Delta_v H_{\text{H}_2O}$

$$\dot{Q}_{\text{Evaporator}} = \left(0.258\frac{\text{kg}}{\text{s}} * 0.87\frac{\text{kJ}}{\text{kg} * °\text{C}} + 1.46\frac{\text{kg}}{\text{s}} * 4.2\frac{\text{kJ}}{\text{kg} * °\text{C}}\right) * (105 - 90)\,°\text{C} + 1.46\frac{\text{kg}}{\text{s}} * 2250\frac{\text{kJ}}{\text{kg}}$$

$$\dot{Q}_{\textbf{Evaporator}} = \mathbf{3380\frac{kJ}{s}}$$

db. $\dot{Q}_{\text{real}} = \frac{\dot{Q}_{\text{Evaporator}}}{\eta_{\text{thermal}}} = \frac{3380\,\text{kJ}}{\text{s}*0.85} = 3976\frac{\text{kJ}}{\text{s}}$

$$\dot{Q}_{\text{real}} = \dot{V}_{H_2} * LHV_{H_2} + \dot{V}_{CH_4} * LHV_{CH_4} \quad \dot{V}_{H_2} = \dot{V}_{\text{total}} * 0.25 \quad \dot{V}_{CH_4} = \dot{V}_{\text{total}} * 0.75$$

$$\dot{Q}_{\text{real}} = \dot{V}_{\text{total}} * \left(0.25 * LHV_{H_2} + 0.75 * LHV_{CH_4}\right)$$

$$\dot{V}_{\text{total-Std-Conditions}} = \frac{\dot{Q}_{\text{real}}}{0.25 * LHV_{H_2} + 0.75 * LHV_{CH_4}} = \frac{3976\,\text{kJ}*\text{m}^3}{\text{s} * (0.25 * 10{,}840 + 0.75 * 36{,}040)\text{kJ}} = 0.134\frac{\text{m}^3}{\text{s}}$$

$$\dot{V}_1 = \frac{\dot{V}_o * p_o * T_1}{T_o * p_1}$$

$$\dot{V}_{\text{total-5 bar, 10 °C}} = \frac{0.134\,\text{m}^3 * 1\,\text{bar} * (273 + 10)\text{K}}{\text{s} * 273\,\text{K} * 5\,\text{bar}} = \mathbf{0.0278\frac{m^3}{s}} = \mathbf{100\frac{m^3}{h}}$$

dc. $p * \dot{V} = \dot{n} * R * T \rightarrow \dot{n} = \frac{p*\dot{V}}{R*T}\dot{m} = \dot{n} * M$

$$\dot{n}_{\text{total}} = \frac{5\,\text{bar} * 0.0278\,\text{m}^3 * \text{mol} * \text{K}}{\text{s} * 8.315 * 10^{-5}\,\text{bar} * \text{m}^3 * 283\,\text{K}} = 5.91\frac{\text{mol}}{\text{s}}$$

$$\dot{n}_{H_2} = 0.25 * 5.91\frac{\text{mol}}{\text{s}} = 1.48\frac{\text{mol}}{\text{s}}$$

$$\dot{m}_{H_2} = 1.48\frac{\text{mol}}{\text{s}} * 0.002\frac{\text{kg}}{\text{mol}} = 0.0030\frac{\text{kg}}{\text{s}} = 10.6\frac{\text{kg}}{\text{h}}$$

$$\dot{n}_{CH_4} = 0.75 * 5.91\frac{\text{mol}}{\text{s}} = 4.43\frac{\text{mol}}{\text{s}}$$

$$\dot{m}_{CH_4} = 4.43\frac{\text{mol}}{\text{s}} * 0.016\frac{\text{kg}}{\text{mol}} = 0.0709\frac{\text{kg}}{\text{s}} = 255.2\frac{\text{kg}}{\text{h}}$$

$$\dot{m}_{\text{total}} = (0.0030 + 0.0709)\frac{\text{kg}}{\text{s}} = \mathbf{0.0739\frac{kg}{s}} = \mathbf{266\frac{kg}{h}}$$

→ ***Result***

a) **22.3 tons of solid salt are produced per day. In the concentration plant, 346 tons of water are evaporated per day in the graduation tower.**
b) **The electrical power consumption of the brine pump is 5.32 kW. About 130 kWh are used for brine production per day.**
c) **A steam stream (105°C) of 0.212 kg/s=763 kg/h is required to heat the 15% brine. This is 14.4% of the steam released when the water content of**

the 15% brine is evaporated. The heat exchanger for preheating the 15% brine must be at least 5.11 m long.

d) **The evaporator requires a heat capacity of 3380 kW. The total volume flow of the fuel gas is 0.0278 $m^3/s \cong 100\ m^3/h$ at 10°C and 5 bar, while the mass flow is 0.0739 kg/s = 266 kg/h.**

Task 109

In a pipe reactor, the etherification of an alcohol (MAlc = 60 g/mol) is carried out at an acidic catalyst bed at 65 °C according to the following reaction equation:

$2\ ROH \rightarrow R\text{-}O\text{-}R + H_2O$

The mean residence time is 15 min. The throughput is 1.1 L/s. The concentration of alcohol in the feed stream is 10.8 mol/L. The density of the feed stream at 65 °C is 850 kg/m^3 and changes only negligibly during the reaction. The mixture leaving the reactor is separated in its components by distillation. The resulting ether (M_{Et} = 102 g/mol) is pumped into a product tank and unreacted alcohol is recycled to the reactor feed stream. In laboratory experiments, the following kinetic parameters were measured at the given concentration of the acidic catalyst:

Maximum rate constant of the second-order reaction $k_o = 1.36 * 10^2$ L/(mol * s); activation energy E_A = 35 kJ/mol.

The reaction enthalpy is -11 kJ/mol alcohol. The entire reactor is kept at 65 ° C by cooling.

a) What is the active volume of the tube reactor?
b) What is the molar composition of the mixture leaving the reactor, if no side reactions occur? It is approximately assumed that the reaction volume and density remain constant.
c) What is the alcohol conversion, the ether yield and the selectivity of the reaction with respect to the alcohol?
d) How much ether is produced per hour (molar and mass flow)?
e) How much heat must be removed from the reactor per unit of time in order to keep the temperature at 65 ° C? A cooling water stream of 3 m ^{3} per hour and 15 ° C is available. The density of the cooling water is 1000 kg/m ^{3}. To what temperature does the cooling water heat up [cp_w = 4.2 kJ/(kg * ° C]?
f) What power of the pump motor (total efficiency of pump & motor = 65%) is required if the delivery height of the cooling water is 5.5 m and a differential pressure cooling water source to heat exchanger feed of 1.5 bar exists, with friction losses being neglected?
   To increase production capacity, the temperature shall be increased to 85 ° C.
g) What composition of the mixture leaving the reactor would be expected if no side reactions occur?
h) What would be the alcohol conversion and the ether yield in this case?
i. How many kg of ether would be produced per hour in this case?

⊗ **Solution**

→ ***Strategy***

a) The active volume of the tubular reactor, that is the total volume minus the volume of the solid catalyst, is calculated by multiplying the residence time by the volume flow entering, which does not change during the reaction.
b) The alcohol concentration of the mixture of substances emerging after the second-order reaction results from Formula 15c. For this purpose, the rate constant at 65 °C is determined by Formula 17a. Since no side reactions occur and the reaction volume does not change, the concentration of ether and water in the final mixture results from the stoichiometric balance as half of the number of alcohol molecules converted, that is, half the concentration difference of the alcohol.
c) Alcohol conversion, ether yield and selectivity with respect to alcohol are calculated using Formulas 9c, 10c and 11c, since no change in volume occurs during the reaction.
d) The molar flow of ether produced is calculated by multiplying the ether concentration by the volume flow. To calculate the mass flow, the molar flow is multiplied by the molar mass of the ether.
e) The heat generation of the reaction is calculated according to Formula 27c. Formula 19 describes the heat removal by the cooling water. It is converted to the temperature difference of the cooling medium.
f) The theoretically necessary pump power results from Formula 39a. By means of the overall efficiency, the actual power consumption of the pump motor is determined. The value of the overall height required for this purpose is calculated by Formula 38a. The calculation of the pressure height is carried out with Formula 38b.

g), h), i) These calculations are carried out at 85 °C instead of 65 °C, as described above.

→ ***Calculation***

a) $\tau = \frac{V_R}{\dot{V}} \quad V_R = \tau * \dot{V} \quad \tau = 15\,\text{min} * \frac{60\,\text{s}}{\text{min}} = 900\,\text{s}$

$$V_R = 900\,\text{s} * 1.1\frac{\text{L}}{\text{s}} = \mathbf{990\,L} = \mathbf{0.99\,m^3}$$

b) $c_{\text{Alc}} = \frac{c_{\text{Alc}_0}}{(1 + c_{\text{Alc}_0} * k * \tau)}$

$$k = k_0 * e^{-E_A/R * T}$$

$$k_{65} = 1.36 * 10^2 \frac{\text{L}}{\text{mol} * \text{s}} * e^{-\frac{35\,\text{kJ} * \text{mol} * \text{K}}{\text{mol} * 0.008315\,\text{kJ} * (273+65)\,\text{K}}} = 5.31 * 10^{-4} \frac{\text{L}}{\text{mol} * \text{s}}$$

$$c_{\mathrm{Alc_o}} * k_{65} * \tau = 10.8\frac{\mathrm{mol}}{\mathrm{L}} * 5.31 * 10^{-4}\frac{\mathrm{L}}{\mathrm{mol} * \mathrm{s}} * 900\,\mathrm{s} = 5.16$$

$$\boldsymbol{c_{\mathrm{Alc}}} = \frac{10.8\frac{\mathrm{mol}}{\mathrm{L}}}{1 + 5.16} = \mathbf{1.75\frac{mol}{L}}$$

$$\boldsymbol{\dot{n}_{\mathrm{H_2O}}} = \frac{\Delta\dot{n}_{\mathrm{Alc}}}{2} \rightarrow c_{\mathrm{H_2O}} = \frac{\Delta c_{\mathrm{Alc}}}{2} = \frac{(10.8 - 1.75)\mathrm{mol}}{2\,\mathrm{L}} = \mathbf{4.525\frac{mol}{L}}$$

$$\boldsymbol{\dot{n}_{\mathrm{Et}}} = \frac{\Delta\dot{n}_{\mathrm{Alc}}}{2} \rightarrow c_{\mathrm{Et}} = \frac{\Delta c_{\mathrm{Alc}}}{2} = \frac{(10.8 - 1.75)\mathrm{mol}}{2\,\mathrm{L}} = \mathbf{4.525}\frac{\mathrm{mol}}{\mathrm{L}}$$

c) $\boldsymbol{X_{\mathrm{Alc}}} = \frac{c_{\mathrm{Alc_o}} - c_{\mathrm{Alc}}}{c_{\mathrm{Alc_o}}} = \frac{(10.8-1.75)\frac{\mathrm{mol}}{\mathrm{L}}}{10.8\frac{\mathrm{mol}}{\mathrm{L}}} = \mathbf{0.843 \rightarrow 84.3\,\%}$

$$\boldsymbol{Y_{\mathrm{Et/Alc}}} = \frac{\upsilon_{\mathrm{Alc}} * (c_{\mathrm{Eto}} - c_{\mathrm{Et}})}{\upsilon_{\mathrm{Et}} * c_{\mathrm{Alc}}} = \frac{-2 * (0 - 4.55)\frac{\mathrm{mol}}{\mathrm{L}}}{+1 * 10.8\frac{\mathrm{mol}}{\mathrm{L}}} = \mathbf{0.838 \rightarrow 83.8\,\%}$$

$$\boldsymbol{S_{\mathrm{Et/Alc}}} = \frac{\upsilon_{\mathrm{Alc}} * \left(c_{\mathrm{Et_o}} - c_{\mathrm{Et}}\right)}{\upsilon_{\mathrm{Et}} * (c_{\mathrm{Alc_o}} - c_{\mathrm{Alc}})} = \frac{-2 * (0 - 4.53)\frac{\mathrm{mol}}{\mathrm{L}}}{+1 * (10.8 - 1.75)\frac{\mathrm{mol}}{\mathrm{L}}} = \mathbf{1.00 \rightarrow 100\,\%}$$

d) $\dot{n}_{\mathrm{Et}} = c_{\mathrm{Et}} * \dot{V} = 4.525\frac{\mathrm{mol}}{\mathrm{L}} * 1.1\frac{\mathrm{L}}{\mathrm{s}} = 4.98\frac{\mathrm{mol}}{\mathrm{s}} = 4.98 * 3600\frac{\mathrm{mol}}{\mathrm{h}}$

$= \mathbf{17,919\frac{mol}{h}}$

$$\boldsymbol{\dot{m}_{\mathrm{Et}}} = \dot{n}_{\mathrm{Et}} * M_{\mathrm{Et}} = 17.919\frac{\mathrm{mol}}{\mathrm{h}} * 0.102\frac{\mathrm{kg}}{\mathrm{mol}} = \mathbf{1.83\frac{t}{h}}$$

e) $\dot{Q} = -\Delta\dot{n}_{\mathrm{Alc}} * \Delta_R H$

$$\Delta\dot{n}_{\mathrm{Alc}} = \Delta c_{\mathrm{Alc}} * \dot{V} = (10.8 - 1.75)\frac{\mathrm{mol}}{\mathrm{L}} * 1.1\frac{\mathrm{L}}{\mathrm{s}} = 9.96\frac{\mathrm{mol}}{\mathrm{s}}$$

$$\dot{Q} = -9.96\frac{\mathrm{mol}}{\mathrm{s}} * \left(-11\frac{\mathrm{kJ}}{\mathrm{mol}}\right) = 109.6\frac{\mathrm{kJ}}{\mathrm{s}}$$

$\dot{Q} = \dot{m}_W * cp_W * \Delta T_W$

$\dot{m}_w = \dot{V}_w * \rho_w = 3\frac{\mathrm{m}^3}{\mathrm{h}} * 1000\frac{\mathrm{kg}}{\mathrm{m}^3} * \frac{\mathrm{h}}{3600\,\mathrm{s}} = 0.833\frac{\mathrm{kg}}{\mathrm{s}}$ (subscript $w \rightarrow$ coolingwater)

$$\Delta T_W = \frac{\dot{Q}}{\dot{m}_W * cp_W} = \frac{109.6\,\mathrm{kJ} * \mathrm{s} * \mathrm{kg} * {}^\circ\mathrm{C}}{\mathrm{s} * 0.833\,\mathrm{kg} * 4.2\,\mathrm{kJ}} = 31.3\,{}^\circ\mathrm{C}$$

$$\boldsymbol{T_{w-\mathrm{ex}}} = T_{w-\mathrm{in}} + \Delta T_w = (15 + 31.3)\,{}^\circ\mathrm{C} = \mathbf{46.3\,{}^\circ C}$$

f) $P = \dot{m}_w * g * H$

$$H = h_{geo} + h_p + h_f$$

$$h_{geo} = 5.5\,\text{m} \quad h_f = 0.0\,\text{m}$$

$$h_p = \frac{\Delta p}{\rho_w * g} = \frac{1.5\,\text{bar} * \text{m}^3 * \text{s}^2}{1000\,\text{kg} * 9.81\,\text{m}} = \frac{1.5 * 10^5\,\text{kg} * \text{m}^3 * \text{s}^2}{\text{m} * \text{s}^2 * 1000\,\text{kg} * 9.81\,\text{m}} = 15.3\,\text{m}$$

$$H = (5.5 + 15.3 + 0)\text{m} = 20.8\,\text{m}$$

$$P_{\text{theor}} = 0.833\frac{\text{kg}}{\text{s}} * 9.81\frac{\text{m}}{\text{s}^2} * 20.8\,\text{m} = 170\frac{\text{kg} * \text{m}}{\text{s}^3} = 170\,\text{W}$$

$$\boldsymbol{P_{\text{real}}} = \frac{P_{\text{theor}}}{\eta} = \frac{170\,\text{W}}{0.65} = \mathbf{261.5\,W \cong 300\,W}$$

g) $c_{\text{Alc}} = \frac{c_{\text{Alc}}}{(1 + c_{\text{Alc}} * k * \tau)}$

$$k = k_0 * e^{-E_A/R * T}$$

$$k_{85} = 1.36 * 10^2 \frac{\text{L}}{\text{mol} * \text{s}} * e^{-\frac{35\,\text{kJ} * \text{mol} * \text{K}}{\text{mol} * 0.008315\,\text{kJ} * (273+85)\text{K}}} = 1.065 * 10^{-3}\frac{\text{L}}{\text{mol} * \text{s}}$$

$$c_{\text{Alc}_0} * k_{85} * \tau = 10.8\frac{\text{mol}}{\text{L}} * 1.065 * 10^{-3}\frac{\text{L}}{\text{mol} * \text{s}} * 900\,\text{s} = 10.35$$

$$\boldsymbol{c_{\text{Alc}}} = \frac{10.8\frac{\text{mol}}{\text{L}}}{1 + 10.35} = \mathbf{0.952\frac{mol}{L}}$$

$$\boldsymbol{\dot{n}_{\text{H}_2\text{O}}} = \frac{\Delta\dot{n}_{\text{Alc}}}{2} \rightarrow c_{\text{H}_2\text{O}} = \frac{\Delta c_{\text{Alc}}}{2} = \frac{(10.8 - 0.952)\text{mol}}{2\,\text{L}} = \mathbf{4.925\frac{mol}{L}}$$

$$\boldsymbol{\dot{n}_{\text{Et}}} = \frac{\Delta\dot{n}_{\text{Alc}}}{2} \rightarrow c_{\text{Et}} = \frac{\Delta c_{\text{Alc}}}{2} = \frac{(10.8 - 0.952)\text{mol}}{2\,\text{L}} = \mathbf{4.925\frac{mol}{L}}$$

h) $\boldsymbol{X_{\text{Alc}}} = \frac{c_{\text{Alc}_0} - c_{\text{Alc}}}{c_{\text{Alc}_0}} = \frac{(10.8 - 0.952)\frac{\text{mol}}{\text{L}}}{10.8\frac{\text{mol}}{\text{L}}} = \mathbf{0.912 \rightarrow 91.2\,\%}$

$$\boldsymbol{Y_{\text{Et/Alc}}} = \frac{\upsilon_{\text{Alc}} * \left(c_{\text{Et}_0} - c_{\text{Et}}\right)}{\upsilon_{\text{Et}} * c_{\text{Alc}_0}} = \frac{-2 * (0 - 4.925)\frac{\text{mol}}{\text{L}}}{+1 * 10.8\frac{\text{mol}}{\text{L}}} = \mathbf{0.912 \rightarrow 91.2\,\%}$$

i. $\dot{n}_{\text{et}} = c_{\text{et}} * \dot{V} = 4.925\frac{\text{mol}}{\text{L}} * 1.1\frac{\text{L}}{\text{s}} = 5.42\frac{\text{mol}}{\text{s}}$

$$\dot{m}_{Et} = \dot{n}_{Et} * M_{Et} = 5.42\frac{mol}{s} * 0.102\frac{kg}{mol} = \mathbf{0.553\frac{kg}{s}} = 0.553\frac{kg}{s} * \frac{3600\,s}{h}$$

$$= \mathbf{1.99\frac{t}{h} \cong 2.0\frac{t}{h}}$$

→ ***Result***

a) The free reactor volume is 0.99 m^3≅1.0 m^3.
b) The reaction mixture leaving the reactor at 65 °C has the following molar composition: 1.75 mol/L alcohol and 4.525 mol/L water and ether each.
c) The alcohol conversion at 65 °C reaction temperature is 0.838 or 83.8 %. The yield of ether relative to alcohol has the same value since no side reactions occur. Since no side reactions occur, the selectivity of the reaction to ether is equal to 1.0 or 100 %.
d) The production rate of ether at 65 °C is 0.511 kg/s or 1.84 t/h.
e) The necessary cooling capacity is 109.6kW. The cooling water is heated to 46.3 °C.
f) The necessary cooling capacity is 109.6 kW. The motor of the cooling water pump has a power consumption of at least 261.5 W, hence about 300 W.
g) The reaction mixture leaving the reactor at 85 °C has the following molar composition: 0.952 mol/L alcohol and 4.925 mol/L water and ether each.
h) The alcohol conversion at 85 °C reaction temperature is 0.912 i.e. 91.2 %. The yield of ether relative to alcohol has the same value since no side reactions occur. Since no side reactions occur, the selectivity of the reaction to ether is equal to 1.0 i.e. 100 %.
i) The production rate of ether at 85 °C is 0.553 kg/s i.e. 2.0 t/h. It is only slightly higher than at 65 °C.

Task 110

An improved process for the production of monoethanolamine (MEA) was tested in the laboratory. For this purpose, a mixture of 1.2 wt% ethylene oxide (EO), 16.0 wt% ammonia (NH_3) and 82.8 wt% hexane (Hex) was reacted in an autoclave at 90 °C for 5 min. The conversion of ethylene oxide was 98.5%, the yield of MEA based on ethylene oxide 85%.

The following material data are known at the given reaction conditions:

	Molar mass g/mol	Density kg/m³	Heat capacity kJ/(kg*°C)	Std Enthalpy of formation kJ/mol
Hexane	66.2	660	2.3	
EO	44.1	890	2.0	-78.0
NH3	17.0	450	6.5	-46.0
MEA	61.4	990	2.95	-507
Water	18.0	1000	4.2	

In a pilot plant, a scale-up in a CSTR of 1m^3 total volume, which may be filled to 75%, is to be carried out at elevated pressure. Due to the relatively small change in concentrations during the reaction and the associated low influence of backmixing, the difference in kinetics between batch operation and CSTR is to be neglected to a first approximation and the conversion and yield data from the laboratory experiment are to be transferred directly to the CSTR. Thus, it should be operated as close as possible to the laboratory conditions, therefore a CSTR residence time of 5 min is chosen. The CSTR is fed with the streams at 20 °C. The temperature in the CSTR should be 90 °C. For this purpose, the CSTR is heated with 130 °C saturated steam (condensation heat = 2300 kJ/kg). The heat transfer coefficient under the mentioned conditions is Kw = 600 W/(m^2 * °C). The heating area of the CSTR is 1.5 m^2.

The flow leaving the reactor is to be cooled in a tube bundle heat exchanger with 50 tubes (outer diameter 2″, wall thickness 4 mm, heat conductivity exchanger wall = 45 W/[m * °C]), estimated heat transfer coefficients: inside = 350 W/[m ^{2} * °C], outside = 750 W/[m ^{2} * °C]) in countercurrent mode with cooling water to 35 °C. The cooling water (cp = 4.2 kJ/[kg * °C]) is supplied to the exchanger at 20 °C. The cooling water outlet temperature must not exceed 40 °C.

a) Formulate a meaningful scale-up factor.
b) How large are the liquid mass and mol feed rates to the pilot reactor?
c) What is the production capacity of the pilot reactor for MEA under the given conditions?
d) How large is the necessary steam flow for heating the CSTR?
e) Is the heating area of the CSTR sufficient for this task?
f) What cooling water flow is necessary to cool the reaction mixture?
g) What exchange area and length must the heat exchanger have at least?

⊗ **Solution**

→ ***Strategy***

a. & b. The volume of the pilot plant reactor used is given, but not the volume of the laboratory reactor. However, the weight-percent composition of the laboratory batch is known. It therefore makes sense to use the volume for a 100 g laboratory batch here, which can be calculated from the sum of the mass individual streams divided by the associated density. From this, a scale-up factor can be formulated by the active pilot plant reactor volume. With this size, the individual feed streams to the pilot plant reactor are then calculated.

c) Equation 10b of the calculation of the MEA yield is switched to the MEA molar flow and this is converted to the MEA mass flow by the MEA molar mass.

d) The total amount of heat required for keeping the reactor content at 90°C is given by equation 21b. It is also the sum of the heat introduced by the steam condensation (equation 24b) and the heat generated by the reaction (equation 27c). This combination of equations is switched to the steam flow. The reaction heat is calculated from the moles of ethylene oxide reacted and the reaction enthalpy according to equation 28.

e) The minimum necessary area for heating with steam is calculated by the appropriately switched equation 30 from the heat flow introduced by the steam condensation.
f) The cooling water flow results from the appropriately switched equation 19a, with the heat flow to be used therein being calculated from equation 19b using the data for ethylene oxide, NH_3 and hexane.
g) The heat flow to be discharged has already been calculated in the previous part of the task. With the appropriately switched equation 30, the necessary exchange area is determined. For this purpose, the heat transfer coefficient is first determined with equation 31a and the mean logarithmic temperature difference according to Sect. 1.2.5. The length of the exchanger is calculated from the exchange area previously calculated, the number of pipes and the appropriately switched equation for the exchange area of a single pipe, using the logarithmic averaging of the pipe diameter according to Sect. 1.2.5.

→ ***Calculation***

a) $V_{\text{total}} = \frac{m}{\rho} = \frac{m_{\text{EO}}}{\rho_{\text{EO}}} + \frac{m_{\text{NH}_3}}{\rho_{\text{NH}_3}} + \frac{m_{\text{Hex}}}{\rho_{\text{Hex}}}$

Laboratory Batch 100 g →

$$m_{\text{total-Lab}} = 100\,\text{g} = 1.2\,\text{g}\,\text{EO} + 16.0\,\text{g}\,\text{NH}_3 + 82.8\,\text{g}\,\text{Hex}$$

$$V_{\text{total-Lab}} = \frac{1.2\,\text{g} * \text{m}^3}{890\,\text{kg}} + \frac{16.0\,\text{g} * \text{m}^3}{450\,\text{kg}} + \frac{82.8\,\text{g} * \text{m}^3}{660\,\text{kg}} = 1.62 * 10^{-4}\,\text{m}^3$$

Pilot Reactor →

$$\dot{V} = \frac{V_{\text{PilotReactor}}}{\tau} \qquad V_{\text{PilotReactor}} = 1.0\,\text{m}^3 * \frac{75\%}{100\%} = 0.75\,\text{m}^3$$

$$\tau = 5\,\text{min} = 300\,\text{s} \rightarrow \dot{V} = \frac{0.75\,\text{m}^3}{300\,\text{s}} = 2.5 * 10^{-3}\frac{\text{m}^3}{\text{s}}$$

$$\boldsymbol{ScF} = \frac{V_{\text{PilotReactor}}}{V_{\text{Lab}}} = \frac{0.75\,\text{m}^3}{1.62 * 10^{-4}\,\text{m}^3} = \mathbf{4630}$$

b) *Pilot Reactor with a Residence Time of* $\tau = 5\,\text{min} = 300\,\text{s}$

$$\dot{m}_{\text{Pilot}} = \frac{m_{\text{Lab}}}{\tau_{\text{Lab}}} * ScF$$

$$\dot{n} = \frac{\dot{m}}{M}$$

$$\dot{\boldsymbol{m}}_{\textbf{EO}} = \frac{0.0012\,\text{kg}}{300\,\text{s}} * 4630 = \mathbf{0.0185\frac{kg}{s}}$$

$$\dot{\boldsymbol{n}}_{\textbf{EO}} = \frac{0.0185\,\text{kg} * \text{mol}}{\text{s} * 0.0441\,\text{kg}} = \mathbf{0.420\frac{mol}{s}}$$

$$\dot{m}_{\mathrm{NH_3}} = \frac{0.016\,\mathrm{kg}}{300\,\mathrm{s}} * 4630 = \mathbf{0.247\frac{kg}{s}}$$
$$\dot{n}_{\mathrm{NH_3}} = \frac{0.247\,\mathrm{kg} * \mathrm{mol}}{\mathrm{s} * 0.017\,\mathrm{kg}} = \mathbf{14.5\frac{mol}{s}}$$

$$\dot{m}_{\mathrm{Hex}} = \frac{0.0828\,\mathrm{kg}}{300\,\mathrm{s}} * 4630 = \mathbf{1.28\frac{kg}{s}}$$
$$\dot{n}_{\mathrm{Hex}} = \frac{1.28\,\mathrm{kg} * \mathrm{mol}}{\mathrm{s} * 0.0662\,\mathrm{kg}} = \mathbf{19.3\frac{mol}{s}}$$

c) $Y_{\mathrm{MEA/EO}} = \frac{\nu_{\mathrm{EO}} * (\dot{n}_{\mathrm{MEA_0}} - \dot{n}_{\mathrm{MEA}})}{\nu_{\mathrm{MEA}} * \dot{n}_{\mathrm{EO_0}}}$

with $\nu_{\mathrm{EO}} = -1 \quad \nu_{\mathrm{MEA}} = +1 \quad \dot{n}_{\mathrm{MEA_0}} = 0$

$$\dot{n}_{\mathrm{MEA}} = Y_{\mathrm{MEA/EO}} * \dot{n}_{\mathrm{EO_0}} = 0.85 * 0.420\frac{\mathrm{mol}}{\mathrm{s}} = \mathbf{0.357\frac{mol}{s}}$$

$$\dot{m}_{\mathrm{MEA}} = 0.357\frac{\mathrm{mol}}{\mathrm{s}} * 0.0611\frac{\mathrm{kg}}{\mathrm{mol}} = \mathbf{0.0218\frac{kg}{s}} = \mathbf{78.5\frac{kg}{h}}$$

d) $\dot{Q}_{\mathrm{total}} = \dot{Q}_{\mathrm{Steam}} + \dot{Q}_{\mathrm{Reaction}}$

$$\dot{Q}_{\mathrm{Steam}} = \dot{m}_{\mathrm{Steam}} * \Delta_v H_{\mathrm{Water}}$$

$$\dot{m}_{\mathrm{Steam}} = \frac{\dot{Q}_{\mathrm{total}} - \dot{Q}_{\mathrm{Reaction}}}{\Delta_v H_{\mathrm{Water}}}$$

$$\dot{Q}_{\mathrm{total}} = \left(\dot{m}_{\mathrm{EO_0}} * cp_{\mathrm{EO}} + \dot{m}_{\mathrm{NH_{3_0}}} * cp_{\mathrm{NH_3}} + \dot{m}_{\mathrm{Hex}} * cp_{\mathrm{Hex}}\right) * (T_{\mathrm{Reactor}} - T_{\mathrm{Feed}})$$

$$\dot{Q}_{\mathrm{total}} = (0.0185 * 2.0 + 0.247 * 6.5 + 1.28 * 2.3)\frac{\mathrm{kg} * \mathrm{kJ}}{\mathrm{s} * \mathrm{kg} * {}^\circ\mathrm{C}} * (90 - 20)\,^\circ\mathrm{C} = 321.1\frac{\mathrm{kJ}}{\mathrm{s}}$$

$$\dot{Q}_{\mathrm{Reaction}} = -\Delta\dot{n}_{\mathrm{EO}} * \Delta_R H$$

$$\Delta\dot{n}_{\mathrm{EO}} = \left(\dot{n}_{\mathrm{EO_0}} - \dot{n}_{\mathrm{EO}}\right) = X_{\mathrm{EO}} * \dot{n}_{\mathrm{EO_0}} = 0.985 * 0.420\frac{\mathrm{mol}}{\mathrm{s}} = 0.414\frac{\mathrm{mol}}{\mathrm{s}}$$

$$\Delta_R H = \nu_{\mathrm{EO}} * \Delta_f H_{\mathrm{EO}} + \nu_{\mathrm{NH_3}} * \Delta_f H_{\mathrm{NH_3}} + \nu_{\mathrm{MEA}} * \Delta_f H_{\mathrm{MEA}}$$
$$\nu_{\mathrm{EO}} = -1 \quad \nu_{\mathrm{NH_3}} = -1 \quad \nu_{\mathrm{MEA}} = +1$$

$$\Delta_R H = \{-1 * (-78) - 1 * (-264) + 1 * (-507)\}\frac{\mathrm{kJ}}{\mathrm{mol}} = -383\frac{\mathrm{kJ}}{\mathrm{mol}}$$

$$\dot{Q}_{\mathrm{Reaction}} = -0.414\frac{\mathrm{mol}}{\mathrm{s}} * \left(-383\frac{\mathrm{kJ}}{\mathrm{mol}}\right) = 158.6\frac{\mathrm{kJ}}{\mathrm{s}}$$

$$\dot{m}_{\mathbf{Steam}} = \frac{(321.1 - 158.6)\text{kJ} * \text{kg}}{\text{s} * 2300\,\text{kJ}} = \mathbf{0.0707\frac{kg}{s}} = \mathbf{254.3\frac{kg}{h}}$$

e) $\dot{Q}_{\text{Steam}} = Kw_{\text{Reactor}} * A_{\text{Reactor}} * (T_{\text{Steam}} - T_{\text{Reactor}})$

$$\dot{Q}_{\text{Steam}} = \dot{Q}_{\text{total}} - \dot{Q}_{\text{Reaction}} = (321.1 - 158.6)\frac{\text{kJ}}{\text{s}} = 162.5\frac{\text{kJ}}{\text{s}}$$

$$A_{\text{Reactor}} = \frac{\dot{Q}_{\text{Steam}}}{Kw_{\text{Reactor}} * (T_{\text{Steam}} - T_{\text{Reactor}})} = \frac{162.5\,\text{kJ} * \text{m}^2 * {}^\circ\text{C}}{\text{s} * 600\,\text{W} * (130 - 90)\,{}^\circ\text{C}} = \frac{162.5\,\text{kJ} * \text{m}^2 *{}^\circ\,\text{C} * \text{s}}{\text{s} * 0.6\,\text{kJ} * (130 - 90)\,{}^\circ\text{C}}$$

$$\boldsymbol{A}_{\mathbf{Reactor}} = \mathbf{6.77\,m^2}$$

f) $\dot{Q}_{\text{CoolWater}} = \dot{m}_{\text{CoolWater}} * cp_{\text{CoolWater}} * (T_{\text{CoolWater-ex}} - T_{\text{CoolWater-in}})$

$$\dot{m}_{\text{CoolWater}} = \frac{\dot{Q}_{\text{CoolWater}}}{cp_{\text{Water}} * (T_{\text{CoolWater-ex}} - T_{\text{CoolWater-in}})}$$

$$\dot{Q}_{\text{CoolWater}} = \Delta\dot{Q}_{\text{ReactionMixture}}$$
$$= \left(\dot{m}_{\text{EO}_o} * cp_{\text{EO}} + \dot{m}_{\text{NH}_{3_o}} * cp_{\text{NH}_3} + \dot{m}_{\text{Hex}} * cp_{\text{Hex}}\right) * (T_{\text{Reactor}} - T_{\text{ReactionMixture-Ex}})$$

$$\dot{Q}_{\text{CoolWater}} = (0.0185 * 2.0 + 0.247 * 6.5 + 1.28 * 2.3)\frac{\text{kg} * \text{kJ}}{\text{s} * \text{kg} * {}^\circ\text{C}} * (90 - 35)\,{}^\circ\text{C} = 252.3\frac{\text{kJ}}{\text{s}}$$

$$\dot{m}_{\mathbf{CoolWater}} = \frac{252.3\,\text{kJ} * \text{kg} * {}^\circ\text{C}}{\text{s} * 4.2\,\text{kJ} * (40 - 20)\,{}^\circ\text{C}} = \mathbf{3.00\frac{kg}{s}} = \mathbf{10.8\frac{t}{h}}$$

g) $\dot{Q}_{\text{CoolWater}} = Kw * A * \overline{\Delta T_M} \rightarrow A = \frac{\dot{Q}_{\text{CoolWater}}}{Kw * \overline{\Delta T_M}}$

$$\frac{1}{Kw} = \frac{1}{\alpha_1} + \frac{s}{\lambda} + \frac{1}{\alpha_2} = \frac{\text{m}^2 * {}^\circ\text{C}}{350\,\text{W}} + \frac{0.004\,\text{m} * \text{m} * {}^\circ\text{C}}{45\,\text{W}} + \frac{\text{m}^2 * {}^\circ\text{C}}{750\,\text{W}} = 0.00437\frac{\text{m}^2 * {}^\circ\text{C}}{\text{W}}$$

$$\rightarrow Kw = 229\frac{\text{W}}{\text{m}^2 * {}^\circ\text{C}}$$

$$\overline{\Delta T_M} = \frac{\Delta T_1 - \Delta T_2}{\ln\frac{\Delta T_1}{\Delta T_2}} \quad \Delta T_1 = (90 - 40)\,{}^\circ\text{C} = 50\,{}^\circ\text{C} \; \Delta T_2 = (35 - 20)\,{}^\circ\text{C} = 15\,{}^\circ\text{C}$$

$$\overline{\Delta T_M} = \frac{(50 - 15)\,{}^\circ\text{C}}{\ln\frac{50}{15}} = 29.2\,{}^\circ\text{C}$$

$$\boldsymbol{A} = \frac{252.3\,\text{kJ} * \text{m}^2 * {}^\circ\text{C}}{\text{s} * 229\,\text{W} * 29.2\,{}^\circ\text{C}} = \frac{252.3\,\text{kJ} * \text{m}^2 *{}^\circ\,\text{C} * \text{s}}{\text{s} * 0.229\,\text{kJ} * 29.2\,{}^\circ\text{C}} = \mathbf{37.7\,m^2}$$

$$A = \overline{d}_M * \pi * L * n_{\text{Pipes}}$$

$$L = \frac{A}{\overline{d}_M * \pi * n_{\text{Pipes}}}$$

$$\overline{d}_M = \frac{d_a - d_i}{\ln \frac{d_a}{d_i}}$$

$$d_a = 2'' * 25.4 \frac{\text{mm}}{''} = 50.8\,\text{mm} \quad d_i = d_a - 2\,\text{s} = (50.8 - 8)\,\text{mm} = 42.8\,\text{mm}$$

$$\overline{d}_M = \frac{(0.0508 - 0.0428)\text{m}}{\ln \frac{0.0508}{0.0428}} = 0.0468\,\text{m}$$

$$L = \frac{37.7\,\text{m}^2}{0.0468\,\text{m} * \pi * 50} = \mathbf{5.13\,m \cong 5.2\,m}$$

→ ***Result***

a) **The scale-up factor was chosen as the ratio of the volumes of the laboratory experiment with the use of 100 g of reaction mixture and the active volume of the pilot plant reactor. This results in a scale-up factor of 4630.**
b) **The feed rates for the pilot plant reactor are as follows:**
Ethylene oxide: 0.0185 kg/s = 0.420 mol/s → 66.6 kg/h = 1512 mol/h
Ammonia: 0.247 kg/s = 14.5 mol/s → 889.2 kg/h = 52,200 mol/h
Hexane: 1.28 kg/s = 19.3 mol/s → 4608 kg/h = 69,480 mol/h
c) **The production capacity of MEA in the pilot plant reactor is 0.357 mol/s, which is 0.0218 kg/s or 78.5 kg/h.**
d) **To heat the reactor, 254.3 kg of 130 °C saturated steam is required per hour.**
e) **The minimum required heat transfer area was calculated to be ~6.8 m^2. Since the pilot plant reactor has 1.5 m^2 of heat exchanger area, this is not sufficient. An external heat exchanger for the CSTR-heating by means of steam is required.**
f) **3 kg/s i.e. 10.8 t/h of cooling water is required.**
g) **A minimum heat exchanger area of 37.7 m^2 is required, which corresponds to a length of the heat exchanger of 5.2 m.**

Task 111

In a production plant, a bromine and sulfur-containing liquid waste stream of 250 kg per hour results, which is to be disposed of by combustion. The plant is to be roughly designed.

The following average values of the waste stream are known from laboratory measurements:

Bromine content = 0.95 wt%, sulfur content = 1.6 wt%, low heating value LHV = 18,000 kJ/kg,

Specific heat capacity $cp_{\text{Waste}} = 1.83$ kJ/(kg * °C)

In order to reduce the viscosity of the liquid waste and thus enable its atomizing in a burner, the stream is warmed-up in a heat exchanger from 30 °C to 85 °C before combustion. The stainless steel heat exchanger is operated with saturated steam at 130 °C. The steam-side heat transfer coefficient was estimated with $\alpha_1 = 5000\ W/(m^2 * °C)$ and the liquid-side with $\alpha_2 = 100\ W/(m^2 * °C)$. The wall thickness of the exchanger tubes is s = 4 mm. The thermal conductivity of the stainless steel is $\lambda = 21\ W/(m * °C)$.

The heat of combustion is to be used for steam generation. A thermal efficiency of 60% is expected during the combustion process itself.

The hydrogen bromide and the sulfur dioxide formed during combustion are removed from the cooled flue gas with neutral water in a washer. The water leaving the washer is to be neutralized with 10% sodium hydroxide solution. Chlorine is introduced for the recovery of bromine and the bromine is extracted from the water phase with a tetrachloromethane stream of 40 L per hour. The following reactions take place:

$2Br^- + Cl_2 \rightarrow Br_2 + 2Cl^-$ and

$SO_3^{2-} + Cl_2 + H_2O \rightarrow SO_4^{2-} + 2H^+ + 2Cl^-$

a) What is the minimum area of the heat exchanger required to heat the waste stream in front of the burner?
b) How much saturated steam at 130 °C can be generated by this plant, if the feed-water is supplied at a temperature of 105 °C (water: cpw = 4.17 kJ/[kg * °C]; $\Delta_v H_w = 2174$ kJ/kg) and the steam consumption for heating the waste stream before its combustion is taken into account?
c) How much 10% sodium hydroxide is required for neutralization of the water exiting the flue gas scrubber? The low carbon dioxide content of the wash water is to be neglected here.
d) What is the flow rate of chlorine at 15 °C and 1.5 bar that must be passed through the neutralized water if a 1.5% chlorine excess is required to achieve complete oxidation of the bromide ions and sulfite ions?
e) How many kg of bromine are recovered per hour if 2.5% losses are to be expected? What are the molar and mass concentrations of bromine in tetrachloromethane ($\rho_{CCl_4} = 1600\ kg/m^3$) in this case?

Molar masses: NaOH: 40 g/mol; S = 32.1 g/tom; Br: 79.9 g/tom

⊗ **Solution**

→ ***Strategy***

a) The necessary heat transfer area is calculated from the corresponding rearranged formula 30. The heat output is calculated from formula 19a with the exit and inlet temperatures of the liquid waste stream. The heat transfer coefficient results from formula 31a. The calculation of the mean logarithmic temperature difference is described in the section Averagings (1.2.5). Here, the differences of the steam temperature to the inlet and outlet temperature of the liquid stream are used.

b) The useful 60% of the heat flow generated by the combustion of the liquid waste (calculated according to the modified formula 27c from the low heating value, see 2.4.4) is used for the heating of the boiler feed water (formula 19a) and the steam generation (formula 24b) as well as the heating of the waste stream. The mass flow of the boiler feed water is equal to that of the generated steam. The formulas are combined accordingly and rearranged to the generated steam flow.

c) The molar flow of hydrogen bromide resulting from the combustion is equal to that of bromine in the waste stream and can be calculated by a stoichiometric balance. The same applies to the resulting sulphurous acid. From this, the necessary molar flow of NaOH is calculated and related to 10% sodium hydroxide.

d) For 2 moles of bromide, 1 mole of chlorine and for 1 mole of sulphite, 1 mole of chlorine is required respectively. From the previously calculated molar flows of bromide and sulphite, the required number of moles of chlorine results. With the help of the rearranged ideal gas law (formula 2), the flow rate of chlorine follows.

e) The molar flow of bromine (Br_2) is half as large as that of bromide. The bromine losses of 2.5 % have to be taken into account. The concentration of bromine in tetrachloromethane results from the division of its mass flow by the volume or mass flow of this solvent.

→ ***Calculation***

a) $\dot{Q} = Kw * A * \Delta T$

$$A = \frac{\dot{Q}}{Kw * \Delta T} \quad \text{with } \Delta T = \Delta \overline{T}_M = \frac{\Delta T_1 - \Delta T_2}{\ln \frac{\Delta T_1}{\Delta T_2}}$$

$$\Delta T_1 = (130 - 30)\,°\text{C} = 100\,°\text{C} \quad \Delta T_2 = (130 - 85)\,°\text{C} = 45\,°\text{C}$$

$$\Delta \overline{T}_M = \frac{(100 - 45)\,°\text{C}}{\ln \frac{100}{45}} = 68.9\,°\text{C}$$

$$\dot{Q}_{WasteHeating} = \dot{m}_{Waste} * cp_{Waste} * (T_{\text{Waste-ex}} - T_{Waste-\text{in}}) = 250 \frac{\text{kg}}{\text{h}} * 1.83 \frac{\text{kJ}}{\text{kg} * °\text{C}} * (85 - 30)\,°\text{C}$$

$$= 25163 \frac{\text{kJ}}{\text{h}} = 7.00 \frac{\text{kJ}}{\text{s}} = 7.00\,\text{kW}$$

$$\frac{1}{Kw} = \frac{1}{\alpha_1} + \frac{s}{\lambda} + \frac{1}{\alpha_2} = \frac{\text{m}^2 * °\text{C}}{5000\,\text{W}} + \frac{0.004\,\text{m} * \text{m} * °\text{C}}{21\,\text{W}} + \frac{\text{m}^2 * °\text{C}}{100\text{W}}$$

$$= 0.0104 \frac{\text{m}^2 * °\text{C}}{\text{W}} \quad K_W = 96.15 \frac{\text{W}}{\text{m}^2 * °\text{C}}$$

$$A = \frac{7000\,\mathrm{W} * \mathrm{m}^2 * {}^\circ\mathrm{C}}{96.15\,\mathrm{W} * 68.9\,{}^\circ\mathrm{C}} = \mathbf{1.06\,m^2} \cong \mathbf{1.1\,m^2}$$

b)
$$\dot{Q}_{\text{Combustion-total}} = \dot{m}_A * LHV = 250\frac{\mathrm{kg}}{\mathrm{h}} * 18{,}000\frac{\mathrm{kJ}}{\mathrm{kg}}$$
$$= 4{,}500{,}000\frac{\mathrm{kJ}}{\mathrm{h}} = 1250\frac{\mathrm{kJ}}{\mathrm{s}}$$

$$\dot{Q}_{\text{Utilized}} = \frac{\eta_{\text{therm}}}{100\%} * \dot{Q}_{\text{Combustion-total}} = \frac{60\%}{100\%} * 1250\frac{\mathrm{kJ}}{\mathrm{s}} = 750\frac{\mathrm{kJ}}{\mathrm{s}}$$

$$\dot{Q}_{\text{Utilized}} = Q_{\text{Waste-Heating}} + \dot{Q}_W + \dot{Q}_{St} \quad \dot{Q}_W = \dot{m}_{St} * cp_W * (T_{St} - T_W) \quad \dot{Q}_{St} = \dot{m}_{St} * \Delta_v H_w$$

$$\dot{Q}_{\text{Utilized}} = \dot{m}_{St} * \left[cp_W * (T_{St} - T_W) + \Delta_v H_w\right] + Q_{\text{WasteHeating}}$$

$$\dot{m}_D = \frac{\dot{Q}_{\text{Utilized}} - \dot{Q}_{\text{WasteHeating}}}{cp_W * (T_{St} - T_W) + \Delta_v H_w} = \frac{(750 - 7)\frac{\mathrm{kJ}}{\mathrm{s}}}{4.17\frac{\mathrm{kJ}}{\mathrm{kg}*{}^\circ\mathrm{C}} * (130 - 105)\,{}^\circ\mathrm{C} + 2174\frac{\mathrm{kJ}}{\mathrm{kg}}}$$
$$= \mathbf{0.326\frac{kg}{s}} = \mathbf{1.17\frac{t}{h}} \cong \mathbf{1.2\frac{t}{h}}$$

c) $$\dot{n}_{\text{HBr}} = \dot{n}_{\text{Br}} = \frac{\dot{m}_{\text{Br}}}{M_{\text{Br}}} = \frac{\dot{m}_A}{M_{\text{Br}}} * \frac{\%\text{Br}}{100\%} = \frac{250\,\mathrm{kg} * \mathrm{mol}}{\mathrm{h} * 0.0799\,\mathrm{kg}} * \frac{0.95\%}{100\%} = 29.7\frac{\mathrm{mol}}{\mathrm{h}}$$

$$\dot{n}_{H_2SO_3} = \dot{n}_S = \frac{\dot{m}_S}{M_S} = \frac{\dot{m}_A}{M_S} * \frac{\%S}{100\%} = \frac{250\,\mathrm{kg} * \mathrm{mol}}{\mathrm{h} * 0.0321\,\mathrm{kg}} * \frac{1.6\%}{100\%} = 124.6\frac{\mathrm{mol}}{\mathrm{h}}$$

$$\dot{n}_{\text{NaOH}} = \dot{n}_{\text{HBr}} + 2 * \dot{n}_{H_2SO_3} = (29.7 + 2 * 124.6)\frac{\mathrm{mol}}{\mathrm{h}} = 278.9\frac{\mathrm{mol}}{\mathrm{h}}$$

$$\dot{m}_{\text{NaOH}} = \dot{n}_{\text{NaOH}} * M_{\text{NaOH}} = 278.9\frac{\mathrm{mol}}{\mathrm{h}} * 0.040\frac{\mathrm{kg}}{\mathrm{mol}} = 11.16\frac{\mathrm{kg}}{\mathrm{h}}$$

$$\dot{m}_{\textbf{Caustic 10\%}} = \dot{m}_{\text{NaOH}} * \frac{100\%}{10\%} = 11.16\frac{\mathrm{kg}}{\mathrm{h}} * 10 = \mathbf{111.6\frac{kg}{h}} = \mathbf{0.031\frac{kg}{s}}$$

d) $$\dot{n}_{Cl_2\text{-theor}} = \frac{\dot{n}_{\text{Br-}}}{2} + \dot{n}_{SO_3-} = \left(\frac{29.7}{2} + 124.6\right)\frac{\mathrm{mol}}{\mathrm{h}} = 139.45\frac{\mathrm{mol}}{\mathrm{h}}$$

$$\dot{n}_{Cl_2\text{-real}} = \dot{n}_{Cl_2\text{-theor}} * 1.015 = 139.45\frac{\mathrm{mol}}{\mathrm{h}} * 1.015 = 141.5\frac{\mathrm{mol}}{\mathrm{h}}$$

$$p * \dot{V}_{Cl_2} = \dot{n}_{Cl_2} * R * T$$

$$\dot{V}_{Cl_2} = \frac{\dot{n}_{Cl_2} * R * T}{p} = \frac{141.5\,\mathrm{mol} * 8.315 * 10^{-5}\,\mathrm{bar} * \mathrm{m}^3 * (273 + 15)\mathrm{K}}{\mathrm{h} * \mathrm{mol} * \mathrm{K} * 1.5\,\mathrm{bar}}$$
$$= \mathbf{2.26\frac{m^3}{h}} = \mathbf{0.628\frac{L}{s}}$$

e) $\dot{n}_{\text{Br2-real}} = \frac{\dot{n}_{\text{Br-}}}{2} * \frac{(100 - 2.5)\%}{100\%} = \frac{29.7\,\text{mol}}{2 * \text{h}} * 0.975$

$= 14.5 \frac{\text{mol}}{\text{h}}$

$$\dot{m}_{\textbf{Br2-real}} = \dot{n}_{\text{Br2-real}} * M_{\text{Br2}} = 14.5 \frac{\text{mol}}{\text{h}} * 2 * 0.0799 \frac{\text{kg}}{\text{mol}} = \mathbf{2.32 \frac{kg}{h}}$$

$$c_{\textbf{Br2}} = \frac{\dot{n}_{\text{Br2-real}}}{\dot{V}_{\text{Tetra}}} = \frac{14.5\,\text{mol} * \text{h}}{\text{h} * 40\,\text{L}} = \mathbf{0.363 \frac{mol}{L}}$$

$$C_{\textbf{Br2}} = \frac{\dot{m}_{\text{Br2-real}}}{\dot{m}_{\text{Tetra}}} * 100\% = \frac{\dot{m}_{\text{Br2-real}}}{\dot{V}_{\text{Tetra}} * \rho_{\text{Tetra}}} * 100\% = \frac{2.32\,\text{kg} * \text{h} * \text{m}^3}{\text{h} * 0.040\,\text{m}^3 * 1600\,\text{kg}} * 100\% = \mathbf{3.63\,wt\%}$$

→ ***Result:***

a) **The minimum required heat exchanger area is 1.1 m^2.**
b) **1.2 t of saturated steam at 130 °C is generated per hour.**
c) **The consumption of 10% caustic soda for neutralization of the wash water is 111.6 kg per hour i.e. 0.031 kg per second.**
d) **A flow rate of 2.26 m^3/h i.e. 0.628 L/s of chlorine at given pressure and temperature must be supplied for release of the bromine and oxidation of the sulfite.**
e) **2.32 kg of bromine is recovered per hour, resulting in a solution in tetrachloromethane with a concentration of 0.363 mol/L i.e. 3.63 wt% bromine.**

"All's well that ends well!"

Printed in the United States
by Baker & Taylor Publisher Services